Primate Conservation in the Tropical Rain Forest

MONOGRAPHS IN PRIMATOLOGY

Volume 1: Child Abuse: The Nonhuman Primate Data
Martin Reite and Nancy G. Caine, *Editors*

Volume 2: Viral and Immunological Diseases in Nonhuman Primates
S.S. Kalter, *Editor*

Volume 3: Blood Groups of Primates: Theory, Practice, Evolutionary Meaning
Wladyslaw W. Socha and Jacques Ruffié

Volume 4: Female Primates: Studies by Women Primatologists
Meredith F. Small, *Editor*

Volume 5: Clinical Management of Infant Great Apes
Charles E. Graham and James A. Bowen, *Editors*

Volume 6: Nonhuman Primate Models for Human Growth and Development
Elizabeth S. Watts, *Editor*

Volume 7: The Lion-Tailed Macaque: Status and Conservation
Paul G. Heltne, *Editor*

Volume 8: Behavior and Pathology of Aging in Rhesus Monkeys
Roger T. Davis and Charles W. Leathers, *Editors*

Volume 9: Primate Conservation in the Tropical Rain Forest
Clive W. Marsh and Russell A. Mittermeier, *Editors*

Primate Conservation in the Tropical Rain Forest

Editors

Clive W. Marsh
Sabah Foundation
Kota Kinabalu, Sabah
Malaysia

Russell A. Mittermeier
World Wildlife Fund
Washington, D.C.
and
Department of Anatomical Sciences, HSC
State University of New York
Stony Brook, New York

ALAN R. LISS, INC., NEW YORK

Address all Inquiries to the Publisher
Alan R. Liss, Inc., 41 East 11th Street, New York, NY 10003

Library of Congress Cataloging-in-Publication Data

Primate conservation in the tropical rain forest.

Includes index.
1. Primates—Tropics—Ecology. 2. Wildlife conservation—Tropics. 3. Rain forest ecology.
I. Marsh, Clive W. II. Mittermeier, Russell A.
QL737.P9P672325 1987 333.95'9 85-10414
ISBN 0-8451-3408-6

Contents

SECTION III: PRIORITIES

Contributors

Rauf Ali, Department of Biology, Faculty of Science, Mahidol University, Bangkok 10400, Thailand; present address: Auroville Center, Aurodam, 605101 South Arcot, India **[23]**

Rosalind J. Aveling, Frankfurt Zoological Society, 6000 Frankfurt/Main 1, Federal Republic of Germany, and Zaire Gorilla Conservation Project, Goma, Kivu, Zaire **[231]**

J. Marcio Ayres, Department of Anatomy, University of Cambridge, Cambridge CB2 1QS, United Kingdom **[83]**

Warren Y. Brockelman, Department of Biology, Faculty of Science, Mahidol University, Bangkok 10400, Thailand **[23]**

Ardith A. Eudey, Department of Anthropology, University of Nevada, Reno, Nevada 89557-0006; present address: Department of Psychology, University of California, Riverside, CA 92521 **[147]**

Nathan R. Flesness, International Species Inventory System, Minnesota Zoological Garden, Apple Valley, MN 55124 **[263]**

Thomas J. Foose, AAZPA Conservation, Minnesota Zoological Garden, Apple Valley, MN 55124 **[263]**

J. Stephen Gartlan, Wisconsin Regional Primate Research Center, University of Wisconsin, Madison, WI 53706 **[321]**

Ruth E. Happel, Department of Anthropology, Peabody Museum, Harvard University, Cambridge, MA 02138 **[63]**

Andrew D. Johns, Department of Anatomy, University of Cambridge, Cambridge CB2 1QS, United Kingdom **[83]**

Michael Kavanagh, Wildlife Trade Monitoring Unit, IUCN Conservation Monitoring Centre, Cambridge CB3 0DL, United Kingdom; present address: World Wildlife Fund Malaysia, Box 10769, 50724 Kuala Lumpur, Malaysia **[147]**

David Mack, TRAFFIC (U.S.A.), World Wildlife Fund–United States, Washington, D.C. 20009; present address: World Wildlife Fund–United States, Washington, D.C. 20037 **[147]**

The numbers in brackets are the opening page numbers of the contributors' articles.

Clive W. Marsh, Sabah Foundation, 88817 Kota Kinabalu, Sabah, Malaysia **[xi,63,83,343]**

Jeffrey A. McNeely, Programme and Policy Division, International Union for Conservation of Nature and Natural Resources, 1196 Gland, Switzerland **[181]**

Kenton R. Miller, International Union for Conservation of Nature and Natural Resouces, 1196 Gland, Switzerland **[181]**

Russell A. Mittermeier, Primate Program, World Wildlife Fund, Washington, DC 20037, and Department of Anatomical Sciences, Health Sciences Center, State University of New York, Stony Brook, NY 11794 **[xi,109,305]**

Norman Myers, Consultant in Environment and Development, Upper Meadow, Old Road, Headington, Oxford OX3 8SZ, United Kingdom **[3]**

John F. Noss, Department of Anthropology, Peabody Museum, Harvard University, Cambridge, MA 02138 **[63]**

John F. Oates, Department of Anthropology, Hunter College of the City University of New York, New York, NY 10021 **[321]**

Alison F. Richard, Department of Anthropology, Yale University, New Haven, CT 06520 **[329]**

Paul Richards, 14 Wooton Way, Cambridge CB3 9LX, United Kingdom **[ix]**

U. S. Seal, IUCN Captive Breeding Specialist Group, V.A. Hospital, Minneapolis, MN 55417 **[263]**

Thomas T. Struhsaker, Wildlife Conservation International, New York Zoological Society, Bronx Zoo, Bronx, NY 10460 **[321]**

Robert W. Sussman, Department of Anthropology, Washington University, St. Louis, MO 63130 **[329]**

James W. Thorsell, Commission on National Parks and Protected Areas, International Union for Conservation of Nature and Natural Resources, 1196 Gland, Switzerland **[181]**

A. William Weber, Institute for Environmental Studies, University of Wisconsin, Madison, WI 53706 present address: Kigali (c), USAID, Washington, DC 20523 **[205]**

Foreword

Most of the world's primates—other than man—are found only in the tropics. Some live in savannas and other open habitats, but the majority inhabit closed forests, either the tropical rain forests, semi-deciduous and deciduous forests of the lowlands, or the mainly evergreen montane forests of the tropical mountains. The fate of the surviving primates and of tropical forests are thus inevitably linked together.

Earlier in this century it was still believed that the tropical forests had survived, probably with little change, since the Cretaceous. It was also assumed that, in spite of the inroads of agriculture and timber exploitation, substantial areas of rain forest would remain into the foreseeable future. Palynologists and biogeographers have since shown that the first belief is far from the truth; during the last 50 or 60 million years the forest-covered area of the tropics has waxed and waned and during the Pleistocene glaciations the tropical rain forest was reduced to small widely separated refugia, with profound effects on their plant and animal populations. But of even greater importance as far as primate conservation is concerned, the growth of human populations and the development of modern technology have resulted in a far more rapid shrinkage of the forest area than was foreseen 50 years ago and it has greatly accelerated in the last two or three decades.

As the earlier chapters of this books show, much more than half the tropical rain forest still existing in 1900 has already been converted to farmland or destroyed and of what remains most has been drastically modified by timber extraction and shifting cultivation. Though the amount of tropical forest which will survive for another 20 years cannot be accurately predicted, it is certain that unless determined and well organized efforts are made to conserve what now exists, the world's tropical forests will have largely disappeared, if not by the end of this century, then certainly long before the end of the next one. Many of the probable consequences of this destruction are well known, but their full significance for the human environment is still not fully realized. One of the most tragic of these consequences is the likely extinction of vast numbers of plant and animal species, many of them still undescribed and hardly known.

Among the animals most affected by the destruction and fragmentation of tropical forests are the apes, monkeys, and lemurs which for many reasons, including their blood relationship to man, are of quite exceptional interest.

The conservation of primates is therefore in itself a very important reason for trying to conserve as much as possible of the remaining tropical forests. If these disappear, so will most of the primates, and future human generations will know them only from illustrations in books, or, at best, from captives in zoos.

To the aim of conserving all the forest-living primates of the world, this book should make a valuable contribution. Research and administrative measures in aid of primate conservation are going on in many countries, but there has been no book before this one which reviews the subject on a world-wide scale and none which examines in depth the effects on wild primates of timber extraction, forest fragmentation, trapping, trade, and other threats to their survival. It also attempts to assess current conservation policies affecting what are necessarily multifarious and intractable problems. I am glad to have this opportunity of welcoming this book and I hope it will have the success it deserves.

Paul Richards

Introduction

The stimulus for this book was a session of the Seventh Congress of the International Primatological Society held in Bangalore, India, in 1979, at which a number of field primatologists spoke informally on the problems facing primates in the countries in which they worked. The approach was comparative and spanned all three main regions of the tropics. Some problems, such as habitat loss, were recurrent in one talk after another, while others, such as the hunting of animals, varied greatly between species and countries. Prospects for effective conservation action also seemed highly variable, depending not only on biological considerations of species abundance, rates of habitat loss and so on, but also on local economic, social and educational conditions.

At the time of the Bangalore meeting, Prince Rainier and Bourne [1977] had recently published a collection of conservation status reviews on selected primate species. Admirable as many of these were, we felt that the approach was limited. To those not immediately involved with a particular species the value of successive case histories lies in the emergence of pattern. We therefore decided to solicit a set of papers that would compare the problems faced by primates in different parts of the world and examine directions for progress in practical conservation efforts.

Since its conception, the project has been seven years in gestation, during which time there has been considerable progress in conservation theory and in some aspects of its practice: for example, in the collation of data on endangered species and protected areas, and in attempts to rationalize criteria for determining conservation priorities. This has been one of the main achievements of the International Union for Conservation of Nature and Natural Resources (IUCN) and has led to clearer perspectives at the international level. The scientific study of conservation problems has also developed greatly. Among useful publications we would single out especially Soulé and Wilcox's *Conservation Biology*, published in 1980, as having done much to define and encourage a science by this name. We believe that the papers in the present book lie within the same tradition and will be of interest to many conservation biologists, not just those concerned with primates. If some of the papers are now more than two years out of date, this is partly because of the field commitments of many of the authors. We hope that the finished product is worth the delay.

Two main themes are expounded in the early part of the book. Firstly, since the great majority of primates, and particularly threatened species, live in the rain forest, the problems of primate conservation are largely a function of the decline of this biome. The destruction of the world's tropical rain forests is one of the great events of our time, perhaps even the greatest in its consequences. We therefore begin with an overview of the issue by its best-known exponent, Norman Myers. His paper describes the methodological problems of assessing trends in forest area statistics and shows how the nature and scale of forest losses vary between countries and continents. More briefly treated are the far-reaching consequences of forest loss for the sustainable use of natural resources, for the maintenance of stable environments and for the world's endowment of species.

In such a wide context, the justification for a focus on primates, rather than other taxonomic groups, must be principally that they make excellent "ambassadors" or "flagships" for the plight of tropical forests. Primates have a wide range of sizes and diets and they have been quite intensively studied over the past twenty-five years. Thus, their problems are better documented than those of many other vertebrate groups. In addition, of course, the complexity of primate social behavior and the inclusion of our own species in this order, give primates a special place of interest and concern to the public at large in both developed and developing countries.

A second theme is the recognition of variation in the threats to primate populations and the need for more research that is directly relevant to their elucidation. This need is clear in each of the six chapters of Section I of the book grouped under the heading "Problems." At present, it is probably fair to say that most field primatology is concerned with essentially academic questions of behavior and function to which conservation problems are incidental. This may to help to explain a curious dualism that otherwise rigorous scientists sometimes lapse into thoroughly vague or emotional pronouncements on conservation, as though their own professional standards could be waived in a good cause. This is an unaffordable luxury which may invite unsympathetic criticism [e.g., Simon, 1986].

A prerequisite to any conservation assessment for a species is some knowledge of the distribution, density, and size of its populations. Estimation of even simple parameters such as these poses severe problems for rain forest primates because visibility in the field is poor and the animals are generally cryptic. As in other sciences, accuracy is largely a function of the time and resources available, the best method of assessment in this case being through intensive, long-term study of neighboring groups at several sites. Quicker appraisal of larger populations requires sampling techniques based on sightings, calls, or nest counts over a wider area. In Chapter 2, Brockelman and Ali describe some of the main techniques available and the biases to which they are liable. An appreciation of these problems breeds some salutory caution in pronouncements on population densities and size. The refinement and testing of survey methods remains an important field for further study in primate conservation biology.

Our third paper, by Happel, Noss, and Marsh, reviews available data on the population density and geographical range area of a large sample of primates, and seeks explanations for patterns of abundance and scarcity in terms of species ecological characteristics. In other words, are some species intrinsically more likely to be rare and prone to endangerment than others? The problem is a complex one and only a few tentative generalizations are possible. For example, big animals tend to occur at lower densities and reproduce more slowly than smaller ones. They are also a more economic proposition for hunters and thus predisposed to scarcity in several ways. More surprisingly, species classified by IUCN as threatened tend to have either longer or shorter than average gestation periods although female body mass does not show this pattern. While fertile ground for further study, academic findings of this kind cannot be expected to yield close predictions of species' endangerment because of the paramount importance of human history for patterns of habitat destruction. Thus, the fact that so many primates are endangered in Madagascar or in the Atlantic forest region of Brazil may be partly a function of, say, high rates of endemism per unit area, specialized diets, or other ecological characters, but the primary causes must be sought in the history of settlement in these areas.

The next three papers examine consequences of human activities on primates. Marsh, Johns, and Ayres offer a classification of different kinds of habitat disturbance and review studies to date on how primates survive them. Although habitat loss is by far the greatest threat to these animals, many species are surprisingly resilient to habitat modification that does not amount to clear-felling. However, this ability varies with species ecology, and especially with dietary specialization. Many species are also threatened in the longer term by fragmentation of remaining habitat, which reduces the size of breeding populations and thus increases the likelihood of local extinctions. A further problem both for conservation efforts and research design is that habitat disturbance is often compounded in its effects by hunting.

Mittermeier examines the occurrence of primate hunting worldwide and shows both the severity of its effects in some areas and the great range of cultural variation between regions and countries. Primates are a significant source of protein for human consumption in some areas, and this constitutes a major conservation problem in three parts of the world: Amazonia, West Africa, and Central Africa. At a local level, there remains fascinating scope for more detailed studies combining ecological and anthropological techniques to examine the interaction of hunters and their prey and the necessary conditions for a sustainable harvest. A promising start in this field has been made recently by Caldecott [1986].

A less widespread but sometimes contentious form of direct exploitation of primates is their trapping and trade, principally for use in biomedical research. Kavanagh, Eudey, and Mack review the available trade statistics and we are very pleased that their research has also contributed to a recent, multi-authored review of the international primate trade [Mack and Mittermeier, 1984]. Although trapping for trade is generally a secondary threat for all but a few species it merits careful attention because it is more amenable to control both within source countries and through the Convention on International Trade in Endangered Species of Wild Fauna and Flora (CITES). There is thus less excuse for defaults. An underlying problem, however, in many cases of possible overexploitation, is that population estimates, and still less scientifically respectable quotas for sustainable yields, are poor or lacking.

After six chapters concerned with the threats to primates and their assessment, Section II of the book considers conservation remedies under the heading "Approaches." The theme intended here is that conservation problems need to be addressed from several directions at once, but that for any long-term hope of success, impetus for action must develop from within the countries concerned. The foreign scientist is here in a difficult position. He may see clearly what he thinks should be done, but in the post-colonial world he can rarely determine or help to implement policies, only suggest and encourage. Hence, qualities of diplomacy and knowledge of the local language, culture, and political situation are essential. Ultimately, the most enduring contribution which the outsider can make is often in catalyzing students, government officers and others to a greater appreciation of their own country's wildlife and of the dilemma of short-term gains based on over-exploiting natural resources versus the longer term benefits of more sustainable patterns of development.

Since habitat loss is the principal threat to primates it follows that habitat protection should be the highest conservation priority. Hence, the most valuable direct means of assisting species conservation is in most cases through the establishment and management of protected areas. McNeely, Miller, and Thorsell, who are all senior staff members of IUCN, have contributed an overview of the coverage of tropical rain forest within national parks and other categories of protected areas. The main aim of a national system of such reserves is to protect representative examples of as many as possible of the habitats to be found within a country and, in addition, species or additional features which are especially at risk. This principle is illustrated by reference to two of the most important countries containing rain forest: Brazil and Indonesia.

While many of the benefits of protected areas, and especially the reasons for conserving rare animals, are more apparent in a national or global perspective, most of the costs of refraining from exploiting protected areas are borne locally. This asymmetry lies at the root of most conservation management problems and demands attention to the problems and welfare of ordinary people whose lives are affected by the creation of protected areas. In Chapter 8, Weber describes a rare and instructive attempt to survey the attitudes of local residents towards reserves in the Zaire-Nile Divide region of Rwanda and Burundi. His study demonstrates clearly the need to view problems of park management in their wider social setting and to find means of returning economic benefits from parks to the local people, if only through tourism. The Mountain Gorilla Project, of which this study is a part, provides an exemplary model for integrated conservation efforts elsewhere.

That education is an essential ingredient in achieving conservation goals is a truism. But what efforts are actually taking place in tropical countries, and how effective are they? In Chapter 9, Aveling contributes a wide-ranging review of current initiatives and programmes in environmental education at all levels. A comparative approach is particularly valuable here as successful programs in one country—for example, the Wildlife Clubs of Kenya scheme—can serve as a model and inspiration for work in others. In many countries, primates have proved an effective focus for a wider educational mission. The muriqui in Brazil, gorillas in Rwanda and Zaire, the lion-tailed macaque in India, and the orang-utan in Indonesia and Malaysia, have all played this role of "flagship" species.

Returning to what might be termed the "hardware" of conservation—i.e. physical actions rather than attempts to change attitudes—the plight of a few species of primates is clearly so desperate that captive breeding programs may offer the only alternative to extinction, or at least provide a prudent insurance policy. This is a field in which zoos and research institutions have a vital role to play. One might also add that the development of a direct role in conservation for zoos has in recent years done much to revitalize their public image and their own sense of purpose. We are therefore pleased to include here a review of captive breeding of primates by Foose, Seal, and Flesness, who are prominent members of IUCN's Captive Breeding Specialist Group.

The last four chapters of the book are grouped in Section III under the title "Priorities." The sequence of the book thus begins in Section I with problems, progresses in Section II to a discussion of practical approaches to conservation, and then returns in Section III to specific problems in the field and the need to determine priorities among them. We have therefore solicited summary reviews of species status and conservation requirements for the four principal biogeographic regions of the world for primates. These are personal assessments and less comprehensive than the "Action Plans" which are periodically revised by the Primate Specialist Group of IUCN's Species Survival Commission, but they serve to show the value of a systematic approach to determining conservation priorities.

Priorities and progress in the Neotropics are reviewed by Mittermeier, who points out that the region has over half of all the world's tropical forest and fully one third of its primate species. This region has also the dubious distinction of having the most species currently recognized as threatened. Oates, Gartlan, and Struhsaker present a paper on Africa, which has lost proportionately more of its forest than tropical Asia or the Neotropics and has the highest rate of population growth. Although much of the Zaire basin is expected to remain little-disturbed until well into the next century, forests and several primate taxa in West and East Africa are more immediately in danger. Madagascar is treated as a separate region from Africa because of its extraordinary diversity of primates, almost all of which are endemic and threatened. Conservation needs for this troubled nation subcontinent are surveyed by Richard and Sussman. Tropical Asia, reviewed by Marsh, has the highest human population density and is today witnessing the most dramatic rate of forest loss, due principally to shifting cultivation and timber extraction. Each continental region thus has distinctive features and a claim on the limited pool of resources and expertise available for species conservation.

Among the four regions, in our view, Madagascar presents indisputably the most urgent challenge to prevent major losses of diversity among primates, followed by the Atlantic forest region of eastern Brazil, which is a subregion of the Neotropics. Within regions and countries, decisions on what *can* be done must inevitably be influenced by pragmatic considerations of where prospects for funding and political support look best. Such decisions, however, should be viewed against increasingly objective criteria for determining what *needs* to be done. Achieving consensus on regional and national priorities is often a difficult task for specialists, but it becomes essential when project proposals must compete for scarce resources. We believe that the IUCN Primate Specialist Group has in recent years become an increasingly effective mechanism for providing balanced advice to funding agencies and governments.

As editors of this book, we have had difficulty in deciding where to limit its scope when passing from technical towards social and political aspects of the subject matter. Our judgement has been to confine the book to scientific and field-related aspects of conservation, and to refrain from including chapters on the broader political and socio-economic constraints to rain forest conservation, although we are well aware of the importance of action at this level.

Since almost all tropical forests occur in developing countries, problems of their conservation must be viewed in the wider context of the economic and social development of the nations involved, and of growth in the human population. This realization invites a review of the roles played by governments, business interests, international aid and loan agencies, non-governmental organizations and even the life styles of ordinary citizens. The field is clearly huge and has been addressed in general terms in the World Conservation Strategy [IUCN-UNEP-WWF, 1980], and its national derivatives, or, in more polemic fashion, by authors such as Myers [1979] and Caufield [1985].

Although the politics of conservation must remain beyond the scope of this book, the role of the field primatologist in the wider scheme of things seems worth stressing. Field biologists, by observing plants and animals directly, collect the raw data on which assessments of conservation problems must be made. They are also sometimes in a position to assist in making and implementing conservation plans and training staff. Thus, in an obvious sense, they are at the leading edge of conservation. But the field biologist is often also in a privileged position to understand why people hunt animals to extinction, cut timber, or clear land in areas which are patently unsuitable for cultivation, because he or she sees these things at first hand. Emotional reactions to this often begin with anger and disdain for "stupid" or "greedy" people, who, in addition, are usually of a very different culture from the scientist, even if he is working within his own country. In consequence, the visitor may feel little sympathy. However, the man or woman who stays longer and looks harder at the situation will usually come away with a deeper, if sadder, understanding of the simple necessities and desires of life which drive us all to abuse our environment in one way or another, and hence of the inevitability of so much of what is changing in the natural world. This is not to suggest that the pursuit of conservation objectives in tropical forests is futile—far from it. The protection of primate species within adequate areas of habitat in the wild is still eminently attainable in nearly all cases. To be an effective catalyst for action, however, the field biologist must not only master his technical subject, but also must come to understand the people, the language, and the cultures involved. This is the real challenge of conservation biology in the field.

In concluding, we would like to record our special thanks to Steve Gartlan, who contributed much to the early stages of planning this book, and to Mike Kavanagh for his advice and encouragement to complete it. The editors would also like to thank the Sabah Foundation and World Wildlife Fund—U.S., without whose support this book would never have been possible.

Clive W. Marsh
Russell A. Mittermeier

REFERENCES

Caldecott JO (1986): "Hunting and Wildlife Management in Sarawak." Kuala Lumpur: World Wildlife Fund Malaysia.

Caufield C (1985): "In the Rainforest." London: Heinemann.

IUCN-UNEP-WWF (1980): "World Conservation Strategy." Gland, Switzerland: IUCN.

Mack D, Mittermeier RA (eds) (1984): "The International Primate Trade, Volume 1, Legislation, Trade and Captive Breeding." Washington, DC: TRAFFIC (U.S.A.), WWF-US and IUCN.

Myers N (1979): "The Sinking Ark.": Oxford: Pergamon.

Rainier III Prince of Monaco, Bourne GH (eds) (1977): "Primate Conservation." New York: Van Nostrand-Reinhold.

Simon JL (1986): Disappearing species, deforestation and data. New Scientist 1508:60–63.

Soulé ME, Wilcox BA (eds) (1980): "Conservation Biology." Sunderland, Massachusetts: Sinauer.

SECTION I: PROBLEMS

Primate Conservation in the Tropical Rain Forest, pages 3–22

1

Trends in the Destruction of Rain Forests

Norman Myers

Consultant in Environment and Development, Oxford OX3 8SZ, United Kingdom

Tropical forests, notably rain forests, but also some less wet forests, are undergoing depletion, both quantitative and qualitative, at progressive rates. If present land-use trends and exploitation patterns persist (and they are likely to accelerate), large sectors of tropical forests will become markedly modified, if not fundamentally transformed or eliminated outright, within the next three to five decades. In some parts of the biome, many forests will have been reduced to degraded remnants, if not destroyed, as soon as the end of this century. This is all the more regrettable in that tropical forests constitute the richest ecosystems on Earth, containing, within their extent of only about 8 percent of the planet's land surface, some 40–50 percent of Earth's estimated 5–10 million species. Yet the biome is less known than any other; we have hitherto identified only an estimated one in six of the forests' species. Of course certain categories of species, such as primates, are fairly well known. Loss of their forest habitats could well prove to be the predominant factor in the decline of tropical forest primates.

INTRODUCTION

Much has been asserted in recent years about the impending demise of tropical forests. All authoritative observers agree that a pattern of disruption and destruction is overtaking the biome, resulting, on the one hand, in complete and permanent elimination of extensive tracts of forest and, on the other hand, in significant disturbance through degradation and impoverishment of forest ecosystems to a degree that will not permit recovery to complete primary status within less than several decades, in some cases even centuries.

The key question is how quickly are these depletive processes affecting remaining forests and is this harmful impact of a primarily qualitative or quantitative nature? In other words, is the principal injury to tropical forest ecosystems of a sort that stops short of outright deforestation, i.e., imposes considerable injury to forest ecosystems while leaving some trees standing? Or, is it more a case of total elimination of forest cover with the land being given over to an entirely different purpose?

It is of central importance to draw a clear distinction between these two forms of depletion. Much of the confusion in recent years about supposed discrepancies between different estimates of depletion rates has arisen through misunderstandings concerning the nature of the depletion—or of "conversion," to use a catch-all term that stands for all types of forest depletion ranging from marked modification to fundamental transformation to outright destruction. For the purposes of this paper, the first set of processes, viz. major disturbance, is considered to result in qualitative depletion leading to pronounced (as opposed to marginal) impoverishment of forest ecosystems. The second set of processes is viewed as deforestation, i.e., complete and permanent removal of all forest cover.

Plainly, both these impacts can have deep-seated repercussions for primate communities, species, and populations. The first set of processes can precipitate a decline in carrying capacity for many species but also have little or no effect for other species and even some increase in carrying capacity for still other species. By contrast, the second set of processes obviously results in an absolute end to primate habitat.

RAIN FORESTS AND MOIST FORESTS

Before addressing the central question of this chapter, viz. the actual rates of depletion, let us consider the nature of the forests at issue. In other words, let us define our use of the term *rain forest* within a context of other forest formations in the humid tropics. As strictly understood by certain ecologists and biogeographers [for example, Brünig, 1977; Golley, 1983; Holdridge, 1967; Prance, 1982], *rain forest* properly refers to forest in areas that generally receive 3,000 mm of annual rainfall (and have a mean annual temperature of 24°C), which, provided that it is spread more or less evenly around the year with no significant dry season, and supposing that other conditions such as soils and nutrient stocks are suitable, leads to evergreen wet forest. In areas receiving less than 3,000 mm but more than 1,500 mm of annual rainfall, but not less than 100 mm of rainfall in any month for 2 of 3 years, we find *moist forest*, a form of forest with the typically luxuriant vegetation of tropical forest as generally understood and with several more or less distinct strata to the forest structure (though with less biomass and less abundance and diversity of species than rain forest).

Were we to accept these criteria, we should find ourselves constrained, for purposes of this chapter, which is ostensibly limited to rain forest, by the fact that much of the zone that is generally accounted "tropical forest" in Asia would fall outside our definition. In fact, as soon as we move a little north of the Malaysia/Thailand border to the Kra Isthmus, located at about 11°N, we should find ourselves in the subzone of seasonal (monsoonal) forest, with only about 200 genera of plants in contrast with 375 genera to the south [van Steenis, 1950]. Moving farther north and northwestward from the subzone of seasonal forest, we should find that semievergreen forest gives way to semideciduous and deciduous forest. In Africa, moreover, we should find next to no rain forest in the strict sense as postulated by, e.g., the Holdridge life-zone system [Holdridge, 1967]. In Latin America, however, we should find that there is more rain forest than any other type of forest.

This chapter, as this book, is intended to review the situation with regard to tropical forests (in the sense of closed-canopy forests with exceptional concentrations of species) throughout the humid tropics, so the writer presumes to adopt a definition of tropical forest that is sometimes known as "moist forest," i.e., with a minimum of 1,500 mm of annual precipitation and no drought beyond a few months. In short, the type of forest we are dealing with is that type, or rather types, of forest that constitutes an integrative community of intrinsically (though not necessarily) broadleaved evergreen hygrophilous trees, generally with several strata of trees and shrubs, and a good number of other plants such as vines and epiphytes. We generally find these forests at altitudes below 1,300 m, though they are sometimes higher in Amazonia and generally lower in Southeast Asia. Above this altitudinal cut-off line, the luxuriance and variety of the forest, notably its abundance and diversity of species, gives way to relatively depauperate form that can be collectively described as *montane forest*. Although montane forests have their own worthy characteristics, notably their unique species assemblies and vegetation formations, and although they are undergoing rapid depletion in several sectors of the biome, they nevertheless constitute only a very small fraction of the total extent of the biome. For the sake of brevity and succinctness, we shall not deal further with montane forests here.

DEPLETION SURVEYS

During the past 5 years, two surveys have been conducted of tropical forest depletion rates, both of a more comprehensive and systematized kind than earlier surveys. They are, first, the one conducted by Norman Myers, for the National Academy of Sciences in Washington, DC, the findings published by the National Research Council under the title *Conversion of Tropical Moist Forests* [Myers, 1980], and, second, the one conducted by a team of experts at the Food and Agriculture Organization and the United Nations Environment Programme, led by Jean-Paul Lanly. The results are published in three volumes under the title *Tropical Forest Resources* [Lanly, 1982].

The main feature that distinguishes these two reports from earlier survey reports lies with their good fortune in being able to call on the results of remote-finding technologies, notably in the form of satellite imagery through Landsats though also through aerial photography with conventional aircraft, side-looking radar, and other techniques. In the case of the first survey cited above, the principal investigator was able to call on remote-sensing findings for almost 45% of the biome and, in the case of the second survey, approximately 58% of the biome. Remote-sensing findings are particularly valuable insofar as they usually yield methodical and objective data concerning human intrusions into natural vegetation formations.

The Myers report has come up with a "bottom line" conclusion to the effect that approximately 200,000 km^2 of tropical forest are being "converted" from its original status each year, i.e., are undergoing depletion of various kinds whether quantitative or qualitative. Although he did not possess substantive data to offer a conclusive assessment, Myers hazarded a best-judgement appraisal to the effect that about half of this total figure for depletion can be characterized as outright destruction, i.e, deforestation, and the other half can be characterized as significant disruption and degradation. The Lanly report concludes that some 76,000 km^2 of tropical forest are being eliminated each year, i.e., deforested. Because the Lanly report is principally interested in commercial timber values rather than bio-ecological and conservation issues it looks only at outright deforestation. The Myers report, by contrast, having been required by the National Academy of Sciences to look at both sets of depletive processes, gives as much attention to degradation as to destruction.

Subsequent to the release of the Lanly report, a number of observers have suggested that there is an apparently gross discrepancy between the figure of 200,000 km^2 of depletion and the figure of 76,000 km^2 of deforestation per year. As we have seen, however, this supposed discrepancy has arisen through a confusion of definitions and criteria. To clarify the situation, Myers, in conjunction with colleagues at the Ecosystems Center in Woods Hole, Massachusetts, has engaged in a fine-grain analysis of the two reports, concentrating on deforestation, viz. outright destruction, as manifested in the predominant form of forest in question, viz. tropical moist lowland forest. They have found that, according to calculations of the first report, Myers reveals a deforestation rate for this forest category of 91,000 km^2 per year, and, in the case of the second report, Lanly reveals 71,000 km^2 [Houghton et al., 1984; Melillo et al., 1983]. Myers and his colleagues have concluded that depletion in the sense of disruption accounts for about 100,000 km^2 of forest per year, meaning that the overall total for depletion, both quantitative and qualitative, is slightly less than the 200,000 km^2 per year originally postulated by Myers [1980]. At the same time, Lanly has joined with one of his UNESCO colleagues to come up with a parallel, though much less detailed analysis: they agree that there is no significant divergence between the two reports [Hadley and Lanly, 1983].

These follow-up publications by Myers and Lanly and their colleagues should help to clear away the confusion once and for all. When we engage in an exercise of "comparing apples with apples," the difference between the two sets of findings amounts to only about 25%. Moreover, this marginal difference can be accounted for by what we can justifiably call a "cautious and conservative" approach that has been necessarily adopted by the two U.N. agencies in question insofar as they are obliged to accept data supplied by member-state governments. In the case of Indonesia, for example, the official figure cited for forest cover is 1.2 million km^2—a figure that the Indonesian government has been promulgating for years and indeed a figure that the Indonesians obtained from the former Dutch colonialists. In 1978, the Indonesian government requested a group of forest inventory experts from Canada to come and check out their forest cover. The experts drew on findings from remote-sensing technology and other systematized methods, concluding that Indonesia possesses about one-third less "real" forest than the government believes, a full 400,000 km^2 less. Yet, for reasons of its own, the Indonesian government prefers to stick with its former figure—and the FAO/UNEP report, being beholden to the Indonesian government, agrees that the 1.2 million figure should stand. The report does not even make reference to the experts' survey and its findings [Oudschoorn, 1979], presumably because the Indonesian government has declared it "sensitive information."

A number of other countries are represented in the FAO/UNEP report as possessing more forest than actually exists on the ground. In Bangladesh, the Madhupur tract of forest in the middle of the country is reported by government officials to cover 337 km^2. However, on-the-ground observers, including field foresters and technical experts, agree that a more realistic figure is only 91 km^2 in total and, when the remnant scraps of undisturbed forest are documented, only 28 km^2 [Green, 1981]. In the neighboring Himalaya foothills of India, where forest cover has recently been estimated at about one-fifth of land area (minimum desirable is one-third), remote-sensing surveys reveal as little as one-seventh in some areas and only one-tenth in the rest [Myers, 1984]. In Burma, the Arakan sector, although officially classified as "forest," has been reduced by forest farmers to bamboo vegetation with only a few scattered trees. In Honduras, the actual amount of forest surviving is reported to be less than half as much as the government asserts.

DIFFERENTIATED DEPLETION RATES

Although the biome is undergoing depletion in many extensive tracts, depletion trends and patterns overall are highly differentiated. Certain sectors of the biome are experiencing widespread depletion at rapid rates, other sectors are undergoing moderate depletion at intermediate rates, and still other areas are encountering little change. Thus depletion of forests is not a phenomenon common to all parts of the biome.

In brief summary, we can say that virtually all lowland forests of the Philippines and Peninsular Malaysia (aside from a few protected areas) seem likely to become logged over by 1990 at the latest, possibly earlier. Much the same applies to most parts of West Africa. Little could remain of Central America's forests within another 10 years, probably less. Almost all of Indonesia's lowland forests have been scheduled for timber exploitation by the year 2000, and at least half by 1990. Extensive areas of Amazonia in Colombia and Peru could be claimed for cattle ranching and various forms of cultivator settlement by the end of the century; something similar is true for much of the southern and eastern sectors of Brazilian Amazonia. In marked contrast, Central Africa, viz. Zaire, Congo, and Gabon, is sparsely inhabited and possesses abundant mineral resources. This means that these countries sense little incentive to liquidate their forest capital in order to generate funding for various forms of economic development. So there could well remain large expanses of little-disturbed forest in the Zaire Basin until the end of the century, conceivably far beyond. Similarly, the western portion of Brazilian Amazonia, because of its remoteness and perhumid climate, could undergo only moderate change for at least 2 decades, possibly longer. In short, the overall outcome is likely to be extremely differentiated in terms of both geographic area and degree of depletion.

It would represent too much of a digression to give a detailed account here of all those areas that are undergoing broadscale depletion at rapid rates. Details are set out in the Appendix, which likewise presents data on areas experiencing moderate depletion at intermediate rates and areas apparently undergoing little change. Suffice it to say here that, in sum, the situation is patchy to a considerable degree both between and within the three main regions and within several individual countries.

The overall situation can be clarified for us if we note that only a few of the more than 50 countries with tropical forests constitute most of the biome. The following possess at least 100,000 km^2 of forest cover, i.e., well over 1% of the biome: Cameroon, Congo, Gabon, and Zaire in Africa; Bolivia, Brazil, Colombia, Ecuador, Guyana, Peru, and Venezuela in Latin America; and Burma, Indonesia, Malaysia, Papua New Guinea, the Philippines, and Thailand in Southeast Asia. These 17 countries contain at least three-quarters of the biome. Let us further note that an additional four countries possess not insignificant tracts of forest, namely, French Guiana, Suriname, Ivory Coast, and Madagascar. If we are to focus on countries that possess sizeable amounts, i.e., 100,000 km^2 or more, of true rain forest, we can confine our attentions to just 10 countries, these being Brazil, Colombia, Ecuador, Gabon, Guyana, Indonesia, Malaysia, Peru, Venezuela, and Zaire: These countries account for some 80–85% of true rain forest in the entire biome.

THREE KEY NATIONS: BRAZIL, INDONESIA, AND ZAIRE

Very significantly for conservation interests, three countries contain more than half of all tropical forests. These are Zaire with almost 1 million km^2, Indonesia with about 800,000 km^2, and Brazil with 2,600,000 km^2, making a total of 4,400,000 km^2. This means that if these three countries do a good job of rational development and conservation of their forests, they will safeguard a very large expanse of forest. Conversely, of course, if these three countries engage in broadscale felling and disturbance of their forests, the outlook will be all the bleaker.

Their record is decidedly mixed, though it is now becoming a good deal better than it was only half a dozen years ago. Zaire, as we have seen, suffers little population pressure, with only 33 million people in a territory of 2.3 million km^2, and a remarkedly high rate of urbanization for a country of black Africa—30%. In addition, the population growth rate of Zaire is "only" 2.9% a year (a good deal less than the average for black Africa), and the total when zero growth rate is reached around the year 2100 is projected to be "only" 156 million. Given its mineral wealth, Zaire might perceive little incentive to "cash in" its forest resources for a good while to come.

By contrast, Indonesia is depleting its forests quite rapidly [Hansen, 1981; Myers, 1984; Ross, 1980]. The government believes that it makes sense to press on with disruptive logging programs, which are being pursued on a scale that could leave little undisturbed forest in Sumatra and Sulawesi by 1990, in Kalimantan (Indonesia's sector of Borneo) and most smaller islands by 1995, and in Irian Jaya (western New Guinea) by the year 2000. During the recent slump in oil prices, and the fall from absolute power of OPEC, Indonesia, dependent on major foreign exchange earnings to pay for its imported food, proclaims that it has "little alternative" to converting its forest resource stocks into liquid capital even if that means that the forests are harvested on far less than a renewable basis.

At the same time, Indonesia has embarked on a vigorous transmigration program in order to relieve exceptional population pressures in Java, Bali, and Madura. After all, the government argues, the average population density in Java is 700 persons per km^2, in contrast with Kalimantan (fewer than 10) and Irian Jaya (only 2). The transmigration effort aims to resettle 2.5 million people during the period 1980–1984, at a cost of $2.5 billion—even though, during the same 5 years, Java will add about 9 million people to its populace. The eventual aim is to translocate 15 million people, occupying 100,000 km^2 of land in the "outer islands" of Borneo, Sumatra, and Sulawesi, virtually all the land in question being forested. Because the cost amounts to more than $1,000 per family, one might well ask whether the same money could not be better spent in upgrading small holder agriculture back in Java, Madura, and Bali.

As for Brazil, the country seems to have abandoned the early-1970s campaign to populate the "demographic void" of Amazonia [Moran, 1982; Myers, 1984; Smith, 1982]. The settlement schemes for which the Trans-Amazon Highway system was being built have been scaled down pronouncedly, or ended altogether, and the government no longer offers much support for cattle-raising activities. However, some sizable logging schemes are being formulated for Amazonia, and the government sees little cause to manage or direct the "spontaneous colonization" of Rondônia and several fringe sectors of southeastern Amazonia—activities which, with their attritive impact, could steadily lead to early loss of substantial sectors of Amazonia [Fearnside, 1986].

Despite this patchy record, let us bear in mind that all three countries have taken some decisive steps in recent years to expand their networks of parks and reserves in forest territories. Indonesia aims to bring at least 10% and possibly 12% of its forest lands under protected status by the end of the 1980s, if not sooner; Zaire talks about 15%. Brazil is not quite so ambitious, but it has accepted the principle of a strategically located network of protected areas, in accord with the theories of Pleistocene refugia and island biogeography [Wetterberg et al., 1976; Prance, 1982; see also McNeely et al., this volume].

AGENTS OF DEPLETION

There are four main agents of depletion at work [Myers, 1980, 1984]. The cattle raiser, confined to Latin America, is causing deforestation of at least 20,000 km^2 per year. The fuelwood gatherer, significant in parts of Africa and very significant in much of Asia, is probably causing the destruction of 25,000 km^2 of forest each year, but his impact is less consequential in the broad picture insofar as most logged forests are soon occupied by small scale cultivators and other farmers, these being the persons who are causing by far the greatest damage across the board—and these also being the main persons who (with the partial exception of the fuelwood gatherer) cannot readily gain their livelihood in alternative ways. True, ways are available, through, e.g., a gene revolution, to make the forestland farmer's agriculture more productive and stable, more intensive rather than extensive, more efficient rather than wasteful, but this is all another story [Myers, 1983, 1984]. Meanwhile, let us note that, in his various guises of traditional shifting cultivator, migrant squatter, and landless "colonizer," all being life-styles that entail slash-and-burn cultivation, the forest farmer can be reckoned to be depleting 160,000 km^2 of forest per year, possibly a good deal more.

To emphasize a central factor, these calculations represent no more than informed estimates. They are certainly not presented as a concise account of what is actually occurring in tropical forests. A great range of reliability must be attached to the statistical details. Indeed many quantified statements should be qualified with phrases such as "so far as can be ascertained . . . " or "best-judgment assessments suggest that . . . ." In other words, this writer tries on the one hand to describe semidocumented situations with an appropriate degree of "precise imprecision." To put it yet another way, the bottom-line reckoning should be amplified to read, "to the extent that we can establish the situation, on the basis of limited appraisals and with a sometimes less-than-satisfactory data base of variable reliability for parts of the biome, tropical forests appear to be undergoing depletion at a significant rate that can be put, for want of any better assessment, at a figure approaching 200,000 km^2 per year."

So much for the main agents of depletion. In addition, there are several other agents, notably the plantation operator, who replaces original forest with monocultures of tree species such as rubber, oil palm, cocoa, and tea, or food crops such as sugar and bananas. According to the World Bank [Spears, 1983], plantations can be estimated to cover 250,000 km^2 of formerly forested land, many of them having been in continuous production for over 50 years—which means that this is a highly sustainable form of forest land use and that only a relatively small amount of forest is converted to plantation agriculture each year. Moreover, plantations are productive in other ways, on the grounds that they offer employment and livelihood to some 50 million small-holders and agricultural workers at income levels well above the average. Were a further 20,000 km^2 of forest to be converted to perennial tree crop plantations between now and the year 2000, they could provide a sustainable livelihood for at least 20 million people, or one in 10 of the 200 million people now pursuing a subsistence-level and extensive form of agriculture, i.e., disruptive agriculture in tropical forests.

At the same time, however, these plantations sometimes cater to marketplace demand in rich nations of the temperate world, supplying ostensibly "cheap" products, some of them less than necessary. An example is cassava, now grown in miniplantations in Thailand for the European Common Market [Myers, 1984]. The main manufacturers of livestock feed in West Germany, the Netherlands, and Belgium started to search in the late 1960s for a calorie-rich crop that could be made available at prices that would help to counter inflationary spirals in the livestock industry in Western Europe. They chose cassava as a sound candidate. Today one-tenth of all cassava grown in the tropics finds its way to the European Common Market, virtually all of it coming from Thailand. In 1973, the Common Market imported 1.5 million tons of cassava from Thailand, an amount that soared by 1982 to almost 6 million tons. Cassava growers in Thailand establish their plots, often on plantation scale (it is a booming commercial crop), at the expense of natural forest, through a type of "shifting cassava cultivation."

We can make similar comments about bananas. The giant food corporations from the United States that originally fostered the banana republics of Central America are now engaged in "bananaization" of sizable forests in the Philippines in order to meet soaring demand for bananas from Japan [Third World Studies Center of the Philippines, 1981].

These instances parallel the "hamburger connection," a phenomenon stimulated by demand from inflation-fearing beef-eaters in North America. Hamburger corporations, together with other enterprises in the fast-food industry, believe that they can find a solution to their problemns of ever-growing prices for beef by persuading the U.S. government to increase imports of so-called reasonably priced beef from Central America. The consequence is an outburst of ranchlands, almost entirely established at the expense of natural forests [Myers, 1981].

FOREST FARMERS: FUTURE TRENDS

Because the forest farmer is by far the most important source of depletion of tropical forests, let us briefly consider the prospect for the future. Today approximately 200 million people gain their subsistence through this type of agriculture. (According to some experts [e.g., Denevan, 1982; UNESCO, 1978; Watters, 1971] the total number of forest farmers could have reached 200 million as early as 1970 and could now be approaching 300 million—although the FAO/UNEP/Report [Lanly, 1982] appears to propose a total of well under 100 million, albeit a total documented in surprisingly inconsistent detail.) The greatest loss to forest farmers is currently thought to be occurring in Southeast Asia, where cultivators clear a minimum of 85,000 km^2 per year, adding to the 1.2 million km^2 of formerly forested croplands in the region [Chandrasekharan, 1978; Kartawinata et al., 1981; Myers, 1980, 1984]. Africa south of the Sahara is believed to have lost 1 million km^2 of forest to these cultivators even before modern development started to gain momentum after World War II; by the mid-1970s the region's loss has been estimated to have reached 40,000 km^2 per year in a zone of 400,000 km^2 of forest lands under that style of agriculture [Braun, 1974; Hauck, 1974; Persson, 1977]. Latin America is likewise believed to have been losing forest to small-scale cultivators at a rate of 40,000 km^2 each year by the mid-1970s, though fewer details are available [Denevan, 1982; Watters, 1971].

However, on-the-ground investigations reveal that census figures for forest farmers in several countries are incomplete. For various political reasons, governments prefer to ignore refugees, guerillas, and other insurgents and general "troublemakers." Thus official estimates for the Philippines, Thailand, and half-a-dozen other countries could well be doubled or even tripled to make a realistic figure; whereas in Ivory Coast and several other countries of West Africa official estimates deserve to be expanded by at least one-half to take account of illegal immigrants fleeing from the Sahel disaster [Myers, 1984].

Precisely because of migration patterns on the part of numbers of landless peasants who believe they can find salvation for themselves in the only other "free," i.e., public, lands left available in their countries, the numbers of forest farmers can be expected to continue to grow at a rate far faster than that of national populations. Whereas population growth in the countries concerned is projected to produce an increase of almost 50% during the last 2 decades of this century, the number of forest farmers could well double. As for the longer-term future, none of the major tropical forest countries is projected to reach zero population growth, even with stepped-up family planning programs, until late in the next century if not early in the twenty-second century, due to the "youthfulness" of their present populations. So we must anticipate a future in which the population of the Philippines grows from its 1980 total of 49.1 million to 125 million, of Indonesia from 146.6

million to 388.4 million, of Brazil from 118.7 million to 281 million, and of coastal West Africa (not counting Nigeria, with its 85 million projected to reach 459 million) from 46 million to 263 million. Of the incremental numbers, a solid proportion will, if current land-use trends persist, become forest farmers, to add to the existing multitudes. At the same time, however, let us not forget a hopeful aspect: Zaire, Gabon, and Congo, with a total forest area of 1.2 million km^2, are projected to grow from a 1980 total of about 35 million to an eventual total of no more than [sic] 168 million. For further details see Table I.

REPERCUSSIONS OF DEFORESTATION

As forests disappear, we lose two sets of outputs: material goods and environmental services.

Loss of Material Goods

The material goods include, notably, commercial timber, fuelwood, and genetic resources among other so-called minor forest products. There is not space here to consider all these items. Let us note that export of forest products from the humid tropics currently generates revenues of more than $7 billion, or more than cotton and as much as rubber and coffee combined. These timber earnings could persist in perpetuity, were sustainable logging to be undertaken; regrettably, this is not the case nearly as often as it should be. Furthermore, these revenues could readily be doubled and even tripled if the nations in question engage in increased processing of their lumber, in order to export sawn timber and semifinished or even finished industrial-wood products instead of logs. Thus processing might (albeit theoretically) relieve some to the headlong urge to fell extensive tracts of tropical forests on the grounds that the countries concerned could derive as much foreign exchange through the export of much smaller amounts of timber.

As a measure of how rapidly excessive logging can overtake even well endowed countries, let us note that Thailand, Nigeria, and Peninsular Malaysia have already become net importers of hardwood logs, with strongly adverse impact on their balance of payments. As for genetic resources, the daily welfare of people around the world, and especially of those in advanced nations (these being the nations with the technology to exploit genetic resources), depends increasingly on start-point materials, especially germplasm, from plants and animals of tropical forests. When we visit our drugstore to purchase a medicine or pharmaceutical, there is roughly one chance in four that the preparation we purchase owes its origin, in some manner or other,

TABLE I. Population Projections for Some Tropical Forest Countries[a]

Country	Population in 1980 (millions)	Rural population (%)	Population in 2000 (millions)	Year of net reproduction rate	Year of stationary population	Total of stationary population (millions)
Latin America						
Brazil	118.7	39	176.5	2010–15	2075	281
Colombia	26.7	40	39.4	2005–10	2065	60.2
Costa Rica	2.2	59	3.3	2000–05	2065	4.8
Ecuador	8	58	13.6	2020–25	2085	28.4
Guatemala	7.2	64	12	2020–25	2085	24.2
Guyana	0.8	60	1.2	2000–05	2065	1.8
Honduras	3.7	69	6.7	2025–30	2090	16
Mexico	69.8	36	115	2010–15	2075	202.6
Nicaragua	2.6	51	4.7	2025–30	2090	10.8
Panama	1.8	50	2.8	2005–10	2070	4.3
Peru	17.4	38	27.3	2015–20	2080	48.8
Suriname	0.4	34	0.6	2005–10	2070	0.9
Venezuela	14.9	25	23.8	2005–10	2075	38.6
Southern/Southeast Asia						
Bangladesh	88.5	91	141	2030–35	2125	338.2
Burma	34.8	78	54	2025–30	2095	89.7
India	673.2		994.1	2015–20	2115	1,621.5
Indonesia	146.6	82	216	2015–20	2110	388.4
Malaysia	13.9	73	20.8	2000–10	2070	30
Papua New Guinea	3.1	87	4.2	2030–35	2125	9.4
Philippines	49.1	68	76.9	2010–15	2075	125
Thailand	46.9	87	68	2000–05	2070	102.5
Vietnam	54.2	78	87.9	2010–15	2075	153.4
Tropical Africa						
Cameroon	8.5	71	14.2	2035–40	2110	40.8
Congo	1.5	60	2.7	2035–40	2100	10.5
Gabon	0.7	68	0.9	2035–40	2130	2
Ivory Coast	8.2	68	14.8	2035–40	2110	47.4
Madagascar	8.7	84	16.1	2035–40	2110	51.1
Zaire	28.3	70	51	2035–40	2110	156.1

[a]Source: Vu and Elwan [1982]: "Short-Term Population Projection 1980–2000, and Long-Term Projection 2000 to Stationary Stage, by Age and Sex for All Countries of the World." Washington, DC: The World Bank.

to wild organisms of tropical forests. Cross-counter sales of these diverse items now amount to some $20 billion per year worldwide [Myers, 1983]. Similar figures, albeit on not quite such a scale, can be adduced for agriculture and industry. Moreover, as the burgeoning new sector of genetic engineering expands, and places a premium on ever-growing supplies of genetic variability, there will be yet greater demand for genetic materials from the ecological zone that possesses almost as much as all other natural environments put together, viz., tropical forests [Myers, 1983, 1984]. Yet tropical forests, with between 40% and 50% of the Earth's 5–10 million species, is the zone where the first great waves of extinctions are already occurring and where we can anticipate the major spasms of extinctions within the coming few decades. When we speak of one species becoming extinct per day right now [Myers, 1979], we refer principally to tropical forests, with their species concentrations, often characterized by high levels of endemism, together with ecological specializations that make them unduly susceptible to summary extinction.

This anticipated loss of biological diversity appears likely to occur on a compressed scale unmatched during the course of life's development on the face of the planet during the past 3.6 billion years. This would therefore represent a biological débâcle of unprecedented order. So profound could the implications be that this could affect even the course of evolution itself. In that tropical forest ecosystems are biologically richer and ecologically more complex than any others on Earth, these forests represent, so to speak, a "powerhouse" of evolutionary processes. Were the forests to be grossly impoverished, or removed outright, within the next few decades, this could markedly affect some critical workings of evolution.

Loss of Environmental Services

Equally to the point, the environmental repercussions of broadscale deforestation could be extensive and serious. Familiar phenomena include loss of topsoil with nutrient stocks, disruption of hydrological and watershed systems, and sedimentation of irrigation channels and hydropower facilities, among many other effects. Let us confine ourselves here to a single category of environmental consequences: climatic effects.

Tropical forests exert significant regulatory effects on climate, whether regional or global. As a regional example, let us take Amazonia, where more than half of all moisture circulating through the basin's ecosystems remains within the forest zone, thanks to the forest's evapotranspiration function: Rainwater is absorbed by plants, before being "breathed out" by them again into the atmosphere. Were a large part of the forest to disappear, the remainder could well become less able to retain so much moisture, meaning that it would become a drier forest [Lettau et al., 1979; Salati et al., 1979]. As more forest was cleared, the desiccating process would become more pronounced, reducing the moisture stock of the entire basin—with deep-seated consequences for the entire biota of this, the region that is biologically richer than any other on Earth.

Tropical forests also appear to help stabilize climate at the global level [Dickinson, 1981; Henderson-Sellers, 1981; Pinker et al., 1980; Potter et al., 1975; Sagan et al., 1979]. Their mass of vegetation absorbs much solar radiation. When the forest cover is cleared, there is an increase in the "shininess" of the planet's surface, leading to an albedo effect. As a result of deforestation-caused increase in albedo, there could well be disruption for convection patterns, wind currents, and rainfall in lands far outside the tropics, and one consequence could be less moisture for North America's grain-growing territories.

Furthermore, tropical forests play an important part in the carbon dioxide budget of the global atmosphere [Houghton et al., 1983; Woodwell et al., 1983]. Forests represent one of Earth's main stocks of stored carbon. When the forests are burned by large-scale ranchers and small-scale cultivators, they release considerable quantities of carbon into the atmosphere—with all that that could imply for a greenhouse effect, possibly bringing on ultimately warmer climates for everyone but drier climates for some, making agriculture a changeable affair at best. The amount of carbon released by burning of tropical forests is a subject of much argument. According to the best estimates to date, the net amount could well have reached as much as 2 gigatons per year, or roughly two-fifths of the amount released through combustion of fossil fuels.

CONCLUSION

Thus the decline of tropical forests. The situation represents a travesty, if not a tragedy, of development and conservation. It is all the more regrettable in that there are plenty of sound reasons why we should view tropical forests as a source of many goods and services to support our welfare into the indefinite future. Were we to undertake a comprehensive evaluation of all that tropical forests have to offer us, we would surely find it in our interest to reverse the present tendency toward elimination of these valuable and unique stocks of natural resources.

Suppose we were to conduct an evaluation of this sort, and suppose it were to persuade us to do a better job with tropical forests. We would surely find that a strong case would be included for enhanced conservation of primates. To this extent, primatologists might consider whether they could not make stronger common cause with those who promote the increased processing of specialist hardwoods, the shift from ever more expensive petrochemicals to phytochemicals, and the cost-benefit articulation of environmental trade-offs arising from widespread deforestation. All these measures would assist the survival outlook of tropical forests.

REFERENCES

Braun H (1974): "Shifting Cultivation in Africa. Report of the FAO/SIDA/ARON Regional Seminar of Shifting Cultivation and Soil Conservation in Africa." Rome, Italy: Food and Agriculture Organization.

Brünig EF (1977): The tropical rain forests—A wasted asset or an essential biospheric resource? Ambio 6(4):187–191.

Chandrasekharan C (ed) (1978): Shifting cultivation. Forest News 2(2):1–25.

Denevan WM (1982): Deforestation in Latin America (Paper prepared for Office of Technology Assessment, Washington, DC). Madison, Wisconsin: Department of Geography.

Dickinson RE (1981): Effects of tropical deforestation on climate. Studies in Third World Societies 14:411–442.

Fearnside PM (1986): "Human Carrying Capacity of The Tropical Rain Forest." New York: Columbia University Press.

Golley FB (ed) (1983): "Tropical Rain Forest Ecosystems." New York: Elsevier Scientific Publishing Company.

Green KM (1981): "Digital Processing of Tropical Forest Habitat in Bangladesh." Washington, DC: National Zoological Park, Smithsonian Institution.

Hadley M, Lanly JP (1983): Tropical forest ecosystems: identifying differences, seeing similarities. Nature and Resources 19(1):2–19.

Hansen GE (ed) (1981): "Agricultural and Rural Development in Indonesia." Boulder, Colorado: Westview Press.

Hauck FW (1974): Shifting cultivation and soil conservation in Africa (Soils Bulletin No. 24). Rome, Italy: Food and Agriculture Organization.

Henderson-Sellers A (1981): The effects of land clearance and agricultural practices on climate. Studies in Third World Societies 14:443–486.

Holdridge LR (1967): "Life Zone Ecology" (revised edition)." San Jose, Costa Rica: Tropical Science Center.

Houghton RA, et al. (1983): Changes in the carbon content of terrestrial biota and soils between 1860 and 1980: A net release of CO_2 to the atmosphere. Ecological Monographs 53:235–326.

Houghton RA, et al. (1984): Current estimates of deforestation in tropical forest: A synthesis. Nature 316:617–620.

Kartawinata K, et al. (1981): The impact of man on a tropical forest in Indonesia. Ambio 10(2-3):115–119.

Lanly J-P (1982): "Tropical Forest Resources." Rome, Italy: Food and Agriculture Organization (in conjunction with Nairobi, Kenya: United Nations Environmental Programme).

Lettau H, et al. (1979): Amazonia's hydrologic cycle and the role of atmosphere recycling in assessing deforestation effects. Monthly Weather Review 107:227–238.

Melillo J, et al. (1983): A comparison of recent estimates of disturbance in tropical forests. Environmental Conservation 12:37–40.
Moran EF (ed) (1982): "The Dilemma of Amazonian Development." Boulder, Colorado: Westview Press.
Myers N (1979): "The Sinking Ark." Oxford, U.K.: Pergamon Press.
Myers N (1980): "Conversion of Tropical Moist Forests." Washington DC: National Research Council.
Myers N (1981): The hamburger connection: How Central America's forests become North America's hamburgers. Ambio 10(1):3–8.
Myers N (1983): "A Wealth of Wild Species." Boulder, Colorado:Westview Press.
Myers N (1984): "The Primary Source: Tropical Forests and the Earth's Future." New York: WW Norton.
Oudschoorn W (1979): "Environmental Control and Forest Development: Report of Government of Indonesia and Canadian International Development Agency." Vancouver, Forestal International Ltd.
Persson R (1977): "Forest Resources of Africa, Part II: Regional Analysis" (Department of Forest survey, research Notes No. 22). Stockholm, Sweden: Royal College of Forestry.
Pinker RT, et al. (1980): The albedo of a tropical evergreen forest. Meteorological Soc 106:551–558.
Potter GL et al. (1975): Possible climatic impact of tropical deforestation. Nature 258:697–698.
Prance GT (ed) (1982): "Biological Diversification in the Tropics." New York: Columbia University Press.
Ross MR (1980): The role of land clearing in Indonesia's transmigration program. Bulletin of Indonesian Economic Studies 16(1):75–87.
Sagan C, et al. (1979): Athropogenic albedo changes and the earth's climate. Science 206:1363–1368.
Salati E, et al. (1979): Recycling of water in the Amazon basin: An isotopic study. Water Resources Research 15(5): 1250–1258.
Smith NJH (1982): "Rainforest Corridors: The Trans-Amazon Colonization Scheme." Berkeley, California: University of California Press.
Spears JS (1983): "Saving the Tropical Forest Eco-System. Discussion Paper." Washington DC: The World Bank.
Third World Studies Center of The Philippines (1981): TNC control of the Philippines banana industry. AMPO Japan—Asia Quarterly Review 13(3):2–61.
UNESCO (1978): "Tropical Forest Ecosystems." Paris, France: UNESCO.
van Steenis OG (1950): The delimitation of Malaysia and its main plant geographical divisions. Flora Malesiani Ser I(1): 70–75.
Watters RF (1971): "Shifting Cultivation in Latin America." Rome, Italy: Food and Agriculture Organization.
Wetterberg GA, et al. (1976): "An Analysis of Nature Conservation Priorities in the Amazon" (NDP/FAO/IBDF/BRA 545, Technical Series No. 8). Brasilia, Brazil: Instituto Brasileiro de Desenvolvimento Florestal.
Woodwell GM, et al. (1983): The earth's biotas and the global carbon cycle (tentative title). Science, in press.

APPENDIX: COUNTRY-BY-COUNTRY REVIEW OF DEPLETION RATES IN TROPICAL FORESTS

This summary review is taken from the 1980 publication, Conversion of Tropical Moist Forests [Myers, 1980]. The author sees no reason to change any of the details now that we have reached 1986.

Areas Undergoing Broad-Scale Depletion at Rapid Rates

1. Most of Australia's lowland tropical forests, both rain forests and seasonal forests, due to timber exploitation and planned agriculture; could be little left by 1990 if not earlier.

2. Most of Bangladesh's forests, both lowland and upland, predominantly rain forests, due to timber exploitation, forest farming, and population pressure; could be little left by 1990 if not earlier.

3. Most of India's forests, predominantly seasonal forests, mainly upland, due to forest farming and population pressure; could be little left by 1990.

4. Much if not most of Indonesia's lowland forest, predominantly rain forests, due to timber exploitation, forest farming, and transmigration programs; could be little left in Sumatra and Sulawesi by 1990, in Kalimantan and most of the smaller islands by 1995, and in Irian Jaya by the year 2000.

5. Much of Sumatra's and Sabah's lowland forests, almost all rain forests, due to timber exploitation; could be little left by the year 2000 if not earlier.

6. Most of Peninsular Malaysia's lowland forests, almost all rain forests, due to timber exploitation and planned agriculture; could be little left by 1990 if not earlier.

7. Much if not most of Melanesia's lowland forests, due to timber exploitation and planned agriculture; could be little left by 1990.

8. Most of the Philippines' lowland forests, predominantly rain forests, because of timber exploitation and forest farming; could be little left by 1990 if not earlier.

9. Much if not most of Sri Lanka's forests, predominantly rain forests, mostly upland, due to timber exploitation and forest farming; could be little left by 1990.

10. Much if not most of Thailand's forests, almost all seasonal forests, both lowland and upland, due to timber exploitation (especially illegal felling) and forest farming; could be little left by 1990 if not earlier.

11. Much of Vietnam's forests, almost all seasonal forests, both lowlands and upland, especially in the south, due to forest farming, timber exploitation, and immigration from the north; could be little left by 1990.

12. Parts of Brazil's southern and eastern sectors of Amazonia, lowland rain forests, notably in Rondônia, Pará, and Mato Grosso, due to cattle raising and colonist settlement; appreciable tracts could be depleted by 1990 and much more by the year 2000.

13. Most if not virtually all of Brazil's Atlantic coast strip of moist forest, due to timber exploitation and cash-crop agriculture, notably sugarcane plantations; could be little left by 1990 if not a good deal earlier.

14. Much if not most of Central America's forests, notably rain forests, both lowland and upland, due to forest farming, cattle raising, and timber exploitation; could be little left by 1990 if not earlier.

15. Parts of Colombia's lowland rain forests on the borders of Amazonia, especially in Caquetá and Putumayo, due to colonist settlement and cattle raising; extensive tracts could be depleted by 1990 and much more by the year 2000.

16. Much of Ecuador's Pacific coast forests, mostly very wet and very rich rain forests, both lowland and upland, due to plantation agriculture and some timber exploitation; could be almost entirely depleted by 1990.

17. Much if not most of Madagascar's forests, especially rain forests, both lowland and upland, due to forest farming and timber exploitation; could be little left by 1990 if not earlier.

18. Much if not most of East Africa's relict montane forests, especially in northern Tanzania, mostly seasonal forests, due to timber exploitation, firewood cutting, and forest farming; could be little left by 1990.

19. Much if not most of West Africa's forests, mainly seasonal forests, due to timber exploitation and forest farming; could be little left by 1990 if not earlier.

Areas Undergoing Moderate Depletion at Intermediate Rates

These areas cannot be so readily listed as those undergoing broadscale depletion at rapid rates, because less is known about their present status and future prospects. The listing is deliberately conservative.

1. Parts of Burma's lowland forests, almost all seasonal due to forest farming and some timber exploitation; appreciable areas could be depleted by the year 2000 if not earlier.

2. Parts of Papua New Guinea's forests, mostly seasonal, both lowland and upland, due to timber exploitation and forest farming; extensive areas could be depleted by the year 2000 if not earlier.

3. Parts of Brazil's Amazonian forests, lowland rain forests, notably in Amapá, Acre, sections of the Trans-Amazonia Highway system and of the várzea floodplains, and areas selected for timber exploitation, e.g., Tapajós River area, due to colonist settlement, forest farming, cattle raising, and timber exploitation; appreciable tracts could be depleted by 1990.

4. Parts of Colombia's Pacific coast forests, very wet and very rich rain forests, both lowland and upland, due to timber exploitation; extensive sectors could become depleted by 1990.

5. Much of Ecuador's Amazonian forests, almost all rain forests, both lowland and upland, due to colonist settlement, forest farming, some planned

agriculture, and also oil exploitation; appreciable areas could be depleted by 1990 and much more by the year 2000.

6. Much of Peru's Amazonian forests, almost all rain forests, both lowland and upland, due to colonist settlement, forest farming, and some planned agriculture; appreciable areas could be depleted by 1990 and much more by the year 2000.

7. Parts of Cameroon's forests, both seasonal and rain forests, both lowland and upland, due to timber exploitation and forest farming; extensive areas could be depleted by the year 2000.

Areas Apparently Undergoing Little Change

Like the areas undergoing moderate depletion, these areas cannot be so readily listed as those areas rapidly undergoing broad-scale depletion, because less is known about their present status and future prospects. The listing is deliberately conservative, especially as concerns the long term.

1. Much of Brazil's western Amazonian lowland rain forests, generally wetter and richer than eastern Amazonia; except for some timber extraction in limited areas, and some cultivation of várzea floodplains, exploitation of this huge zone could prove difficult in view of its unusually wet climate and distance from markets; it is reasonable to anticipate—so far as can be ascertained, and the point is stressed—that much of this vast tract of lowland rain forest could remain little changed for some time, possibly until the year 2000, even for a time thereafter.

2. Much of the forests of French Guiana, Guyana, and Suriname, almost all rain forests, both lowland and upland; timber exploitation, at present very limited, might expand, but, because population pressures are low, there is little likelihood of widespread colonist settlement and forest farming. It is reasonable to anticipate—with caveat as for Brazil above—that large areas may remain little changed for a good while to come, possibly until the year 2000, conceivably for a time thereafter.

3. Much of the Zaire basin, comprising Congo, Gabon, and Zaire; some rain forest in Gabon, remainder mainly seasonal almost entirely lowland; population pressures are low, and there are abundant mineral resources on which to base national economic development; timber exploitation, primarily limited to northern Congo and to Gabon, could expand, but, in the main, it is reasonable to anticipate—with caveat as for Brazil above—that large areas will remain little changed for a good while to come, possibly until the year 2000, even for a time thereafter.

Primate Conservation in the Tropical Rain Forest, pages 23–62

2

Methods of Surveying and Sampling Forest Primate Populations

Warren Y. Brockelman and Rauf Ali

Center for Wildlife Research, Department of Biology, Faculty of Science, Mahidol University, Bangkok 10400, Thailand

INTRODUCTION

In this chapter, we discuss the objectives and general methods for estimating the distribution and abundance of primates in tropical forest habitats. The literature on sampling forest animals is large, and we will not attempt to review it but we will select only those methods that we believe are presently most useful and defensible and discuss problems that we believe deserve the most emphasis at present or that have not been previously resolvable. Forest primates are difficult animals to sample for a number of obvious reasons, and there are still outstanding problems in estimating their numbers even crudely. The subject will require updating periodically as new experiences are gained and new needs are formulated.

The chapter is divided into three sections: distributional surveys, population sampling, and intensive study areas. The activities have distinct objectives but reinforce each other: No one should attempt a distributional survey without becoming intimately familiar with the habitat type, troop structure, typical densities, and vocalizations of the animals. No one should attempt to sample densities without understanding these aspects plus dispersion patterns of individuals and groups, activity patterns, reactions to humans, average calling frequencies, etc. Thus, although it seems logical to begin a primate field study with a broad survey, in practical terms it is advisable to begin with an intensive study of a year's duration or more or at least work with a more experienced person for several months to become familiar with the species.

DISTRIBUTIONAL SURVEY

Objectives

The importance of data on the distribution of primates has been emphasized by Heltne and Thorington [1976] and others. The importance of publishing reliable and detailed data on distribution has not been fully appreciated by many primate field workers, and it serves several major objectives.

Conservation. Extensive data on distribution perhaps accompanied by crude but objectively defined relative estimates of abundance may be more important to the evaluation of the conservation status of a species than an estimate of overall numbers alone or density estimates for a few areas. The survival prospects of a species are highly dependent on the number of areas containing self-sustaining populations and on trends in the numbers of areas remaining. Trends in distribution and abundance can be assessed only if exact location data are published along with time of survey and the methods used. Normally, species populations are not reduced uniformly everywhere by hunting, trapping, or habitat destruction; the occupied range tends to be "rolled back" along frontiers of destructive activity surrounding more undisturbed enclaves. Reliable data on distribution and relative abundance are sufficient to stimulate conservation action in the form of creating new protected areas or improving protection of existing ones if adequate publicity is created.

Taxonomy and evolution. Primates are a rich and still mostly untapped source of data relevant to the problem of how genetic differentiation and speciation occurs. Although some evolutionary biologists have proposed theories of primate speciation that involve macrogenetic changes in sympatry [Bush, 1975; White, 1978; Cronin et al., 1984; Shafer et al., 1984], very closely related forms of primates are virtually all allopatric or parapatric, which supports the more widely accepted theory of allopatric speciation. Detailed studies of contact zones or hybrid zones between closely related forms are urgently needed. Species or subspecies in parapatric contact, in many cases with hybrids, occur in most primate genera. Protection of these areas is extremely important. Research in systematics is of direct importance to conservation. We need to answer the questions: What are we conserving? How many forms do we need to conserve?

Ecology and zoogeography. These subjects overlap the first two, but ask additional questions about distributions: What are the species' range limits in terms of geographical barriers and ecological tolerances? What relation do species distributions have to the inferred process of speciation? Extirpation of species over broad areas, and lack of reliable historical records, will make it very difficult to reconstruct the original ranges of most species. For example, the ranges and habitats of hoolock, concolor, and lar gibbons (each having several subspecies) in much of Burma, Thailand, Laos, Viet Nam, and China cannot now be ascertained because of forest destruction and political restrictions. We have little knowledge concerning the extent to which gibbons occupied drier lowland forests in the north because few such forests remain, and such areas are more accessible to hunters than are upland areas.

Methods

As pointed out above, one should not initiate a distributional, or "broad survey" [NRC, 1981], before obtaining an understanding of the species' behavior and ecology. After that, the initial preparations for a survey should include: 1) obtaining the most detailed maps available; 2) obtaining current information on forest types and cover from vegetation maps, aerial photographs, and satellite images; 3) communicating with persons who have previously worked in the areas to be visited; 4) assembling all previous records and reports; and 5) selecting survey routes and sample locations. Preference might be given to potential conservation areas, borders of ranges, and locations with few or no previous reports. Surveying can involve traveling by vehicle or by boat, extensive trekking in back country, or all of these. It is always necessary to hike extensively into the forest on foot (at least several kilometers) at each sample location. The lazy surveyor who observes and listens only from roads or well worn trails will miss many good records; well visited areas are usually the first areas to be hunted out. Needless to say, one must become proficient with map and compass to survey properly. It is often advantageous (and safest) to hire local rural farmers and other residents to help in getting around an unfamiliar area, but one should never allow local people to "select" the locations to be checked. We have found that local people usually do not like to penetrate unfamiliar or difficult forest areas and frequently do not know what is in them.

Methods of conducting the survey vary with species and the location. The information recorded will depend partly on the purpose of the survey but ordinarily will include the following: 1) exact location and time; 2) group size and age-sex composition (especially as these may provide evidence of reproduction or mortality); 3) descriptions of pelage and other characters of taxonomic importance; 4) photographs and tape recordings if possible; 5) habitat, especially forest type and condition; and 6) threats to survival, such as human encroachment and hunting.

Two semiquantitative methods of sampling abundance that might be used are as follows: 1) number of encounters of groups or individuals while walking through the forest; distance to center of group at detection should be recorded; such data might be calibrated with line transects through known populations and converted to crude density estimates (see line transect sampling below); 2) for primates such as gibbons giving loud calls or songs, one full morning of listening from a prominent location in each sample area will yield some indication of abundance; for gibbons, this method is more productive than relying on visual encounters provided that the weather is good [e.g., Chivers, 1974].

Relative abundance at each sample site should be scored at least as: dense population at or near carrying capacity, moderately common, rare (one or two sightings or calls only), or absent. The categories should be objectively defined, and one should attempt to spend the same amount of effort searching at each location.

Tape recordings of vocal primates will become increasingly important to studies of taxonomy, racial variation, and speciation. Although we are accustomed to thinking of primates as behaviorally flexible animals with high learning abilities, certain vocal characters are among the easiest ways of distinguishing geographically differentiated species and subspecies [e.g., Marshall et al., 1984; Marshall and Sugardjito, 1986; Struhsaker, 1970, 1981]. Vocalizations may become as useful to taxonomy as "hard" characteristics and can be permanently stored, objectively measured, and published graphically as sonograms. They are, in fact, less variable within many populations and more easily quantified than pelage features. The vocalizations of greatest interest are the ones that are most stereotyped within populations, usually the loud calls and songs. Ideally, tape recordings of vocalizations should be carefully made, catalogued, and stored with all the relevant documentation so that they become part of the permanent scientific record, available for future researchers to study. Because of conservation considerations, the collection of primate specimens is no longer feasible and increases in information for studies of evolution and systematics will have to come mainly from tape recordings and accurate field descriptions [Marshall et al., 1984].

The making of good field recordings requires just as much skill and field experience as does the taking of good photographs, and persons planning to record sounds should first consult more experienced persons. Gulledge [1976] provides a good introduction for prospective recordists. Obtaining acceptable recordings requires the use of a parabolic reflector or a directional microphone; the latter is more convenient to carry through the forest but eliminates less background noise. Loud calls must still be recorded less than 50 m from the animals. Problems of attenuation, echoing, reverberation from foliage, and background noise, often unnoticed at the time of recording, render more distant recordings unsuitable for sonagraphing.

Field Notes

Field notes are the primary repository of data for the field biologist and they should be maintained with the care and permanence that they deserve. Normally, field notes are taken with black indelible ink in a durable notebook in the form of a terse diary. Necessary information for every place visited includes date, location, names of accompanying persons, weather, routes taken, tasks performed, important observations and events, etc. Location data should include more than names of towns and highways; relations to permanent geographic features and map coordinates should also be given. Copies of maps should be bound in the notebook. Special check sheets and data forms can be designed for routine repeated observations and cataloguing of plant specimens, sound recordings, etc. If any sheets are loose, each must have the complete data and identifying information at the top. It is wise to store duplicate copies of important records.

The exact style of maintaining field records is a personal matter, but a good rule of thumb is never to trust anything to your memory. Hard experience frequently reinforces this lesson. Also, ask yourself this question: If I or someone else examines my notes 50 years from now, will they be able to understand what I wrote and be able to relocate easily all the sites visited? Remember that most roads and towns will be changed within 50 years.

Complete identifying data should be taken down for each tape-recorded cut. Gulledge [1976] illustrates the check sheet recommended by the Laboratory of Ornithology of Cornell University, although many of the items listed can be condensed without loss of important information. Information catalogued should include:

Species names
Tape number and cut
Exact location with map coordinates
Time begun
Habitat
Height and location of animals
Recording equipment
Vocalizations recorded and behaviors observed:
Tape time (sec)

Date
Primate group designation
Weather
Terrain and elevation
Disturbed?
Recordist
Background sounds
Recording distance
Description of contents

Under the last two items the recordist should, while relistening to the cut with a stopwatch, note down with comments what sounds are on the tape, what animals made them, etc. This greatly facilitates relocating sounds during future analysis and prevents confusion. At the minimum, the tape number and cut, date, and time must be announced on the tape at the beginning or end of each cut.

There are several bioacoustic archives that store good recordings of animal sounds (either the originals or good copies on half-inch tape) after the owner has extracted the information he needs. Among the best known are the Laboratory of Ornithology, Cornell University, Ithaca, New York; Florida State Museum, University of Florida, Gainesville; and the British Library of Wildlife Sounds, British Library National Sound Archives, London. Copies of all documentation must also be supplied, without which any recording is worthless.

Publication

No study is complete until it is published. Our only purpose in making this point is to urge primate field workers to publish distributional data in detail. Few primate workers other than those in systematic zoology seem to be in the habit of doing this satisfactorily. Maps showing distributional records should be detailed enough to show the locations of the animals within 1 km or less. Frequently, distribution data are shown on countrywide maps with large circles or triangles that do not indicate the sites clearly at all. Such maps are of limited use to anyone, for several reasons. First, one cannot return to the sites in the future to check trends in abundance or distribution unless the locations are accurately shown or described. Second, sites not exactly described are of limited use for systematic purposes, particularly for primates that do not cross rivers or streams. Many primate specimens in museums are of limited value because the collector did not specify on which side of an important stream the specimen was taken. Third, precise location data may be needed to stimulate efforts to bring the populations under better protection.

The use of shading on maps to show species ranges is not an acceptable substitute for actual sites. Only suitable habitat area or the suspected range should be shaded in an original report; otherwise, a misleading impression can be created.

It is sometimes argued that the divulging of exact sites can lead to increased threats from poachers, trappers, tourists, etc. Every case must of course be evaluated for this possibility, but we know of very few instances where this might be a problem. This belief assumes that the species is adequately secure now and that the poachers, etc., both need and are capable of using the published information. These assumptions are nearly always wrong; poachers are not limited by available information—only conservationists and responsible officials are. Virtually every primate field worker finds that to do nothing or restrict information will surely result in the gradual loss of most populations from existing threats. It also seems to be true in every country in the world that agencies responsible for conservation measures are seldom able to act promptly unless they are supplied with convincing documentation and rewarded with publicity.

POPULATION SAMPLING

Population sampling seeks to obtain an estimate of the total population in a region through counts of individual animals or groups of animals in replicate sampling areas selected throughout the region. Most likely, each sample area will yield only an estimate of density; normally in primate studies a researcher has time enough to obtain an accurate census in only a single area, the "intensive study area." Data from this study area may be used to test or calibrate more rapid but less precise methods to be used in the other sample areas.

The first step in sampling a population is to delineate the "region" (the "universe" in the statistician's terminology) to be sampled. This need not be a continuous area but may consist of isolated patches or long strips of a particular habitat type or elevation. The region is best defined from the maps used to carry out an extensive distributional survey. If the region consists of more than one habitat type, these should be sampled separately in a stratified design.

In the next section, we review and comment on several general methods of sampling forest primates and discuss briefly the problem of sampling design (the shape, number, and placement of samples) and estimation of the regional population.

Strip Transect and Quadrat Methods

These methods involve censusing the number of animals (or their signs) in known sampling areas. We use the word *census* to mean a complete count of animals or groups in an area. A strip is simply an elongated quadrat; its main advantages are that 1) habitat heterogeneity will increase the sample error, or variation among samples, less if samples are long and thin and 2) censusing a long strip is simpler and more efficient than a broad quadrat one cannot see across.

Broad quadrats can be censused by the "sweep survey" method [Marsh and Wilson, 1981] in which several persons spaced a regular interval apart

(50–100 m) walk across the quadrat in a coordinated fashion. This method gave rather accurate estimates of primate densities in Kuala Lompat, Malaysia, when tested by Marsh and Wilson [1981]. Good coordination between the persons participating is required to avoid missing groups or counting them twice, as is a sample area that is relatively level and easy to walk across. This last requirement is not often met in tropical forest.

Use of the strip transect (or "strip census") requires that all, or nearly all, primates are detectable when one walks down the middle of the strip. Although many primate surveyors have attempted to use this method assuming that primates can be detected within 40 or 50 m of the center line, very few persons have reported data on the distribution of animal-path distances that would allow testing of this assumption. We recommend instead that surveyors use the line transect method (described below), which makes no assumptions about the limit of detectability.

The density of nests of great apes has been censused using the strip transect [Schaller, 1963; MacKinnon, 1974; Rijksen, 1978; and others cited in NRC, 1981]. The nest density must be divided by the mean life of the nests and by the average number of nests built per individual per day to yield an estimate of animal density. MacKinnon [1974] found that an intensively searched quadrat yielded a much higher orangutan nest density than a long strip transect 60 m wide. We therefore recommend that either replicated, intensively searched quadrats or line transect methods be used for nests. If nest-to-transect line distances are recorded, it will be readily apparent if most or all nests are detectable within any given distance.

Line Transect Methods

The line transect is a plotless method in which one walks along a straight line and records for each animal seen the distance from the observer to the animal (r), the shortest or perpendicular distance from the animal to the path (x), and the angle of observation between the animal and the trail (θ); Figure 1. Any one of these quantities can be calculated from the other two, but noting all three enables the surveyor to check the accuracy of the data gathered. At the outset, nothing is assumed about the detectability of the animals except that an animal directly above the trail has a 100% chance of being detected. The essence of the analysis is to convert the trail-animal distances, or the observer-animal distances, into an estimate of density along the trail. This usually requires making further assumptions about the distributions of x or r.

A perplexing array of methods for analyzing line transect data have been developed in the wildlife management literature [among the most useful

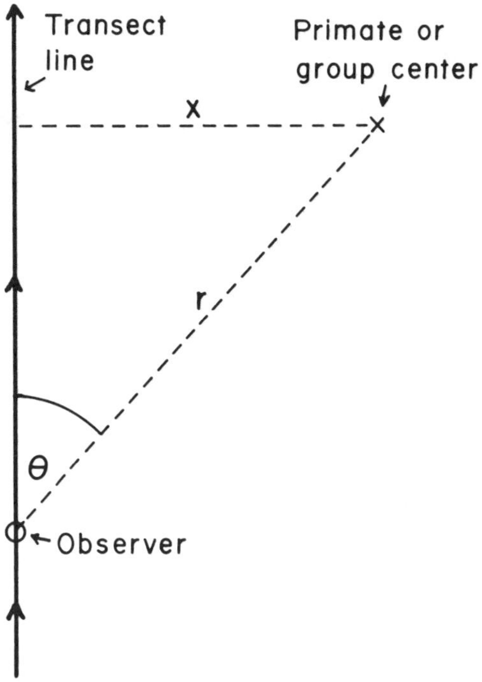

Fig. 1. Illustration of quantities recorded in the line transect sampling method: path-animal distance (x), observer-animal distance (r), and angle of observation (θ).

general references are Hayne, 1949; Gates, 1969; Emlen, 1971; Seber, 1973; Robinette et al., 1974; Eberhardt, 1978; Burnham et al., 1980]. Attempts to test line transect methods on primate populations have been reported by Janson and Terborgh (unpublished ms.), NRC [1981], Marsh and Wilson [1981], and Koster and Butynski [1985]. The monograph by Burnhan et al. now appears to be the definitive work on the subject; it reviews all other important literature and proposes a method that is robust and virtually assumption-free. It contains comprehensive practical suggestions for field workers as well as the background mathematical theory and analysis. Most of the important suggestions and explanations are also accessible in earlier papers [Anderson et al., 1979; Burnham and Anderson, 1976; Burnham et al., 1979; Crain et al., 1978].

Four basic assumptions necessary to the line transect method are as follows [Burnham et al., 1980].

Animals directly over the transect line are not missed. This assumption could be violated in primate surveys if the forest is tall and dense, in which case the method would underestimate density by an unknown amount.

Animals or social groups are seen before they move away or flee. This assumption is sometimes violated, and this can be tested by checking if the mean sighting angle is much larger than 40°. The theoretical mean sighting angle should be around 32.7°. To avoid this the observer must use considerable stealth while walking the transect.

Distances and angles are measured accurately. Distances should be measured to the nearest meter using a tape measure. Pacing is less satisfactory, especially in dense forest with heavy understory on uneven terrain. Marsh and Wilson [1981] have found that distances can be measured by calibrating the focussing ring on binoculars, but distances over 30 m are not measured with the desired accuracy in this way. Transect lines should be as straight as possible. Distance measures should not be pooled during data collection because the most important methods of analysis require exact individual distances. Many observers also have a tendency to "heap" or "dump" data, i.e., round it off to the nearest 5 m while collecting it [see Marsh and Wilson, 1981]. The observer must guard against this tendency.

Sightings are independent events. This assumption will hold reasonably well if social groups are used as detection events. In this case, the measurements must be to the estimated centroid, or geometric center, of the group, and not to the nearest animal seen. If individual animals are tallied, the assumption is violated for group-living species, and the consequence will be a highly increased variance in the density estimate. However, there seems to be little that can be done about this.

An animal at a given distance, x, from the transect line has a certain probability, g(x), of being detected. Assumption 1 above is equivalent to saying that g(0) = 1. Examples of two detection functions are given in Figure 2.

We might ask, if the probability of detecting an animal over or very near the transect is 1, why not simply use a density estimate from animals within, say, 1 m of the transect using a strip transect of width 2 m. This would be quite satisfactory in principle except that there will be too few animals in such a narrow strip from which to obtain an estimate. Animals seen farther from the line give us useful information that helps us in estimating the density over the line. If we know the function g(x), we can fit it to our observation data to estimate the true density.

The various different line transect methods make different assumptions about how g(x) declines with increasing x or else infer the shape of g(x) by noting the distances from the observer at which the animals flush.

Following the terminology of Burnham et al. [1980] closely, we suppose that the maximum distance at which an animal is detected from the transect

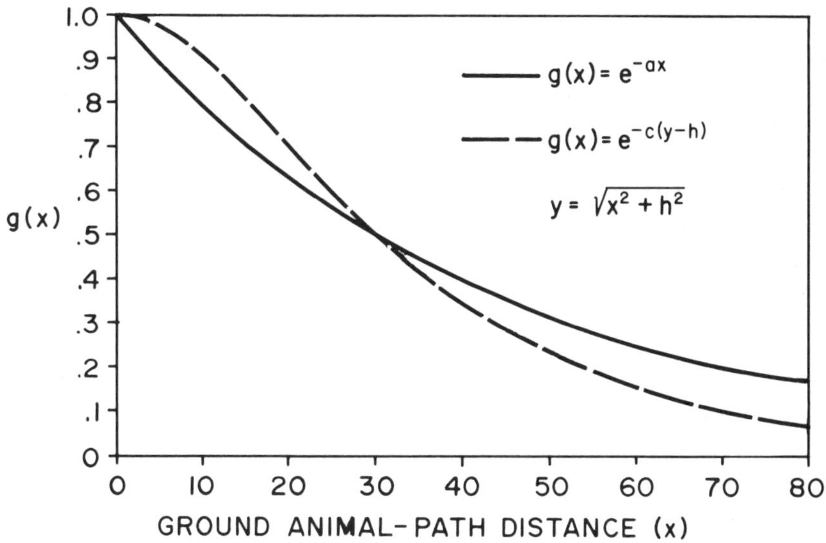

Fig. 2. Two a priori models of detectability in relation to perpendicular (path-animal) distance. The upper model (solid line) is the simple negative exponential curve. The lower model (dashed line) assumes that detectability of primates averaging h meters up in the canopy declines at a constant rate with respect to increasing perpendicular distance through the air, y. Curves are drawn so that g(0) = 1, g(30) = 0.5, h = 20 m.

is w, and the true number of animals in a strip of width 2w and length L is N (the animals seen on both sides are pooled). We actually detect only n of these animals, which represents a proportion, p = n/N. The proportion detected would actually represent the true density within a strip only 2â m wide, where â = pw. The unknown parameter â is defined as the integral of the detection function

$$\hat{a} = \int_0^w g(x)\,dx.$$

By definition, the function, truncated at w, encloses proportion p of the total area in a rectangle of height 1 and width w (Fig. 2).

The total density of animals in the strip of width 2w is the total number, N, divided by the total area, 2Lw. But we estimate N as n/p or, since p = â/w, as nw/â. In the formulas below, $\hat{D}$ is the estimate of the true density D:

$$D = \frac{N}{2Lw}, \quad \hat{D} = \frac{nw}{2Lw\hat{a}} = \frac{n}{2L\hat{a}}.$$

Since w cancels and no longer appears on the right, it is not necessary to consider its value or truncate the distances in the forest.

The basic problem now is to estimate $\hat{a}$ or its reciprocal $1/\hat{a}$, defined as $\hat{f}(0)$, from the distribution of x or r values. Some of the methods for doing this are discussed below. The quantity $\hat{a}$ has been interpreted as one-half "effective strip width" (ESW): the width of a strip such that the animals detected beyond are equal in number to the animals missed inside. The ESW is not known beforehand; it can only be estimated from the data during analysis. It is a purely abstract quantity that has no relevance to the actual conduct of the line transect, but its use has unfortunately resulted in some confusion between strip census and line transect methods. The concept of ESW is in fact not essential to line transect analysis, and, in the method of analysis recommended by Burnham et al. [1980], $1/\hat{a} = \hat{f}(0)$ is the quantity that is directly estimated. Their monograph should be consulted for a more comprehensive explanation.

Data collected while walking the transect should include the following:

Species	Number of individuals
Time sighted	Transect-animal/group distance
Observer-animal distance (r)	Angle of observation (θ)
Location of each sighting along transect	Height of animals
	Group age and sex composition
Method of detection	Behavior when detected

Some of this information might not seem essential in estimating population densities with current methods, but we recommend its collection and publication so that census methods can be evaluated and improved in the future.

The raw data for estimating $\hat{a}$ consist of sets of values for r, x, and θ obtained while walking the transect. The distribution of x and r are shown for some actual data from C.W. Marsh (personal communication) in Figure 3 after pooling into 9-m-wide intervals. We discuss below some important methods of analysis, first those that utilize r values and then those that utilize the perpendicular distance x.

Haynes' method. In the method developed by Haynes [1949], it is assumed that each animal has a constant "flushing radius," which may vary from one animal to the next. When the observer-animal distance declines to the flushing radius for a particular animal, it flushes and is detected. It is assumed that this distance is not affected by the angle of observation. Thus all animals with flushing radius r_i or more that lie within $x_i = r_i$ meters from the trail will be detected. Haynes noted that a greater proportion of animals

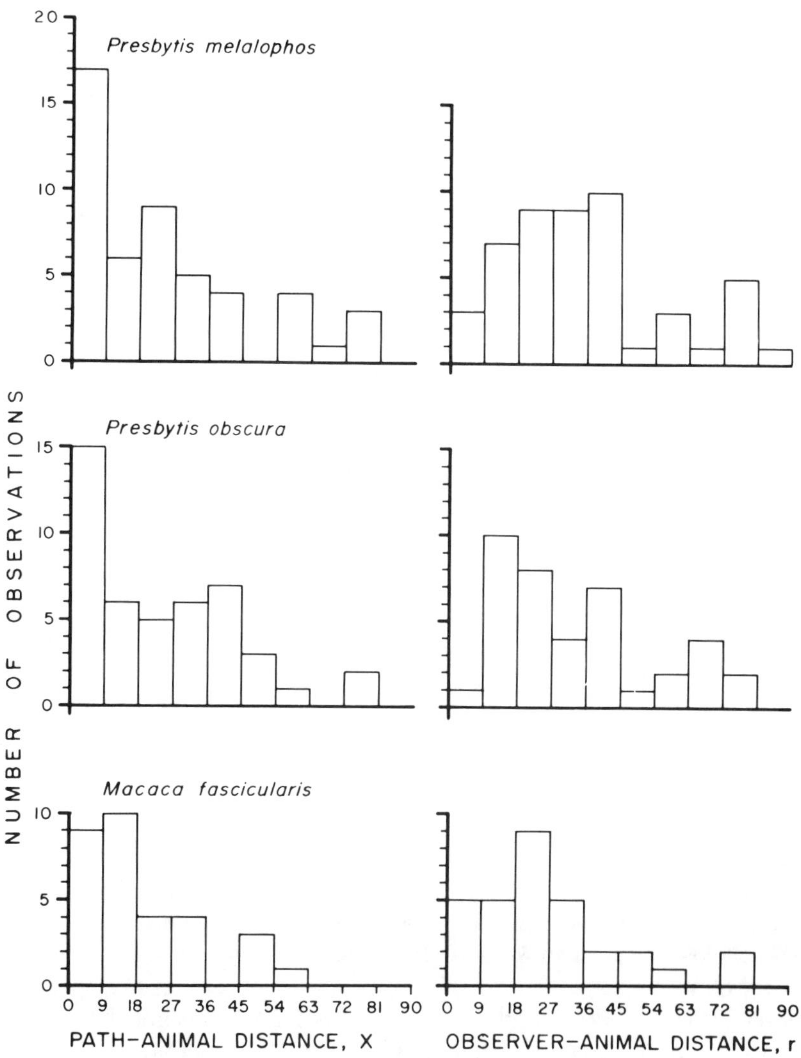

Fig. 3. Distribution of perpendicular path-animal distances (left) and observer-animal distances (right) for the banded and dusky langurs *Presbytis melalophos* and *P. obscura* and the long-tailed macaque *Macaca fascicularis* obtained while walking 19 km of transects in the Krau Game Reserve, Kuala Lompat, Malaysia. Data from C.W. Marsh (personal communication).

with large flushing radii will be detected than animals with small flushing radii. He proposed that the distribution of g(x) thus be modelled by considering the distribution of all observed r_i. It follows mathematically from these assumptions that

$$\frac{1}{\hat{a}} = \frac{1}{n} \sum_{i=1}^{n} \frac{1}{r_i}; \quad \hat{D} = \frac{1}{2L} \sum_{i=1}^{n} \frac{1}{r_i}.$$

A further result is that the average angle of detection should approach 32.7°.

Unfortunately, this elegant method usually underestimates ESW and overestimates density, both with stationary and with moving objects [Robinette et al., 1974; NRC, 1981] (also Janson and Terborgh, unpublished ms.). The problem appears to be that the detection distance does ordinarily vary with the sighting angle. Overestimation generally results because of animals near the transect being detected too close to the observer, which biases the r_i downwards. This could have several causes. Animals behind foliage or cover would be more difficult to detect in front than to the side when the observer's viewing angle is changing rapidly and the distance slowly. Also, observers might tend to concentrate more attention to the sides for fear of missing passing animals. Moving animals that are fleeing from the observer would also affect the r_i values. These factors would increase the mean sighting angle, θ. Calculation of mean θ will thus provide a test of these assumptions. Data collected by Marsh and Wilson [1981] on primates in Pasoh, Malaysia, yield a mean sighting angle of 42.3°, significantly larger than 32.7°, indicating that the Haynes method might not be appropriate. An improvement on the Haynes method called the "modified Haynes method" [Burnham et al., 1980], which takes into account the deviations of the sighting angle from 32.7°, is discussed with an example in the Appendix.

King's method. This method is described by Leopold [1933], Overton [1971], and Eberhardt [1978]. It simply uses the mean r as $\hat{a}$. When tested, this method has often given fairly accurate results [Robinette et al., 1974; Gates, 1969; Marsh and Wilson, 1981] (also Janson and Terborgh, unpublished ms.), in spite of its somewhat uncertain justification. In spite of that, Burnham et al. [1980] do not recommend it because its assumptions concerning r, θ, and g(x) are not explicit. We feel that it sometimes gives a close estimate for the wrong reasons (see below). Under the assumptions of the Haynes model, it should underestimate density [Gates, 1969].

In general, methods that utilize perpendicular path-animal distances are preferable to those that use sighting distances, because the latter require extra assumptions about how the sighting distances can be converted into the perpendicular detection functions. It is the perpendicular distances that are of direct relevance to the estimation of density.

One situation in which use of the observer-animal distances might be preferable is when one is sampling density in a new area where transects have not been cut, perhaps during a distributional survey. A crude estimate of density could be obtained if one recorded sighting distances for each individual or group encountered, which might be improved through calibration on a transect. The distance walked must of course be known.

The following methods utilize the path-animal distances (x_i).

Truncated distance methods. If visibility in the field is good and the assumption that animals are seen in place (before flushing) is well satisfied, it is possible that the perpendicular distances will show a more or less level plateau extending from $x = 0$. It could then be assumed that the animals within the edge of the plateau (distance w_1), or n_1, represent nearly the true density, which could then be estimated as $\hat{D} = n_1/2Lw/_1$. This is the method of Kelker [see Eberhard, 1978; Gates, 1979; Robinette et al., 1974; Emlen, 1971, 1977] and has been applied by Cant [1978] to spider monkeys. Anderson and Pospahala [1970] have a more objective, related method of estimating true density by fitting a polynomial to the frequencies of observation, $n_{\hat{1}}$, in truncated distance classes. Truncated distance methods are inefficient because they sacrifice all observations beyond w_1. Moreover, tropical forest primate data will not often yield a clear plateau, leading to too much subjectivity in determining w_1. Also, strictly speaking, the selection of a method of analysis such as this only after inspection of the data is statistically not justified, unless the data are very extensive and convincing [Burnham et al., 1980]. The method can be used occasionally as a crude first option. We recommend the use of the Fourier series (FS) estimator (see below) whether a plateau is present or not.

Exponential models. One of the best studied methods is that developed by Gates et al [1968] and Gates [1969]. This assumes that detectability drops off at a constant rate with changes in distance, leading to a negative exponential form, $g(x) = e^{-ax}$. A graph of such a detectability function is shown in Figure 2. Although the model has worked with grouse sitting on the ground, it is not so realistic in most situations because the detectability function is peaked on the line instead of flattened.

A modification of the exponential function that might be more appropriate for canopy-dwelling species can be made from a three-dimensional model by assuming that detectability declines exponentially with distance y, where y is the actual distance between the transect at head height and the animal in the tree. A graph of this function is also shown in Figure 2.

This model might be more realistic than the simple exponential, but because of the great variety of shapes of distributions of x we have seen in primate studies we recommend the more robust FS estimator.

Fourier series estimator. This method uses a Fourier series expansion utilizing the values of x_i. The formulas used and a worked example are given in the Appendix. Burnham et al. [1980] believe that it has a number of desirable qualities, the most important of which are:

1. Model robustness: The model is relatively insensitive to those factors that might affect the shape of the detection function.
2. Pooling robustness: The model is not sensitive to variation in the detection probability during data collection. Many data sets can therefore be pooled together for analysis for roughly comparable habitat types.
3. Shape criterion: One important characteristic is a "shoulder" in the detection function near $x = 0$. However, if the distribution of x peaks at some distance from the transect line, as it might do if animals are moving away from the observer, the FS model may be inappropriate, and the use of alternate models such as the exponential polynomial is indicated [Burnham et al., 1980, p. 125].
4. Estimator efficiency: The parameter estimated as $\hat{f}(0)$ must have as small a variance as possible.

Line Transect: Special Problems and Suggestions

Grouping. Nearly all primates occur in well defined social groups. This creates two major problems in recording line transect data: determining the number of animals per sighting and determining their location with respect to the line.

All attempts to apply line transect methods to primates have encountered difficulty in determining group size within the short time available (Janson and Terborgh, unpublished ms.) [Marsh and Wilson, 1981; NRC, 1981]. Primate groups are commonly dispersed over distances exceeding the limits of visibility, and the chances of some animals escaping detection are great. Janson and Terborgh initially tried to count all individuals, but T.T. Struhsaker [NRC, 1981] and Marsh and Wilson used a prior estimate of mean specific group size when the presence of a group was evident. This practice might lead to an overestimate of density for those species in which groups tend to subdivide. However, the most serious problem arises in the estimation of group location; to avoid bias, the geometric center of the *entire* group must be determined.

Both Marsh and Wilson [1981] and Struhsaker used as the center of the group the location of the first animal detected. In retrospect, this practice appears to result in considerable bias and causes an overestimate of $\hat{D}$ [cf. Koster and Butysnki, 1985]. Janson and Terborgh (unpublished ms.) recommend adding one-fourth to one-third of the average troop diameter to the ESW to compensate for the bias.

Because these biases may be the most difficult problems in applying line transects to primate sampling, we offer the following guidelines to be tried by future workers. There is obviously no single solution that is best for all situations.

1. If groups are tight and easily identifiable as such, the observer should quickly place a marker at the point of observation on the transect line and immediately sneak up to the group and count the animals and determine its center. If some animals are often missed, a mean group size estimate can still be used.

2. If the groups tend to be somewhat more fragmented, estimate the center of the individuals actually seen, without necessarily leaving the transect line, and use only the animals seen in estimation. If the number of individuals seen varies considerably, each individual should be treated as a separate sighting for purposes of estimation. This will violate the assumption of independent sightings; this should not bias the estimate of density but will increase its variance.

3. For species that tend to be well dispersed or often in small parties, such as pigtailed macaques or chimpanzees usually are, each individual should be treated as a sighting and its distance, etc., estimated separately. For this alternative, the estimate will be in terms of individuals and not groups.

It is possible that different tactics of data collection will have to be used for different species during the same survey. If both groups and solitary individuals are common in the same species (for instance, in *Presbytis johnii*), then the analysis should be stratified (carried out separately for each category).

If group size has an influence on detection probability (which can easily be tested by plotting group size against x), then density estimates will be biased. To overcome this problem Janson and Terborgh (unpublished ms.) suggest weighting the x_i (or r_i) by group size; so do Burnham et al. [1980], but an explicit method is not suggested. Perhaps the reciprocals of group sizes might be used as weighting factors.

Excessive sightings over transect line. In Struhsaker's census of red colobus populations in Uganda [Struhsaker, 1975; NRC, 1981, p. 61], "nearly 40% of the 166 sightings of red colobus were over the census transect and were scored as zero meters from the trail, although usually they were more than 20 m from the observer." The same problem occurred in Malaysian primate surveys, though to a lesser degree [Marsh and Wilson, 1981]. It is obvious that this result is an artifact of observation conditions or technique (and not a difficulty intrinsic to the use of x). The result is gross underestimation of ESW and overestimation of density by all standard methods.

Problems like this are often not evident until later analysis, and, not having been present during the surveys, we can only offer some possible explanations using the information published. It is possible that visibility down the trail was much better than to the sides (perhaps owing to thick understory or trailside vegetation) and that when a group was encountered the individual most directly ahead was usually the one first seen and scored. An attempt to locate the group center would at least partly have overcome this difficulty.

Another difficult problem to avoid is "dumping" of distance estimates on multiples such as 0, 5, 10, 15, 20 m, etc. It is important that distances are objectively measured and not subjectively estimated, especially those near the transect line.

Most operational problems with estimating primate densities using the line transect method usually lead to overestimation of density. This appears to explain why King's method, now regarded as analytically incorrect and biased in several respects [Burnham et al., 1980], paradoxically leads to the most accurate estimates. The development of improved field methods hopefully will correct this situation.

Velocity of troop movement. If groups range about with an average speed approaching that of the observer, the probability of encounters will increase. Janson and Terborgh (unpublished ms.) have developed a correction for the bias, which is only about 12% if the group velocity is as high as half that of the observer. For most species this problem is negligible.

Sampling design. We will briefly cover some problems concerning placement of transects and population estimation from the standpoint of primate censuses.

To estimate population density over a region, it is necessary that transects be replicated. Statisticians generally agree that the sampling error of density cannot usually be calculated for single transects when assumptions concerning data independence are rarely fulfilled [Burnham et al., 1980; Eberhardt, 1978]. Methods for estimating the variance of density from replicated transects using the Fourier series estimator are given in Burnham et al. [1980; pp. 51–54, 111–113].

For small animal species, short transects may be randomly or regularly placed over an area, but this would be inefficient in most primate surveys because time would be wasted in moving between transects. A severe limiting factor in primate surveys is time to obtain sufficient sightings for analysis; at least 40 sightings per species are recommended for each analysis. The length of transect one can walk in a morning is quite critical. Struhsaker walked 4-km transects and saw an average of 1.7 groups per km of three species while walking in the Kibale Forest, Uganda. Marsh and Wilson [1981] used 2–5-km transects (average 3.3 km) and averaged 0.8 groups per km of four species in Malaysian forests. Janson and Terborgh (unpublished ms.) used 2–3-km transects (but had three or four observers walking at the same time) and had an average of 1.3 encounters per km of eight species in Peru. All observers walked slowly (about 1 km per hour) and paused frequently to look around and listen. Thus primate workers tend to encounter on average between 0.2 and 0.5 groups of a given species per kilometer, and 80–200 km of census effort may be required to obtain enough sightings of even the commoner species. This translates to between 20 and 50 days of effort for one observer walking 4 km per day.

On the basis of these experiences, we recommend making transects somewhat longer and walking without extended pauses at about 2 km per hour. The Marsh and Wilson [1981] data for 28 different transects show that more primates were seen on long transects than on short ones during the same number of days on average. The goal must be to maximize the number of sightings on each day and not the thoroughness with which the forest is searched on both sides. We reiterate that line transect theory takes into account the fact that groups or individuals will be missed. Moreover, sightings far away from the transect line have much less effect on the density estimate than those near the line, and distant sightings (say, $x > 60$ m) may simply be discarded before analysis. Moving faster might also permit primates to be observed more often before they start fleeing, which is very critical to accurate estimation.

The transect can be arranged as the sides of a rectangle 3×1 km in size for efficient sampling. This would result in a total length of 8 km. Transects should be straight, of course, and should not conform to existing roads, ridges, or other terrain features unless absolutely unavoidable.

Several transects should be placed over the region to achieve replication, perhaps in six to twelve different areas. Ideally, they should be randomly or regularly placed, but this will seldom be completely achieved. Accessibility, security, water, food, and other practical considerations will often dictate locations of sample areas in tropical forests. In practice, areas believed to be representative of different parts of the region will most likely be selected; this should be borne in mind during analysis and subsequent interpretation of the data. The goal of transect sampling may be to compare representative ecological communities and not statistically estimate the regional population.

In such a case, it might be worthwhile to sample fewer areas with greater care rather than many areas poorly. In any event, the goal of transect sampling should be clearly defined beforehand so that appropriate measures can be taken to minimize bias in transect placement.

Janson and Terborgh (unpublished ms.) and NRC [1981] suggest placing confidence limits on the number of troop encounters per walk or per day to allow comparisons between censuses. Counts are thus treated as relative density estimates; it must be remembered that this practice assumes that observations are made under similar conditions and that the detectabilities are the same. It would be improper to compare different habitat types or different species in this fashion and risky even to compare results of different observers [Marsh and Wilson, 1981].

Publication and evaluation. Finally, we strongly urge that surveyors publish their raw data, if sufficient for analysis, including the list of x, r, and θ for each observation. We also urge editors to allow it to be published in full. Our task would have been more illuminating and progress in developing methodology would be far more rapid if we and others could examine the assumptions of various methods of analysis using the data. At present, there are virtually no sets of published data that allow testing of the surveyors' assumptions or conclusions for any species of primate.

Use of Loud Calls

Loud calls provide information useful in the localization of groups for a large number of primates. In species that call quite regularly and that can be heard from far away, loud calls can provide a useful indication of the crude or relative density during distributional surveys [e.g., Brockelman et al., 1977; Chivers, 1974; Ellefson, 1974; Kappeler, 1984a]. Under certain conditions, however, loud calls can be used to estimate absolute density (Brockelman and Srikosamatara, unpublished ms.) [Chivers and Davies, 1979; Marsh and Wilson, 1981; World Wildlife Fund, 1982].

Line transect methods tend to underestimate gibbon group densities because such groups are small and are often difficult to detect in tropical rain forest [Marsh and Wilson, 1981]. We will describe methods for estimating density from loud songs developed during studies of lar and pileated gibbons in Thailand and Malaysia; it is hoped that they will be adaptable to surveys of other vocal primates also. Such methods appear to be most suitable for primates that live in relatively small territorial groups.

Singing behavior. Before attempting any census using loud calls, one must become thoroughly familiar with the vocalizations and singing patterns of the species. The most dependable vocalizations for use in censusing gibbons are the morning duets, which commonly carry for 2 km, are given on most mornings in most species, and are sufficiently stereotyped within pairs to allow recognition of individuals frequently. Duets are sung only by mated pairs. Within the duet, the female's "great call" is the most useful and diagnostic sound. In two species that do not sing duets, Kloss's and Javan gibbons [Tenaza, 1976; Whitten, 1982; Kappeler, 1984b], the female still sings great calls regularly in the company of her mate. Unmated juvenile or subadult females of all species of gibbons sing great calls only in close accompaniment with their mothers on the parental territory. They are recognizable as such and thus a useful clue in identifying groups.

General procedure. To sample a population, a listener sits at a "listening post" on a prominent terrain feature with a compass, digital wristwatch, and data forms and notes down the direction, estimated distance, and beginning times of each loud call sequence of each group heard and any vocal features that might aid in group identification (Fig. 4). We find that the compass that can be read most easily and rapidly is the U.S. Army pocket compass. Gibbon duets commonly last up to 10–20 min; groups calling simultaneously are easily distinguished with a knowledge of song organization. Those sing-

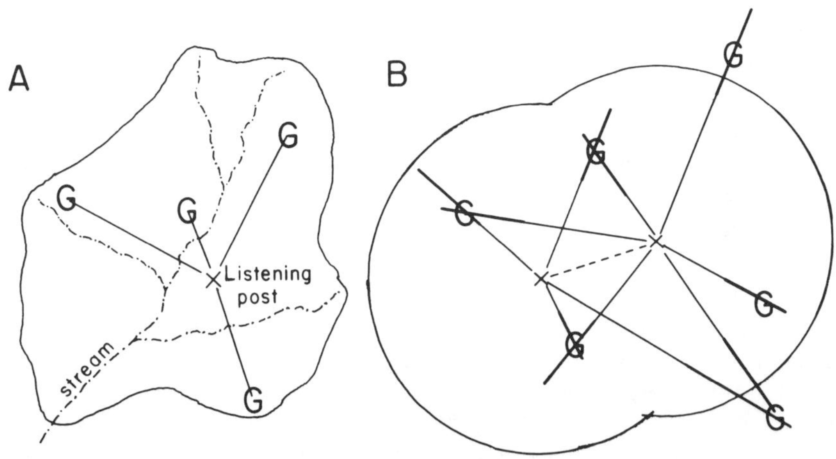

Fig. 4. Two methods of measuring listening area for auditory sampling of groups. A, a single listener sits at a listening post and notes the direction and estimated distance of each group heard. The listening area is delineated from a topographic map. B, method of triangulation in which two listeners occupy posts a measured distance apart and record exact times of calls, directions, and estimated distances (darkened portions of radii). The effective listening radius is selected beforehand, and only groups mapping inside this maximum radius are included in the density estimate. One group mapping within the area was heard by only one listener; it may be included if the listener has confidence in his distance estimate.

ing at different times are distinguished with a combination of directional and distance information if not from individual characteristics. As a rule of thumb, duets that map more than 500 m apart (the approximate diameter of a territory) are considered to be given by separate groups. Gibbons sing from anywhere in their territories, so groups usually cannot be inferred from clumping of singing locations on successive days.

The general formula for estimation of group density is

$$\hat{D} = fn/A, \text{ or } \hat{D} = \frac{n}{p(m)A},$$

where $\hat{D}$ is estimated density, n is the number of groups heard during the sample period (one whole morning or several mornings), A is the listening area, and f is a factor that corrects for the fact that not all groups are expected to sing during a given sample period. Its reciprocal, $1/f = p(m)$, represents the estimated proportion of groups that are expected to sing during a sample period of m days. It is assumed that if a group sings it will be heard; otherwise, the listening area has not been correctly demarcated (see below).

The major problems in sampling populations from songs are the estimation of p(m) and the determination of the listening area, A.

Singing frequency. The use of songs in population sampling assumes that the frequency of singing is approximately uniform throughout the region to be sampled. Although a number of important variables affect singing frequency, the assumption may hold sufficiently well if certain precautions are taken.

The estimated proportion of days on which groups sing, on average, is shown in Table I for several species. For lar gibbons, the range is about 60–85% of days per group. Audibility in some study areas has been found to fall off with increasing distance [e.g., Gittins, 1984] resulting primarily from uneven terrain or unsuitable listening posts.

One of the most serious unresolved problems in the estimation of density accurately from songs is the possible effect of density on singing frequency. In lar, pileated, and agile gibbons, most duets appear to be spontaneous and not directly stimulated by neighbors. However, Raemaekers and Raemaekers [1985], who timed song bouts in a population of lar gibbons over many days, found a tendency for the singing times of neighbors to be more closely correlated than those of nonneighbors. It is evident that duets are sometimes stimulated by neighbors [see also Brockelman et al., 1974]. In *Hylobates hoolock* [Tilson, 1979], *H. klossii* [Tenaza, 1976], *H. moloch* [Kappeler, 1984b], and *H. syndactylus* [Chivers, 1974], singing appears to be more

TABLE I. Estimated Proportion of Days on Which Groups Sing in Study Populations of Various Gibbon Species

Species	No. of groups	$\bar{p}$	Study location	Reference
lar	4	.85	Tanjong Triang, W. Malaysia	Ellefson, 1974
lar	5	.57	Ulu Gombak, W. Malaysia	Chivers, 1974
lar	2[a]	.75	Kuala Lompat, W. Malaysia	Chivers, 1974
lar	3[a]	.60	Ulu Sempam, W. Malaysia	Chivers, 1974
lar	5	.71	Khao Yai, C. Thailand	Kessawai and Brockelman, unpublished
pileatus	30	.47[b]	Khao Soi Dao, S.E. Thailand	Brockelman and Srikosamatara, unpublished
moloch	5	.45	Turalak, W. Java	Kappeler, 1984b
syndactylus	4	.58[c]	Ulu Gombak, W. Malaysia	Chivers, 1974
syndactylus	4	.25[d]	Ulu Gombak, W. Malaysia	Chivers, 1974
syndactylus	6	.25[e]	Kuala Lompat, W. Malaysia	Chivers, 1974
syndactylus	8	.20[e]	Ulu Sempam, W. Malaysia	Chivers, 1974

[a]Only nearest groups included.
[b]Some rainy and windy days excluded. For all 31 days, $\bar{p} = .43$.
[c]Data from 49 days in June and July, 1968.
[d]Data from 36 days throughout year, May, 1969 to April, 1970.
[e]Data from 130–140 days throughout year.

contagious; songs are reported to be passed around the local population (the remaining species have been inadequately studied in the wild). If groups in low-density populations are stimulated to sing less, this will result in their being underestimated. Some suggestions for dealing with this problem are given below.

Weather has been found to affect singing frequency in all species of gibbons that have been studied. Gibbons seldom sing when it is raining, although they may sing just before or after rain showers. Windy or breezy weather also suppresses singing. Bad-weather days should be objectively defined, if possible, and not used in the estimation of p or in the sample periods.

Seasonal variation in singing frequency has been reported in the siamang, lar gibbon [Chivers, 1974], agile gibbon [Gittins, 1984], Javan gibbon [Kappeler, 1984b], and pileated gibbon [Srikosamatara and Brockelman, 1983]. In the pileated gibbon, however, singing frequency did not vary significantly with season when "bad-weather" days were eliminated. In other species, singing frequency was found to correlate with food availability.

Brockelman and Srikosamatara (unpublished ms.) carried out an intensive study of singing frequency and its variability in a population of pileated gibbons in Khao Soi Dao Wildlife Sanctuary, Southeast Thailand. The study area was in a large, deep valley and had ideal terrain for listening. The density was about 6 groups per km^2, and 30 groups within audible range were studied during all seasons of the year, including a continuous 31-day period, which was used to analyze variation in singing frequency between groups and between days.

On average, groups sang on 47% of days of good weather, but the variation among groups (range 23–58%) was very great. This means that an estimate of p must be based on a generous sample of groups and not just on one or two groups no matter how well studied.

Even when bad-weather days were eliminated, there was extremely great intrinsic variation in singing frequency between successive days. The variance of the percentage of groups singing per day was very significantly greater than expected random binomial variance. The unfortunate implication of this is that confidence limits on p cannot be estimated using the binomial model, which would be the logical method.

Because of the large day-to-day variation, a reliable sample of the population cannot be obtained from a single day of listening per sample site. The only way around this problem is to base p and n on a sample period of several days.

There are two ways to obtain an estimate of the number of groups singing in a sample period of m days. One is simply to set n equal to the highest number of groups singing on any single day. The other, preferred method is to consider n as the cumulative number of groups that sing. This requires that groups be individually identified and mapped with the aid of acoustical and directional cues and generally requires more skill and experience.

There are thus many variables influencing the variability of $\hat{p}(m)$—number of sample days, season, weather, method of recording n—which make the variability of f and of $\hat{D}$ very difficult to predict. The standard error of the estimate of regional density will also depend on the number of samples, or listening areas, used, which must be traded off against sample size m.

To determine the optimal combination of sample size, number of samples, and method of recording n, we tabulated (from the 31-day period of listening) empirical distributions of p(m) for $m = 1–5$ days for both techniques of data recording. We then carried out Monte Carlo simulations by computer by repeatedly drawing s random samples ($s = 1–15$) from the distributions of p(m) a large number of times to produce distributions of f for different combinations of s and m. Confidence limits were then read off the simulated distributions of f.

The basic conclusions from this somewhat complex exercise are that 1) counting the number of groups that sing cumulatively results in a narrower confidence interval than does simply using the largest number of groups that sing on a single day and 2) listening at each sample location for at least 2 or 3 days greatly reduces the confidence interval. We have found that most of the variance in a density estimate is attributable to variation in singing between days and not to spatial variation in density or singing behavior.

When counting groups cumulatively, p(m) approaches 1.0 as m becomes large. How fast it approaches 1.0 was examined both empirically, by calculating mean p(m) for all possible successive samples of each size (m = 1–5) in the 31-day period of listening in the pileated gibbon study area, and mathematically, from p(1). For example, if p(1) = 0.47, then the predicted proportion of groups that do not sing during 3 days is $(0.53)^3$, and the predicted proportion that do sing is $1 - (0.53)^3 = 0.85$. The general mathematical formula for predicting the cumulative proportion of groups that sing during an m-day sample period is

$$\hat{p}(m) = 1 - [1 - p(1)]^m.$$

The predicted proportions agree rather well with the empirically determined proportions for the pileated gibbon study area (Fig. 5). If p(1) = 0.5, listening for 3 days per site is predicted to yield a count of 87.5% of the groups in the area. With such a high percentage, one scarcely need apply any correction factors at all. The same is true for 2-day sample periods, providing p(1) is greater than 0.66, as it is for most lar gibbon populations.

In conclusion, the following recommendations are made for determining the appropriate p for use as a correction factor for each species in each region to be sampled.

1. The mean proportion of groups singing per day must be estimated from an intensive study area containing approximately six groups or more. The habitat of the study area should be typical of the region to be sampled and should have a density of groups higher than average for the region. This will tend to prevent underestimation of p and overestimation of average density in the sample areas resulting from a possible correlation between p and density.

2. Weather should be recorded during the listening hours and an arbitrary rule should be formulated for eliminating "bad-weather" days.

3. Seasonal trends in p should be examined if relevant.

4. The ability to recognize groups from a combination of acoustical, directional, and distance information should be developed.

5. A sample size, m (number of days of listening per site), that increases $\hat{p}(m)$ to approximately 0.90 or above should be estimated. Sample size must

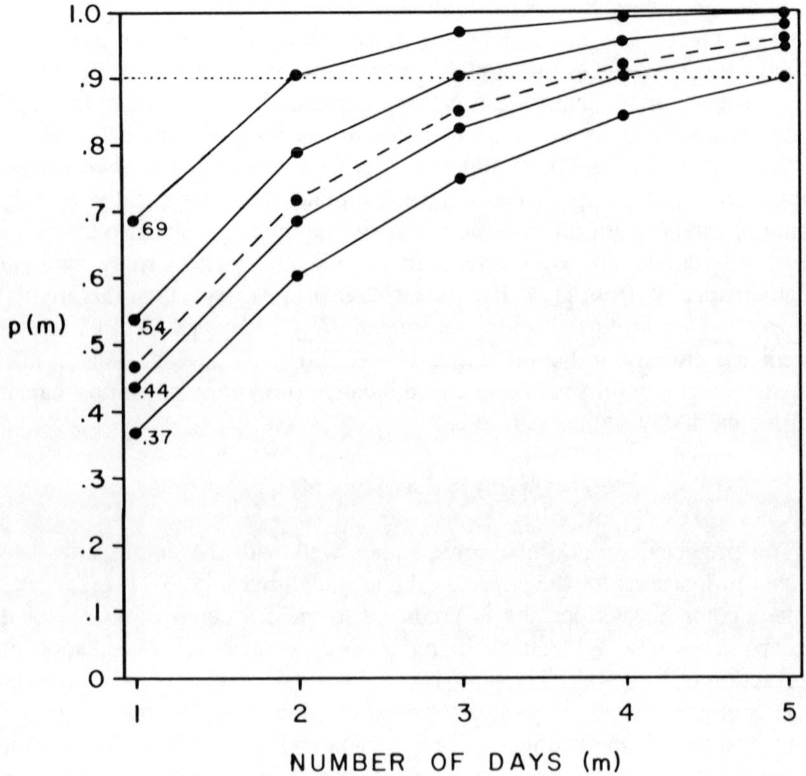

Fig. 5. Cumulative proportion of groups heard during sample period of m days calculated from p(1) (see text; solid lines) and measured empirically (dashed line). Solid lines show the minimum values of p(1) that will achieve approximately 90% of groups singing (dotted line) for sample sizes (m) of 2, 3, 4, and 5 days.

be held constant throughout any given survey, or the use of any particular $\hat{p}(m)$ is not valid.

6. There is no simple way of predicting the sample variance of $\hat{p}(m)$ or of density. A method that predicts variation between successive samples at the same site involves tabulating p(m) for all possible samples within an extended period at a study site. Frequency distributions of p(m) should be examined for m = 1, 2, and 3, at least, and more if p(1) is less than 0.5. For gibbons, it is found that day-to-day fluctuation in singing frequency is the main source of error in density estimates. The standard error of a regional density estimate should be estimated from replicated samples in the region.

Area estimation. Estimation of the listening area is easiest in mountainous terrain where ridges and valleys create natural boundaries. By selecting a prominent location for listening, one can often be sure of hearing all groups that sing in a valley and be certain of not hearing those outside the valley (Fig. 4A). For this method of area estimation, a topographic map is essential. Areas of up to about 10 km^2 can be sampled in this way, depending on the distance that loud calls carry over the forest. Since nearly all remaining gibbon habitat is mountainous, this method usually suffices.

In undulating or level terrain, it cannot easily be determined from a topographic map how far groups can be heard. In such situations, one should rely on triangulation to map the locations of groups; two or more listeners stationed a measured distance apart (~300–600 m for gibbon sampling) note down the exact times, directions, and estimated distances of all groups heard (Fig. 4B). The listeners' watches must be synchronized to the nearest second. It is surprising how often groups are heard by one listener but not by another.

Drawing the outer bounds of a listening area without well defined ridges can be difficult. The maximum audible distance of groups varies so much with terrain and atmospheric conditions that it is not advisable to make arbitrary assumptions about it. In Khao Yai Park, Thailand, a known gibbon group (*Hylobates lar*) was once heard from 3.5 km away across a canyon, but ordinarily audibility is not much more than 1 km through level forest. In undulating terrain, listening "shadows" can occur behind nearby hills. A topographic map may be of help on some sides, but on flat terrain one should place an arbitrary boundary beforehand well within the known limits of audibility. Centers of group ranges mapping outside the boundary should then be excluded. It is difficult to place the boundary after the groups have been mapped without introducing bias.

Loud calls as an aid to censusing quadrats. An accurate but more time-consuming alternative to auditory sampling from stationary posts is simply to map groups by approaching them quickly when they sing and observing them long enough to be able to identify them individually from age composition, pelage, song structure, etc. [Freese, 1976; Neville, 1976]. This method should also work best with primates that live in small territorial groups. Within 10 days of good weather, for example, an experienced observer can crudely map the ranges of all gibbon groups within an area of 1–2 km^2. The method will not work easily with primates whose ranges are large and overlapping.

Inevitably, some group ranges will be only partly inside the census area. Only groups whose ranges appear to be more than half in should be included. No great problem will be caused so long as the census area is substantially larger (at least five times) than the average group range size. One must do some mapping of ranges across the border of the census area to estimate the number of groups within the area.

INTENSIVE STUDY AREAS

An intensive study area is a relatively small area (sometimes less than 100 ha) where one carries out an extended study of the ecology and behavior of a species or community of primates. A grid of trails is cut to permit easy following and accurate mapping of groups, and one group or more is often habituated to observers to allow closer observation. The area is usually selected because of relatively high population density, and the densities and ranges of groups are usually accurately known after a year or two of study. The problem of determining the effective ecological density in species with large overlapping ranges that may be larger than the study area itself has been discussed by Kavanagh [1978] and Struhsaker [NRC, 1981; pp. 74–78].

The density of groups in a study area may thus provide an approximate upper limit for the density estimate for the whole region, but it is not legitimate to extrapolate from a study area to a region; study areas are usually selected because of their favorable habitat and relative security from hunting and other threats.

Work in an intensive study area is normally an essential adjunct to the survey and sampling of primate populations. Below we summarize some important types of information and skills that should be obtained from intensive studies.

Developing detection and observation skills. Observers need to learn and practice detecting and following wild groups and to learn well the species characteristics including vocalizations, pelage, age-class sizes, etc., for survey and sampling purposes. Surveyors need to develop a feel for recognizing the species' habitat. Beginners seldom have even the ability to walk quietly in the forest and may require several weeks to see groups that an experienced observer can find and observe within a few days. One must be able to estimate roughly the distance of groups from loud calls if they are to be used in sampling. For gibbons, beginners tend to underestimate the distance to singing groups by as much as two- or threefold; gibbon songs carry much farther than one might expect from their loudness.

Studying ranging behavior and dispersion within groups. Such information will be useful in deciding which sampling method to use or which method of analysis to use for the line transect. Information on mean group diameter can be of use in crudely estimating the width of the detection radius for a crude application of the strip transect census method. It is of interest to know mean territory size in mapping groups from auditory information.

Testing or calibrating line transect methods. Line transect estimates may be adjustable if the required assumptions are not fully met; however, it is not easy to determine whether the assumptions of the strip census and line transect methods are met. The critical test is thus to apply the methods to an area of known density. The detectability assumptions so important in the strip transect method will critically depend on habitat type and forest density. Line transect methods are more robust to detectability assumptions except that detectability must be 100% over the transect line. The time required for preparation and walking of transect lines should be estimated before designing the sampling program.

Estimating loud calling frequency. If loud calls are to be used in sampling, the average frequency of calling, its variability, and variables affecting it must be studied in a population of known groups. As stated above, such estimates must be based on a sufficient number of groups, not just one or two. It would also be of great interest to know how calling or singing frequency is affected by population density; to our knowledge, this is not known quantitatively for any species. We also have little information about how habitat type or quality can affect singing for any species. The collection of adequate data on singing frequency is time-consuming and sometimes boring, and such work is profitably combined with basic research on vocal behavior and its role in social interactions.

CONCLUSIONS

1. Distributional surveys attempt to map the geographic or regional distribution and habitats and determine crude or relative abundance. They are important to conservation, evolutionary studies, taxonomy, and ecology. For maximum value, careful field records should be maintained and distributional data should be published in detail.

2. Population sampling seeks to estimate the absolute density of a primate species from random (or representative) samples in a region. The most important methods used for sampling primate populations are the quadrat census, strip transect, line transect, and auditory sampling methods.

3. In the quadrat method, an attempt is made to census all primates, groups, or signs such as nests in a defined sample area. The "sweep census" involves a coordinated search by many surveyors. Loud calls can be a useful aid in locating groups that vocalize regularly during census of quadrats.

4. The strip transect or strip census method usually assumes that all groups can be detected a certain distance from the transect path walked. This

assumption has rarely been tested. The line transect technique, a plotless method that usually makes no assumptions about lateral visibility, is preferable. This method requires recording exact perpendicular path-group distances, observer-group distances, and observation angles. It is assumed that detection probability is 100% over the line itself and that animals or groups are seen before fleeing.

5. Analysis of line transect data involves fitting a mathematical or statistical detectability-vs.-distance function to the perpendicular distances or else inferring such a function from the distribution of sighting distances. The recommended method at present is the Fourier series estimator, which uses perpendicular distances.

6. In primate sampling, grouping, fleeing before detection, and a dependence of sighting probability on sighting angle all create problems for accurate line transect estimation of density. Measures must be taken to minimize these problems. If primate groups are the sighting units, distance to the geometric center of each group must be measured.

7. The sample variance of density from transects should be estimated from replicated transects. Obtaining enough sightings per species is a time-limiting factor in transect sampling.

8. Virtually no complete sets of data from line transect sampling have been published for any species of primate. Such data would allow testing many of the assumptions involved and should be published along with final estimates and conclusions.

9. Sampling density from loud calls or songs appears to be most useful for vocal primates that live in relatively small, sedentary groups such as gibbons. Such sampling requires a knowledge of the proportion of days on which the average group sings and a method of estimating the "listening area."

10. Singing frequency in gibbons has been found to vary with weather and season, variables that must be controlled. Even in good weather, there is great unexplained day-to-day variation in the proportion of groups that sing. This variation can be reduced by listening for two or three mornings at each sample site, which has the desirable effect of increasing the proportion of groups that sing (counted cumulatively) per sample period.

11. The listening area can often be estimated easily from a topographic map in mountainous terrain. Otherwise, it should be arbitrarily set well within the audibility limits of the species being sampled and groups mapped by the method of triangulation using two or more listeners.

12. In all distributional survey and population sampling efforts, it is necessary to collect reference information or gain proficiency in observation in an intensive study area beforehand. Some important uses of intensive study

areas are to develop detection skills and learn the characteristics of the species, study group ranging behavior and dispersion, test or calibrate transect methods, and estimate loud calling parameters and their variability.

ACKNOWLEDGMENTS

We thank Clive Marsh for allowing us to use his data and for his encouragement and Charles Janson for allowing us to see his unpublished work. Ken Green and John Oates provided useful comments on a draft analyzing the method of sampling using loud songs. Field support for sampling pileated gibbons was provided by a grant from the Charles A. Lindbergh Fund, Inc. Above all, we thank Ole Hong Thong, who has accompanied us most places we have gone.

REFERENCES

Anderson DR, Laake JL, Crain BR, Burnham KP (1979): Guidelines for line transect sampling of biological populations. J Wildlife Management 43:70–78.

Anderson DR, Pospahala S (1970): Correction of bias in belt transects of immotile objects. J Wildlife Management 34:141–146.

Brockelman WY, Damman D, Thongsuk P, Srikosamatara S (1977): Pileated gibbons survey at Khao Soi Dao Sanctuary–Thailand. Tigerpaper (FAO, Bangkok) 4(4):13–15.

Brockelman WY, Ross BA, Pantuwatana S (1974): Social interactions of adult gibbons (*Hylobates lar*) in an experimental colony. In Rumbaugh DM (ed): "Gibbon and Siamang, 3." Basel: S. Karger, pp 137–156.

Brockelman WY, Srikosamatara S: Estimation of density of gibbons using loud songs. Unpublished manuscript.

Burnham KP, Anderson DR (1976): Mathematical models for nonparametric inference from line transects. Biometrics 32:325–336.

Burnham KP, Anderson DR, Laake JL (1979): Robust estimation from line transect data. J Wildlife Management 43:992–996.

Burnham KP, Anderson DR, Laake JL (1980): "Estimation of Density From Line Transect Sampling of Biological Populations (Wildlife Monogr No. 72)." Washington, DC: The Wildlife Society.

Bush GL (1975): Modes of animal speciation. Annu Rev Ecol Syst 6:339–364.

Bush GL, Case SM, Wilson AC, Patton JL (1977): Rapid speciation and chromosomal evolution in mammals. Proc Natl Acad Sci USA 74:3942.

Cant JGH (1978): Population survey of the spider monkey *Ateles geoffroyi* at Tikal, Guatemala. Primates 19:525–535.

Chivers DJ (1974): "The Siamang in Malaya: A Field Study of a Primate in Tropical Rain Forest (Contr Primatol Vol 4)." Basel: S. Karger.

Chivers DJ, Davies AG (1979): Abundance of primates in the Krau Game Reserve, peninsular Malaysia. In Marshall AG (ed): "The Abundance of Animals in Malesian Rain Forests (Trans Aberdeen-Hull Symp Malesian Ecol No. 6)." Hull: Department of Geography, University of Hull, pp 9–36.

Crain BR, Burnham KP, Anderson DR, Laake JL (1978): "A Fourier Series Estimator of Population Density for Line Transect Sampling." Logan, UT: Utah State University Press.

Cronin JE, Sarich VM, Ryder O (1984): Molecular evolution and speciation in the lesser apes. In Preuschoft H, Chivers DJ, Brockelman WY, Creel N (eds): "The Lesser Apes: Evolutionary and Behavioural Biology." Edinburgh: Edinburgh University Press, pp 467–485.

Eberhardt LL (1978): Transect methods for population studies. J Wildlife Management 42: 1–31.

Ellefson JO (1974): A natural history of white-handed gibbons in the Malayan peninsula. In Rumbaugh DM (ed): "Gibbon and Siamang, 3." Basel: S. Karger, pp 1–136.

Emlen JT (1971): Population densities of birds derived from transect counts. Auk 88:323–342.

Emlen JT (1977): Estimating breeding season bird densities from transect counts. Auk 94:455–468.

Freese C (1976): Censusing *Alouatta palliata, Ateles geoffroyi*, and *Cebus capucinus* in the Costa Rican dry forest. In Thorington RW Jr, Heltne PG (eds): "Neotropical Primates: Field Studies and Conservation." Washington, DC: National Academy of Sciences, pp 4–9.

Gates CE (1969): Simulation study of estimators for the line transect sampling method. Biometrics 25:317–328.

Gates CE, Marshall WH, Olson DP (1968): Line transect method of estimating grouse population densities. Biometrics 24:135–145.

Gittins SP (1984): The vocal repertoire and song of the agile gibbon. In Preuschoft H, Chivers DJ, Brockelman WY, Creel N (eds): "The Lesser Apes: Evolutionary and Behavioural Biology." Edinburgh: Edinburgh University Press, pp 354–375.

Gulledge JL (1976): Recording bird sounds. Living Bird 15:183–203.

Hayne DW (1949): An examination of the strip census method for estimating animal populations. J Wildlife Management 13:145–157.

Heltne PG, Thorington RW Jr (1976): Problems and potentials for primate biology and conservation in the New World. In Thorington RW Jr, Heltne PG (eds): "Neotropical Primates: Field Studies and Conservation." Washington, DC: National Academy of Sciences, pp 110–124.

Janson C, Terborgh J: Censusing primates in rainforest with reference to the primate community at Cocha Cashu Biological Station, Manu National Park, Peru. Unpublished manuscript.

Kappeler M (1984a): The gibbon in Java. In Preuschoft H, Chivers DJ, Brockelman WY, Creel N (eds): "The Lesser Apes: Evolutionary and Behavioural Biology." Edinburgh: Edinburgh University Press, pp 19–31.

Kappeler M (1984b): Vocal bouts and territorial maintenance in the moloch gibbon. In Preuschoft H, Chivers DJ, Brockelman WY, Creel N (eds): "The Lesser Apes: Evolutionary and Behavioural Biology." Edinburgh: Edinburgh University Press, pp 376–389.

Kavanagh M (1978): "Some Inter-Population Variation in the Behavioural Ecology of *Cerpithecus aethiops tantalus*." PhD dissertation, University of Sussex.

Koster S, Butynski T (1985): An evaluation of census methods for estimating abundance. Am J Primatol 8:348 (abstract).

Leopold A (1933): "Game Management." New York: Charles Scribner's Sons.

MacKinnon J (1974): The behaviour and ecology of wild orang-utans (*Pongo pygmaeus*). Anim Behav 22:3–74.

Marsh CW, Wilson WL (1981): "A Survey of Primates in Peninsular Malaysian Forests." Cambridge: Universiti Kebangsaan Malaysia and Cambridge University.

Marshall JT, Sugardjito J (1986): Gibbon systematics. In Swindler D (ed): "Comparative Primate Biology, Vol I. Systematics, Evolution, and Anatomy." New York: Alan R. Liss, Inc, in press.

Marshall JT, Sugardjito J, Markaya M (1984): Gibbons of the lar group: Relationships based on voice. In Preuschoft H, Chivers DJ, Brockelman WY, Creel N (eds): "The Lesser Apes: Evolutionary and Behavioural Biology." Edinburgh: Edinburgh University Press, pp 533–541.

National Research Council (NRC), Subcommittee on Conservation of Natural Populations, Committee on Nonhuman Primates (1981): "Techniques for the Study of Primate Population Ecology." Washington, DC: National Academy Press.

Neville M (1976): The population and conservation of Howler monkeys in Venezuela and Trinadad. In Thorington RW Jr, Heltne PG (eds): "Neotropical Primates: Field Studies and Conservation." Washington, DC: National Academy of Sciences, pp 101–109.

Overton WS (1971): Estimating the numbers of animals in wildlife populations. In Giles RH (ed): "Wildlife Management Techniques, 3rd ed." Washington, DC: The Wildlife Society, pp 403–445.

Raemaekers JJ, Raemaekers PM (1985): Long-range vocal interactions between groups of gibbons (*Hylobates lar*). Behaviour 95:26–44.

Rijksen HD (1978): "A Field Study on Sumatran Orang-Utans (*Pongo pygmaeus abelii* Lesson 1827): Ecology, Behaviour and Conservation." Wageningen, The Netherlands: H. Veeman and Zonen BV.

Robinette WL, Loveless CM, Jones DA (1974): Field tests of strip census methods. J Wildlife Management 38:81–96.

Schaller GB (1963): "The Mountain Gorilla: Ecology and Behavior." Chicago: University of Chicago Press.

Seber GAF (1973): "The Estimation of Animal Abundance." New York: Hafner Press.

Shafer DA, Myers RH, Saltzman D (1984): Biogenetics of the siabon (gibbon-siamang hybrids). In Preuschoft H, Chivers DJ, Brockelman WY, Creel N (eds): "The Lesser Apes: Evolutionary and Behavioural Biology." Edinburgh: Edinburgh University Press, pp 486–497.

Srikosamatara S, Brockelman WY (1983): Patterns of territorial vocalization in the pileated gibbon (*Hylobates pileatus*). In Seth PK (ed): "Perspectives in Primate Biology." New Delhi: T.T.P.P. Publishers, pp 19–22.

Struhsaker TT (1970): Phylogenetic implications of some vocalizations of *Cerpithecus* monkeys. In Napier JR, Napier PH (eds): "Old World Monkeys: Evolution, Systematics and Behavior." New York: Academic Press, pp 365–444.

Struhsaker TT (1975): "The Red Colobus Monkey." Chicago: University of Chicago Press.

Struhsaker TT (1981): Vocalizations, phylogeny and paleogeography of red colobus monkeys (*Colobus badius*). African J Ecol 19:265–283.

Tenaza RR (1976): Songs, choruses and countersinging of Kloss' gibbons (*Hylobates klossii*) in Siberut Island, Indonesia. Z Tierpsychol 40:37–52.

Tilson RL (1979): Behaviour of hoolock gibbons (*Hylobates hoolock*) during different seasons in Assam, India. J Bombay Nat Hist Soc 76:1–16.

White MJD (1978): "Modes of Speciation." San Francisco: W.H. Freeman and Co.

Whitten AJ (1982): The ecology of singing in Kloss gibbons (*Hylobates klossii*) on Siberut Island, Indonesia. Int J Primatol 3:33–51.

World Wildlife Fund (1982): "A Faunal Survey of Sabah (IUCN/WWF project No. 1692)." Kuala Lumpur: World Wildlife Fund Malaysia.

APPENDIX

The data used in this example are unpublished data provided by Clive Marsh and were collected as part of a census of primates at Kuala Lompat, Malaysia, during October, 1979. Detailed calculations are given here for *Presbytis melalophos*, and a summary is given for various estimators for three species. The raw data are presented in Tables II–IV and are presented as histograms in Figure 3. The histograms demonstrate that perpendicular sighting distances for the two *Presbytis* species can be truncated at 80 m and at 63 m for *Macaca fascicularis*. The transect length is measured as 19 km, based on a map given in Marsh and Wilson [1980]. For *Presbytis melalophos*, the truncation distance chosen is thus w = 80 m, and the sample size is n = 49.

TABLE II. Sighting Data (x, r, and θ) for *Presbytis melalophos* in the Krau Game Reserve, Kuala Lompat, Peninsular Malaysia, Taken in October, 1979

Obs. no.	x	r	θ
1	70.0	70.0	90.0
2	20.0	20.0	90.0
3	0.0	35.0	0.0
4	40.0	40.0	90.0
5	10.0	35.0	16.6
6	0.0	25.0	0.0
7	5.0	27.0	10.7
8	20.0	20.0	90.0
9	0.0	15.0	0.0
10	9.0	12.0	48.6
11	55.0	55.0	90.0
12	30.0	30.0	90.0
13	31.0	40.0	50.8
14	20.0	25.0	53.1
15	75.0	75.0	90.0
16	1.0	45.0	1.3
17	20.0	30.0	41.8
18	5.0	5.0	90.0
19	15.0	15.0	90.0
20	11.0	12.0	66.4

(continued)

TABLE II. Sighting Data (x, r, and θ) for *Presbytis melalophos* in the Krau Game Reserve, Kuala Lompat, Peninsular Malaysia, Taken in October, 1979 (Continued)

Obs. no.	x	r	θ
21	0.0	33.0	0.0
22	30.0	30.0	90.0
23	20.0	30.0	41.8
24	0.0	10.0	0.0
25	60.0	60.0	90.0
26	120.0	120.0	90.0
27	40.0	75.0	32.2
28	80.0	85.0	70.3
29	60.0	75.0	53.1
30	175.0	175.0	90.0
31	125.0	125.0	90.0
32	100.0	100.0	90.0
33	20.0	50.0	23.6
34	17.0	30.0	34.5
35	25.0	25.0	90.0
36	4.0	38.0	6.0
37	15.0	35.0	25.4
38	0.0	15.0	0.0
39	10.0	40.0	14.5
40	25.0	45.0	33.7
41	9.0	21.0	25.4
42	55.0	60.0	66.4
43	2.0	6.0	19.5
44	30.0	40.0	48.6
45	5.0	25.0	11.5
46	5.0	45.0	6.4
47	35.0	45.0	51.1
48	25.0	25.0	90.0
49	45.0	80.0	34.2
50	0.0	10.0	0.0
51	7.0	7.0	90.0
52	75.0	75.0	90.0
53	40.0	40.0	90.0

The Fourier series estimator of density is calculated as

$$\hat{D} = \frac{n\hat{f}(0)}{2L}, \text{ where } \hat{f}(0) = \frac{1}{w} + \sum_{k=1}^{m} a_k$$

and L is the total length of the transects walked.

The a_k terms are calculated using the following formula, where x_i are the perpendicular distances:

TABLE III. Sighting Data for *Presbytis obscura*, Krau Game Reserve, 1979

Obs. no.	x	r	θ
1	20.0	25.0	53.1
2	0.0	20.0	0.0
3	50.0	50.0	90.0
4	5.0	10.0	30.0
5	5.0	5.0	90.0
6	50.0	70.0	45.6
7	7.0	45.0	8.9
8	16.0	20.0	53.1
9	50.0	70.0	45.5
10	15.0	22.0	43.0
11	100.0	100.0	90.0
12	0.0	15.0	0.0
13	60.0	60.0	90.0
14	35.0	40.0	61.0
15	25.0	70.0	20.9
16	40.0	45.0	62.7
17	0.0	45.0	0.0
18	40.0	60.0	41.8
19	4.0	10.0	23.6
20	2.0	13.0	8.8
21	0.0	12.0	0.0
22	0.0	70.0	0.0
23	80.0	80.0	90.0
24	17.0	17.0	90.0
25	5.0	300.0	1.0
26	75.0	75.0	90.0
27	8.0	10.0	53.1
28	30.0	30.0	90.0
29	6.0	20.0	17.5
30	15.0	20.0	48.6
31	25.0	30.0	56.4
32	20.0	20.0	90.0
33	4.0	12.0	19.5
34	20.0	20.0	90.0
35	0.0	15.0	0.0
36	15.0	30.0	30.0
37	30.0	30.0	90.0
38	45.0	45.0	90.0
39	18.0	18.0	90.0
40	40.0	40.0	90.0
41	35.0	40.0	61.0

$$a_k = \frac{2}{nw} \sum_{i=1}^{n} \cos(k\pi x_i/w)$$

A total of m terms for a_k are calculated until the stopping rule,

$$a_{m+1} < \frac{1}{w}\left(\frac{2}{n+1}\right)^{0.5},$$

is satisfied. For this example, this calculates to $0.0125(2/50)^{0.5} = 0.0025$. Now, for example,

$$a_2 = \frac{2}{80 \times 49} [\cos(2\pi x_1/55) + \ldots + \cos(2\pi x_{49}/55)] = 0.004734.$$

The following terms were calculated: $a_1 = 0.011304$, $a_2 = 0.004734$, $a_3 = 0.001614$, and $a_4 = 0.0014677$. The stopping rule is satisfied for $a_3 = 0.001614 < 0.0025$. Therefore, the first two terms are used, and we have, adding them up, $\hat{f}(0) = 0.02854$.

To calculate the variance of the estimator, we need to calculate the covariance matrix, where each term is

$$\text{cov}(a_i, a_j) = \frac{1}{n-1}\left[\frac{1}{w}(a_{i+j} + a_{i-j}) - a_i a_j\right].$$

This gives us (each value is $\times 10^{-6}$)

	1	2
1	1.8259	2.2494
2	2.2494	3.1705.

Summing up all these terms gives us var $(\hat{f}(0) = 9.4952 \times 10^{-6}$.

Given that the census path walked was 19 km, we can now calculate the density from the formulas given above as 3.68×10^{-5} sightings/m^2 or 36.8 sightings/km^2. The variance is calculated as

$$\text{var}(\hat{D}) = \hat{D}^2 \left\{\frac{1}{n} + \frac{\text{var}[\hat{f}(0)]}{[\hat{f}(0)]^2}\right\}.$$

TABLE IV. Sighting Data for *Macaca fascicularis* in the Krau Game Reserve, 1979

Obs. no.	x	r	θ
1	10.0	40.0	14.5
2	5.0	10.0	30.0
3	25.0	25.0	90.0
4	15.0	20.0	48.6
5	3.0	5.0	36.9
6	5.0	30.0	9.6
7	1.0	10.0	5.7
8	120.0	120.0	90.0
9	14.0	15.0	69.0
10	30.0	30.0	90.0
11	10.0	20.0	30.0
12	100.0	100.0	90.0
13	10.0	25.0	23.6
14	60.0	60.0	90.0
15	5.0	5.0	90.0
16	20.0	25.0	53.1
17	10.0	10.0	90.0
18	20.0	80.0	14.5
19	7.0	8.0	61.0
20	0.0	20.0	0.0
21	50.0	50.0	90.0
22	20.0	20.0	90.0
23	15.0	20.0	48.6
24	14.0	15.0	69.0
25	15.0	40.0	22.0
26	30.0	30.0	90.0
27	50.0	80.0	38.7
28	50.0	50.0	90.0
29	1.0	7.0	8.2
30	3.0	5.0	36.9
31	30.0	30.0	90.0
32	30.0	30.0	90.0

This calculates out to 43.42 (in km), yielding a standard deviation of 6.59 animals/km^2.

The King method gives a "1/2 ESW" of 43.5 m, yielding a population density estimate of 32.06 animals/km^2.

The Haynes estimator is calculated by the formula

$$\hat{D}_H = \frac{n}{2L}\left(\frac{1}{n}\sum_{i=1}^{n}\frac{1}{r_i}\right),$$

where the r_i are the animal-observer distances. Very small values of r_i are treated as outliers. A "modified Haynes estimator," which is more conservative than the Haynes estimator, corrects for deviations of the mean sighting angle from 32.7° and could be used if the sighting angle is significantly greater than this. It is estimated as $\hat{D}_{MH} = \hat{c}\hat{D}_H$, where $\hat{c} = 1.9661 - 0.02954\ (\bar{\theta})$ and $\bar{\theta}$ is the mean sighting angle. The variance of the modified Haynes estimator is

$$\mathrm{var}(\hat{D}_{MH}) = \hat{D}^2_{MH}\left[\frac{\mathrm{var}(n)}{n^2} + \frac{0.000876}{\hat{c}^2}\cdot\frac{\Sigma(\theta - \bar{\theta})^2}{n(n-1)} + \frac{\Sigma\left(\frac{1}{r_i} - \bar{r}_h\right)^2}{\bar{r}_h^2\, n(n-1)}\right]$$

$$\text{and } \bar{r}_h = \frac{1}{n}\sum_{i=1}^{n}\frac{1}{r_i}.$$

Var(n) = n is the Poisson assumption, although it is probably better to use the more conservative var(n) = 2n. For the data set described above, calculation yields $\hat{D}_H = 57.33$ sightings/km^2 and $\hat{D}_{MH} = 34.21$ sightings/km^2 and the standard deviation as 10.31. Since the mean sighting angle is 46.36°, $\hat{c} = 0.597$. So the modified Haynes estimator gives a result about 40% less than the Haynes estimator. We would advise observers who census regularly to obtain the TRANSECT computer program [see Burnham et al., 1980, for details]. It is advisable to do these calculations on at least a 16-bit computer, since round-off errors in calculation can become very large.

A summary for all the species is given in Table V. The Haynes method gives the highest estimate for all species. These are in all likelihood overestimates. The King and modified Haynes estimators are both conservative and

TABLE V. Summary of Density Estimates Using all Methods. Data from Tables II-IV[a]

Method	*Presbytis melalophos*	*Presbytis obscura*	*Macaca fascicularis*
Fourier series	36.79	44.59	33.69
	(6.59)	(10.38)	(6.82)
Haynes	57.33	47.98	53.21
Modified Haynes	34.21	26.74	24.51
King	32.06	25.58	24.41
Kelker	—	—	27.78

[a]The estimates for the densities given are in terms of sightings/km^2, since the raw data did not indicate whether the sightings were of individuals or groups. Figures in parentheses are the standard deviations.

give similar results, with the Fourier series estimator giving intermediate results. Only *M. fascicularis* yielded x_i suggesting a plateau, with w at 18 m (see Fig. 3). The FS estimator could be an overestimate for this species; only more data would provide an answer. Absolute density estimates are not available because average group sizes have not been given.

Primate Conservation in the Tropical Rain Forest, pages 63–82

3
Distribution, Abundance, and Endangerment of Primates

Ruth E. Happel, John F. Noss, and Clive W. Marsh

Department of Anthropology, Peabody Museum, Harvard University, Cambridge, Massachusetts 02138 (R.E.H., J.F.N.), and Sabah Foundation, Kota Kinabalu, Sabah, Malaysia (C.W.M.)

INTRODUCTION

Although the plight of endangered primates is a cause for widespread concern, little attention has been given to explaining patterns of abundance, rarity, and endangerment across the group as a whole [but see Clutton-Brock and Harvey, 1978]. In other words, are there certain traits that render species likely to be either abundant or rare? In the context of conservation, the question is obscured by the seemingly overwhelming proximate importance of human cultural traditions (e.g., hunting), patterns of colonization, and economic development, which suggests that species endangerment might be entirely a function of human behavior. However, it is known that similar human impacts on a habitat can affect species differently and that some species are hunted or trapped in preference to others (see other papers in this volume). Therefore, it can usefully be asked whether a species' rarity or vulnerability to extinction is at least partly a function of its own biology.

Some valuable insight into the process of extinction has come from studies of island biogeography showing that species diversity and extinction rates are related to habitat area and isolation [e.g., Diamond, 1975]. Several authors have also pointed out that larger species are among the first to drop out of island faunas [e.g., Brown, 1971; Willis, 1974]. Terborgh and Winter [1980] have shown that yet another predictor of extinction probability is the initial population density: Uncommon species are more vulnerable than common ones. Finally, there is a growing recognition that the extinction of one species can precipitate that of others because of the complex web of interdependency among plant and animal species in rain forest.

The aim of this paper is to examine variables that could have some predictive value for primate species' rarity and endangerment. For this purpose, *endangerment* is taken to mean vulnerability to extinction resulting from a declining trend in, or anticipated threats to, an already rare population. *Rarity* is a composite term with two components, which are here considered separately. One component is population density, and the other is species' geographic range area. Thus a primate will be considered rare if it has either a low population density or a small geographic range. We have therefore first conducted a statistical review of some species' characteristics that might correlate with either population density or geographic range.

After seeking possible causes of rarity, we consider additional factors that might contribute to a rare species becoming endangered. These ideas are examined in relation to the current list of primate taxa in the Red Data Rook of the International Union for Conservation of Nature and Natural Resources [IUCN, 1978, 1984].

METHODS

Data for the analyses were taken from the literature, principally the compilations of Clutton-Brock and Harvey [1977a,b, 1978, 1979, 1980], Harvey and Clutton-Brock [1985], Kavanagh [1984], Napier and Napier [1967], and Wolfheim [1983]. Since consideration of only well studied species might introduce a bias towards certain kinds of species, we have included data from as many species as possible. This data set is given in the Appendix.

A number of factors have been linked to rarity. The variables tested here include 1) body size; 2) demography; 3) diet; 4) home range size; 5) degree of arboreality; 6) habitat type; 7) group size; 8) activity cycle (diurnal or nocturnal); and 9) geological age of taxon. The primary analytical techniques used are multiple regression and bivariate correlation. Multiple regression estimates a linear relationship between several predictor variables and a variable of interest by the method of least squares. Correlation analysis uses standard Pearson parametric correlation coefficients [Sokal and Rohlf, 1969]. Normalizing logarithmic transformations were performed for all variables except ordinal and dummy variables. These procedures have been used by Clutton-Brock and Harvey [1977a] and by other workers in quantifying ecological relationships [e.g., Peters and Raelson, 1984].

The unit of analysis used is the species, since this is the least arbitrary of all taxonomic distinctions and the most suitable for examining patterns of rarity and extinction. However, there is great variation in the quality of available data. For some species, such as red colobus (*Colobus badius*) and rhesus macaque (*Macaca mulatta*), information is available for a variety of local populations in different habitats; for other species, much less information is available.

Measures of geographic distribution (in units of 1,000 km^2) were taken primarily from Wolfheim [1983]. This measure is dependent on taxonomy, and, as noted by Crockett [1984], Wolfheim has calculated aggregate geographic ranges for some primates that might represent a number of separate species. Since generally the same taxonomic distinctions are followed here, this is not a major problem for this study.

Population density is a particularly difficult measure to compare because of differences in definition and methods of measurement. Moreover, estimates may be especially hard to obtain for endangered species precisely because they often exist at very low densities. Three population density measures were therefore utilized: average population density and, when available, minimum and maximum recorded population densities per km^2.

Due to difficulties in comparability of measures, species' characteristics are in some cases aggregated into ordinal categories, often with four or five grades. This procedure was followed because parametric statistics are relatively robust to the substitution of monotonic ranks for interval variables [Anderberg, 1973]. This technique permits written descriptions to be coded as quantitative variables. For example, arboreality was defined to quantify heights above the ground of primary activity as follows [modified from Gautier-Hion, 1978]: 0, terrestrial; 1, 0–10 m; 2, 5–20 m; 3, 10–30 m; and 4, 20+ m.

Another procedure utilized here is the use of dummy variables for presence-absence categories (mathematically equivalent to ANOVA). Habitat classification provides an example. If a species has been recorded in a given habitat, it is scored with a 1 value; if never recorded in that habitat, it is scored with a 0 value. Following Wolfheim [1983], the six habitat types utilized here: 1) open country; 2) woodland/bush/scrub; 3) canopy forest; 4) riparian swamp or other wet forest; 5) urban areas/human settlements; and 6) wet habitats only.

Geological age of primate taxa could not be assessed at the species level, because current paleontological data do not permit determination of the initial occurrence of most species, and the divergence dates of even genera and families are the subject of considerable controversy. Assessment of taxonomic age is therefore restricted to the level of the family, with one exception. Although both cercopithecines and colobines are commonly classified together in the family Cercopithecidae, their fundamental dietary and anatomical differences are sufficient to warrant separate treatment in this analysis. The dates used for the Old World monkeys are therefore the estimated date of divergence of the omnivorous cercopithecines from the primarily folivorous colobines. For purposes of internal consistency, all dates of familial divergence are taken from Szalay and Delson [1979].

Diet may be an important factor in the relative survival of primate species in the wild. To maximize comparability of diet measures, all information on primate diets has been taken from a single source [Harding, 1981]. Each species is characterized on an ordinal scale from high to low quality using a nine-grade scale based on its primary sources: 1) animal foods including vertebrate and crustacean flesh plus eggs; 2) invertebrates; 3) human crops; 4) tree parts including bark gums and decayed wood; 5) seeds and nuts; 6) fruits and berries; 7) soft plant parts including immature leaves, buds, shoots, and flowers; 8) roots and grasses; 9) mature leaves (mature and immature leaves both counted when references did not distinguish between them). When more than one food source represented a primary dietary item for a primate, values of the preferred foods were averaged.

One variable that has not been included in this analysis is biomass. In ecological studies of trophic-level interactions it is a useful measure. However, for evolutionary and population questions, numbers of surviving or breeding individuals are the relevant parameters. Biomass is effectively a combined measure of body size and population density, which are here considered separately.

In addition to analyzing the trends within the order as a whole, analyses were conducted by geographic region (Old World, New World, and Madagascar) and by taxonomic family for the larger families (Cebidae, Callitrichidae, and Ceropithecidae). To avoid repetition, the analyses of these data subsets will be discussed only when they differ from trends within the order as a whole.

RESULTS

Population Density and Geographic Range

Using all available information for the 176 primate species in the sample, the highest correlation found with population density was that with home range size ($r = -.58$, $n = 88$; see Table I): Species with high average population densities tend to occur in small home ranges.

Both female body weight ($r = -.36$, $n = 111$) and female head-body length ($r = -.31$, $n = 103$) were also inversely correlated with population density, as might be expected from the relationship between body size and nutritional requirements. However, higher correlations were found with demographic variables. Gestation length ($r = -.53$, $n = 89$) and age of female sexual maturation ($r = -.45$, $n = 63$) are both better predictors of population density than either head-body length or body weight. When multiple regression is performed using both body weight and gestation lengths

TABLE I. Correlation Coefficients (r) for the Population Density and Geographical Range With Selected Other Variables[a]

	Population density		Geographic range	
Variable	n	r	n	r
Female body weight	111	−.36***	131	.04
Female head-body length	103	−.31**	136	.02
Gestation length	89	−.53***	106	.21*
Age of female sexual maturation	63	−.45***	74	.00
Diet	72	−.01	81	−.18
Arboreality	81	.04	102	−.09
Activity period	115	.26**	148	−.01
Home range	88	−.58***	88	.08
Group size	116	−.08	137	.17
Geological age of family	113	.08	113	−.04
Open habitats	106	−.09	142	.20*
Woodland/shrub habitat	106	.18	142	.17
Canopy forest habitat	106	.13	142	−.16
Swamp/riparian habitat	106	.09	142	.06
Urban/developed habitats	106	.07	142	.13
Restricted to wet habitat	106	−.11	142	.05

[a]n: Number of sample species for which data are available for each variable. *,**, and *** indicate correlation coefficients significant at the $P < .05$, $P < .01$, and $P < .001$ levels, respectively.

as predictors of population density, the standardized partial regression coefficients indicate that gestation length is a significantly better predictor of rarity than body weight (beta coefficients of −.49 and −.05, respectively). When both variables are used to predict population density across all primate species, the multiple correlation coefficient (r = −.53) is the same as the bivariate correlation between gestation length and population density alone.

One further significant relationship was that nocturnal forms tend to have higher population densities than diurnal forms (r = +.26, n = 115). This is probably due to prosimians, which form the bulk of the nocturnal primates, having higher population densities than other primates. None of the other tested variables showed a significant relationship with population density.

Geographic range area of different primate species was tested against the same set of variables but with few significant results. However, geographic range area is positively correlated with ability to live in open country (r = +.20, n = 142) and also with female gestation length (r = +.21, n = 106).

Endangerment

If low population density or small geographical range predispose species to endangerment, then variables that correlate significantly with these measures of rarity might also be distributed differently among endangered and nonendangered species. This possibility was examined for six variables using the IUCN Red Data Book as the authority for endangered status. This list is by no means definitive (see regional reviews of conservation priorities in this volume), and some of the forms listed are only endangered at the subspecies level, whereas our data set applies to species as a whole. Of four categories of listing, we have included only those species listed as endangered, vulnerable, or rare, because the category "indeterminate" indicates a lack of information rather than confirmed rarity.

Tested variables were confined to the two earlier measures of rarity, three particularly strong correlates of population density, and one additional variable relating to species breeding systems. In each case, the values were cast into three classes in an attempt to reveal skewing towards either extreme. The results are shown in Table II together with χ^2 values for the difference between listed and unlisted species.

Consider first the taxonomic and biogeographical distribution of listed species (Table III). Three of the five Malagasy families have more than half their members listed. This amounts to 10 of 22 extant Malagasy species. None of the 11 lorises is listed, but one of the three tarsiers is listed. Both main South American families have more than a third of their members threatened, whereas only 18% of Old World Cercopithecidae are listed. A further contrast is seen if the colobine and cercopithecine subfamilies are treated separately. 32% of 28 colobine species are listed compared to only 9% of 45 cercopithecine species. Three of nine lesser apes and all four great

TABLE II. Summary of χ^2 Tests on Distribution of Variables Between Species Listed as Rare, Vulnerable, or Endangered in the Red Data Book and Non-listed Species

			Classes tested					
Variable	n	Units	Lower	Middle	Upper	χ^2	df	Prob
Female body weight	144	kg	<1	1–7.4	>7.4	0.87	2	
Gestation length	120	Days	<141	141–187	>187	11.65	2	<.01
Group home range area	91	km^2	<0.1	0.1–1.0	>1.0	1.66	2	
Geographic range area	148	1,000 km^2	<100	100–1000	<1,000	11.51	2	<.01
Max. population density	104	Indiv/km^2	<20	20–100	>100	3.57	2	
Monogamy	176	Monogamous		—	Other	0.73	1	

TABLE III. Numbers of Species in Each Family Currently Listed in the IUCN Red Data Book[a]

Family	No. species listed	Total No. species	Listed species (%)
Cheirogaleidae	2	7	29
Lemuridae	4	9	44
Lepilemuridae	1	1	100
Indriidae	2	4	50
Daubentoniidae	1	1	100
Lorisidae	0	11	0
Tarsiidae	1	3	33
Callitrichidae	9	22	41
Callimiconidae	1	1	100
Cebidae	12	31	39
Cercopithecidae	13	73	18
Colobinae	9	28	32
Cercopithecinae	4	45	9
Hylobatidae	3	9	33
Pongidae	4	4	100
Total	53	176	30

[a]A series of listed subspecies (e.g., of *Colobus badius*) is here treated as one species, and species listed as "indeterminate" are omitted.

apes are considered threatened. Thus, although all families except the Lorisidae include some listed species, Old World cercopithecines appear possibly less vulnerable to date than the other main groupings. Four small families, the Lepilemuridae, Daubentoniidae, Callimiconidae, and Pongidae, have all their members listed and thus give the gravest cause for concern.

Among possible correlates of endangered status, female body weight, group home range area, and maximum population density showed no significant difference in distribution tested between listed and nonlisted species. However, significant differences were found for geographical range area ($\chi^2 = 11.5$, $P < .01$, df = 2) and gestation length ($\chi^2 = 11.7$, $P < 0.1$, df = 1). Geographical range areas of less than 100 km^2 were disproportionately more frequent among listed species than nonlisted species. In other words, many threatened species have very small geographical ranges, and relatively few have very large ranges. Given the value of gestation length as a predictor of both population density and geographic range in this study, it was interesting to find that listed primates tend to be species with either longer- or shorter- than-average gestation lengths, whereas female body weight shows no such recognizable patern. This suggests that reproductive variables are important components of both rarity and vulnerability to extinction in primates.

Finally, Mackinnon and Mackinnon [1984] have postulated from a comparison of monogamy in gibbons and tarsiers that monogamy is evolutionarily linked to other characteristics, such as strict territoriality and specialized diet, which make such species inflexible in relation to environmental change and thus vulnerable to extinction. We therefore examined the distribution of monogamy vs. other breeding systems [after Kleiman, 1977] among listed and nonlisted species. We found no significant difference in vulnerability related to monogamy.

DISCUSSION

The study demonstrates statistically significant relationships between two measures of rarity and several features of species biology. However, the interpretation of these results is complicated by covariance among some of the tested variables. For example, body size is known to be positively related to gestation length and to home range size, both of which also correlate negatively with population density. One direct consequence of larger body size is increased vulnerability to human predation, because larger animals are easier to see and shoot, and make a bigger meal (Mittermeier, chapter 5, this volume). This will tend to depress already low densities of larger species and increase vulnerability to extinction in some cases.

One of the most consistent trends observed in the analysis is the negative association of gestation length and population density. Results of the multiple regression analysis suggest that this is not principally a function of covariance with body size and that demographic variables have important consequences in their own right. As a further test of this proposition, the data were analyzed separately in subsets for Madagascar, Old World, and New World species. Both the Old and New World primate communities follow a pattern in which most species fall into one of two groups: species with relatively short gestation lengths living at high densities and species with relatively long gestation lengths living at low densities. In Madagascar, this pattern does not occur, perhaps because of extensive recent extinctions of large-bodied lemurs.

Demographic variables such as gestation length appear to be the best predictors of population density available at present. Age of female sexual maturity in the wild might be a still better predictor, but available data are limited. A possible explanation for this is that rapidly reproducing, "r-selected" primates (i.e., with short gestation lengths and/or fast sexual maturation) living at higher densities are generally less specialized than slower reproducing, "K-selected" species which occur at lower densities. This argument could apply, for example, to the large cercopithecids and cebids, among which several congeners often coexist on subtly differing dietary proportions or modes of collecting food [e.g., Struhsaker and Leland, 1979]. Niches are therefore restricted by interspecific competition (i.e., more specialized), causing population densities to remain lower than for some faster breeding and more generalist species.

A third strong correlate of population density is home range size. This reflects the fact that many open country species, for example, *Erythrocebus patas*, have large home ranges but occur at low density as a result of sparcely distributed resources. Species with high densities, such as many *Presbytis* and *Colobus* spp., tend to live in small home ranges and are able to subsist on relatively common and homogeneously distributed resources, such as leaves.

The analysis provides fewer clues to the determinants of geographic range than population density. Perhaps most interesting is the positive association between ability to live in open country and large geographic range. Large geographic ranges among open country species may be due to the very large areas of relatively homogeneous savannah country, found particularly in Africa, and the lack of effective barriers to dispersal.

None of the other tested variables showed significant relationships with either population density or geographic range. This need not mean that they lack influence in particular cases, but in a large sample these might be masked by counteracting effects in different taxonomic groups. For example, one might expect more folivorous species to occur at higher densities than frugivorous ones, as is indeed the case in some communities [e.g., Struhsaker and Leland, 1979; Raemaekers and Chivers, 1980] because their staple food is more abundant. However, when analyzed across the whole order, this trend may be negated by low population densities in certain other herbivorous species, such as *Gorilla gorilla*, and locally high densities of some frugivores, such as some macaques. In short, analysis at this level will detect only very broad tendencies.

A comparison of Red Data Book species with nonlisted species suggests that rarity, in the sense of small geographic range, contributes strongly to species vulnerability to extinction, as might be expected from studies of island biogeography. Population density may be less significant in this regard, first, because species with low population densities often have large geographic ranges and, second, because highly endangered populations sometimes occur at locally high densities in relict habitat patches (e.g., Tana River subspecies of *Colobus badius* and *Cercocebus galeritus*, *Cheirogaleus medius*, or *Propithecus verreauxi*). In some cases, extreme arboreality might also contribute to compression of ranges by hindering colonization of isolated habitat fragments [Marsh, 1978].

Although limits to geographic ranges, especially among generalist species, are sometimes set by gradual habitat changes (e.g., the southern limits of the Sahara) or by competitive replacement along such ecotones, limits are more often set by abrupt physical barriers. The Amazon and Zaire river basins and the Southeast Asian archipelago provide numerous examples of closely related species occupying similar habitats but separated by rivers or seas. Species diversity is then largely a function of the habitat island size and time since its isolation. The ultimate causes of predisposition to endangerment through small range size in such cases are thus, ironically, the same geological and climatic changes that permitted the divergence of species in the first place by dividing formerly continuous geographic ranges.

Among other factors with possible influence, the analysis provided no evidence for the proposition that species with monogamous breeding systems are more likely to be endangered than those with other social arrangements. Nor is the original argument consistent with observations of the effects of habitat disturbance on gibbons, which are some of the animals for which it was originally proposed. Marsh and Wilson [1981], for *Hylobates lar* in Peninsular Malaysia, and Davies and Payne [1982], for *H. muelleri* in Sabah, found that although gibbon densities in logged forest were generally slightly lower than in primary forests the reduction was less than among sympatric *Presbytis* spp. and *Pongo pygmaeus*, which are species with quite different mating systems.

The least-expected finding of our study was that listed species tend to have either shorter- or longer-than-average gestation periods, whereas female body weight shows no such recognizable pattern. This suggests that reproductive variables influence vulnerability to extinction, as well as population density independently of any direct effects of body size. Long-lived species, such as *Pan troglodytes* and *Gorilla gorilla*, tend to be relatively K-selected [Wilson, 1975] with low reproductive rates and high parental investment in each offspring. These animals might be more prone to extinction because losses from hunting or capture will be made up more slowly. Species such as *Callithrix jacchus* and *Microcebus murinus*, on the other hand, with two or more young per year, are better able to replace any losses that do not result from a reduction in habitat carrying capacity.

Perhaps there is an optimal demographic pattern for primates faced with the worldwide modification of habitat and hunting by man. On the one hand, the reproductive interval should be sufficiently short to replace losses rapidly and allow for expansion to available habitat. On the other hand, gestation length must be long enough for the development of a complex neural system to permit the behavioral complexity that can facilitate adaptation to new or changed environments. Those primates falling above or below the norm may be particularly vulnerable to extinction.

ACKNOWLEDGMENTS

The authors gratefully acknowledge support received from the Fulbright Foundation, the National Science Foundation, Harvard University, and the Sabah Foundation while conducting the research reported here. Drs. J. Caldecott, T. Clutton-Brock, J. Fleagle, and J. Oates provided helpful comments on the manuscript.

REFERENCES

Anderberg MR (1973): "Cluster Analysis for Applications." New York: Academic Press.

Brown JH (1971): Mammals on mountaintops: Nonequilibrium insular biogeography. Am Naturalist 105:467–478.

Clutton-Brock TH, Harvey PH (1977a): Species differences in feeding and ranging behaviour in primates. In Clutton-Brock TH (ed): "Primate Ecology." London: Academic Press, pp 557–584.

Clutton-Brock TH, Harvey PH (1977b): Primate ecology and social organization. J Zool 183:1–39.

Clutton-Brock TH, Harvey PH (1978): Home range size, population density and phylogeny in primates. In Bernstein IS, Smith EO (eds): "Primate Ecology and Human Origins: Ecological Influences on Social Organization." New York: Garland STPM Press, pp 201–214.

Clutton-Brock TH, Harvey PH (1979): Comparison and adaptation. Proc R Soc London Series B 205:547–565.

Clutton-Brock TH, Harvey PH (1980): Primates, brains and ecology. J Zool 190:309–323.

Crockett CM (1984): The decline and fall of the world's primates. Am J Primatol 6:115–119.

Davies G, Payne J (1982): "A Faunal Survey of Sabah." Kuala Lumpur: World Wildlife Fund Malaysia.

Diamond JM (1975): The island dilemma: Lessons of modern biogeographic studies for the design of natural reserves. Biol Conserv 7:129–146.

Gautier-Hion A (1978): Food niches and coexistence in sympatric primates in Gabon. In Chivers DJ, Herbert J (eds): "Recent Advances in Primatology. Vol 1. Behaviour." London: Academic Press, pp 269–286.

Harding RSO (1981): An order of omnivores: nonhuman primate diets in the wild. In Harding RSO, Teleki G (eds): "Omnivorous Primates." New York: Columbia University Press, pp 191–214.

Harvey PH, Clutton-Brock TH (1985): Life history variation in primates. Evolution 39:359–581.

International Union for Conservation of Nature and Natural Resources (1978): "Red Data Book Volume I: Mammalia." Morges: IUCN.

International Union for Conservation of Nature and Natural Resources (1984): "Primate Specialist Group Newsletter. No. 4." Stony Brook, NY: IUCN.

Kavanagh M (1984): "A Complete Guide to Monkeys, Apes and Other Primates." New York: Viking Press.

Kleiman DG (1977): Monogamy in mammals. Quart Rev Biol 52:39–69.

Mackinnon JR, Mackinnon KS (1984): Territoriality, monogamy and song in gibbons and tarsiers. In Preuschoft H, Chivers DJ, Brockelman W, Creel N (eds): "The Lesser Apes." Edinburgh: Edinburgh University Press, pp 291–297.

Marsh CW (1978): Problems of primate conservation in a patchy environment along the lower Tana River, Kenya. In Chivers DJ, Lane-Petter W (eds): "Recent Advances in Primatology, Vol. II." New York: Academic Press, pp 85–86.

Marsh CW, Wilson WL (1981): "A Survey of Primates in Peninsular Malaysian Forests." Kuala Lumpur: Universiti Kebangsaan Malaysia.

Napier JR, Napier PH (1967): "A Handbook of Living Primates." London: Academic Press.

Peters RH, Raelson JV (1984): Relations between individual size and mammalian population density. Am Naturalist 124:498–517.

Raemaekers JJ, Chivers DJ (1980): Socio-ecology of Malayan forest primates. In Chivers DJ (ed): "Malayan Forest Primates." New York: Plenum Press, pp 279–316.

Sokal RR, Rohlf FJ (1969): "Biometry." San Francisco: W.H. Freeman and Company.

Struhsaker TT, Leland L (1979): Socioecology of five sympatric monkey species in the Kibale Forest, Uganda. Adv Study of Behav 9:159–228.

Szalay FS, Delson E (1979): "Evolutionary History of the Primates." New York: Academic Press.

Terborgh J, Winter B (1980): Some causes of extinction. In Soulé ME, Wilcox BA (eds): "Conservation Biology." Sunderland, MA: Sinauer Associates, pp 119–133.

Willis EO (1974): Populations and local extinctions of birds on Barro Colorado Island, Panama. Ecol Monogr 44:153–169.

Wilson EO (1975): "Sociobiology." Cambridge: Harvard University Press.

Wolfheim JH (1983): "Primates of the World." Seattle: University of Washington Press.

APPENDIX. Biological and Ecological Features of Primate Species

																		19					
1	2	3	4	5	6	7	8	9	10	11	12	13	14	15	16	17	18	a	b	c	d	e	f
Arctocebus	*calabarensis*	N	0.20	1	2	2.0	—	960	244	1	1	25	134	10	—	02.00	18.00	0	0	1	0	0	0
Galago	*elegantulus*	N	0.30	1	—	15.0	—	740	200	3	1	25	135	—	—	03.00	30.00	0	0	1	1	0	0
	inustus	N	0.26	1	—	—	—	110	144	3	1	25	—	—	1600	—	—	0	0	1	0	0	0
	alleni	N	0.26	1	6	15.0	—	850	200	1	1	25	—	8	—	3.00	25.00	0	1	1	1	0	1
	crassicaudatus	N	1.00	4	4	100.0	0.07	3810	315	—	1	25	135	12	3500	72.00	200.00	1	1	1	1	0	0
	demidovii	N	0.06	1	2	50.0	0.01	5080	123	2	1	25	111	8	2000	32.00	300.00	0	1	1	1	0	0
	senegalensis	N	0.26	2	2	216.0	0.03	10650	152	2	1	25	124	7	2000	0.77	500.00	1	1	1	0	0	0
Loris	*tardigradus*	N	0.30	1	2	100.0	0.01	740	224	2	1	25	163	13	1524	0.01	0.01	0	1	1	1	0	0
Nycticebus	*coucang*	N	1.40	1	2	—	—	4030	302	2	1	25	193	—	1646	—	—	0	0	1	0	0	0
	pygmaeus	N	0.70	1	—	—	—	450	—	2	1	25	—	—	—	—	—	—	—	—	—	—	—
Perodicticus	*potto*	N	1.30	1	6	8.0	0.10	3280	327	2	1	25	193	18	3500	5.00	28.00	0	1	1	1	0	0
Tarsius	*bancanus*	I	0.12	4	6	98.0	0.02	440	127	2	1	30	180	—	1000	80.00	116.00	0	0	1	0	0	0
	spectrum	N	0.13	4	2	—	—	60	127	2	1	30	157	14	1400	—	—	—	—	—	—	—	—
	syrichta	E	0.12	4	—	—	—	50	127	2	1	30	180	—	—	—	—	0	0	1	0	0	0
Allocebus	*trichotis*	R	—	—	—	—	—	—	—	—	1	40	—	—	—	—	—	0	0	1	1	0	1
Cheirogaleus	*major*	N	0.45	1	6	—	—	80	—	—	1	40	70	—	—	—	—	0	0	1	0	0	0
	medius	I	0.25	1	6	350.0	0.04	80	228	1	1	40	—	—	—	300.00	400.00	0	0	1	0	0	0
Microcebus	*coquereli*	E	0.38	1	2	50.0	0.07	40	250	1	1	40	89	9	—	30.00	300.00	0	0	1	0	0	0
	murinus	N	0.08	1	2	400.0	0.02	330	130	1	1	40	62	9	—	250.00	400.00	0	1	1	1	0	1
	rufus	I	0.06	—	—	—	—	—	—	—	—	—	—	—	—	—	—	—	—	—	—	—	—
Phaner	*furcifer*	I	0.30	2	—	250.0	0.04	60	263	2	1	40	—	—	1000	100.00	850.00	0	1	1	1	0	0
Hapalemur	*griseus*	V	1.00	4	9	54.0	—	70	295	2	0	45	140	—	1000	47.00	62.00	0	1	1	1	0	1
	simus	R	2.50	—	—	—	—	—	375	2	0	45	—	—	—	—	—	—	—	—	—	—	—

(continued)

APPENDIX. Biological and Ecological Features of Primate Species (Continued)

																		19					
1	2	3	4	5	6	7	8	9	10	11	12	13	14	15	16	17	18	a	b	c	d	e	f
Lemur	*catta*	N	2.50	18	7	233.0	0.07	100	380	1	0	45	135	—	—	74.00	300.00	0	1	1	1	0	0
	coronatus	N	—	—	—	—	—	7	—	2	0	45	118	—	—	—	—	1	1	1	0	0	0
	fulvus	N	2.10	10	—	1125.0	0.01	200	—	2	0	45	128	10	—	900.00	1222.00	0	0	1	0	0	0
	macaco	E	2.00	10	8	58.0	—	10	375	2	0	45	128	24	—	—	—	0	0	1	1	0	1
	mongoz	V	1.80	3	9	350.0	0.01	20	375	2	0	45	128	—	—	350.00	350.00	0	0	1	1	0	0
	rubriventer	N	2.00	4	9	—	—	40	375	2	0	45	128	—	—	—	—	0	0	1	1	0	1
	variegatus	N	3.00	4	—	—	—	50	600	2	0	45	102	5	—	—	—	0	0	1	1	0	1
Lepilemur	*mustelinus*	E	0.64	1	9	250.0	0.01	150	316	2	1	45	135	21	—	100.00	810.00	0	1	1	1	0	0
Avahi	*laniger*	I	1.30	3	8	—	—	90	320	—	1	15	150	—	—	—	—	0	0	1	1	0	0
Indri	*indri*	E	6.25	3	7	12.5	0.23	30	700	—	0	15	160	—	—	3.00	16.00	0	0	1	1	0	1
Propithecus	*diadema*	E	6.00	4	7	3.0	—	70	495	—	0	15	130	—	—	3.00	3.00	0	0	1	1	0	1
	verreauxi	I	3.50	7	6	103.00	0.08	140	495	—	0	15	140	30	—	30.00	530.00	0	0	1	1	0	0
Daubentonia	*madagascariensis*	E	2.00	1	2	—	0.05	—	400	—	1	15	—	29	—	—	—	0	0	1	1	0	1
Callimico	*goeldii*	R	0.44	6	—	1.5	0.50	1070	190	—	0	34	154	9	—	—	—	0	0	1	0	0	0
Callithrix	*argentata*	V	0.27	6	—	5.5	—	1280	200	—	0	34	140	—	—	—	—	0	0	1	0	0	0
	humeralifer	V	0.27	11	—	30.0	—	350	220	—	0	34	140	—	—	30.00	30.00	—	—	—	—	—	—
	jacchus	N	0.38	11	—	700.0	—	2590	200	—	0	34	148	12	—	700.00	700.00	0	1	1	0	0	0
	aurita	E	0.43	6	—	—	0.11	—	240	1	0	34	140	—	—	—	—	—	—	—	—	—	—
	flaviceps	E	—	—	—	—	—	—	248	—	0	34	140	—	—	—	—	—	—	—	—	—	—
	geoffroyi	N	—	—	—	—	—	—	198	—	0	34	140	—	—	—	—	—	—	—	—	—	—
	penicillata	N	0.18	7	—	—	—	—	220	—	0	34	140	—	—	—	—	—	—	—	—	—	—

Cebuella	*pygmaea*	N	0.15	7	2	250.0	0.01	1470	138	1	0	34	136	24	—	34.00	560.00	0	1	1	1	0	0
Leontopithecus	*rosalia*	E	0.50	4	2	—	—	50	261	1	0	34	129	18	700	—	—	0	0	1	1	0	1
	chrysomelas	E	—	4	—	—	0.40	—	—	1	—	—	129	—	—	—	—	—	—	—	—	—	—
	chrysopygus	E	—	4	—	—	—	—	—	1	—	—	129	—	—	—	—	—	—	—	—	—	—
Saguinus	*bicolor*	I	—	—	—	—	—	160	230	—	0	34	143	—	—	—	—	0	1	1	1	0	0
	fuscicollis	N	0.37	5	4	14.0	0.33	1820	218	—	0	34	149	—	1370	2.40	30.00	0	1	1	1	0	0
	geoffroyi	N	—	6	6	30.0	0.29	—	—	—	0	34	143	—	800	27.00	36.00	—	—	—	—	—	—
	imperator	I	0.33	3	—	12.0	0.35	250	218	—	0	34	143	—	300	2.00	12.00	0	0	1	0	0	0
	inustus	N	—	—	—	—	—	480	230	—	0	34	143	—	—	—	—	0	0	1	1	0	0
	labiatus	N	0.33	6	—	16.0	—	250	240	—	0	34	143	—	—	3.80	27.60	0	0	1	0	0	0
	leucopus	V	—	4	—	9.0	—	40	230	—	0	34	143	—	1500	2.50	15.20	0	1	1	1	0	0
	midas	N	0.33	5	—	15.0	0.09	1570	218	2	0	34	127	20	—	2.10	32.00	0	1	1	1	0	0
	mystax	N	0.63	5	—	—	—	990	218	—	0	34	143	—	610	—	—	0	0	1	1	0	0
	nigricollis	N	0.49	7	—	12.0	0.04	490	218	—	0	34	145	12	914	10.00	19.20	0	1	1	1	0	0
	oedipus	E	0.44	6	2	30.0	0.09	140	240	—	0	34	145	18	1500	20.00	180.00	0	1	1	1	0	0
Alouatta	*belzebul*	N	5.50	20	—	13.0	—	1490	483	—	0	40	187	—	—	—	—	0	0	1	0	0	0
	caraya	N	4.40	8	—	130.0	0.01	2570	600	—	0	40	187	—	—	0.80	0.80	0	0	1	1	0	0
	fusca	I	5.50	—	—	—	—	850	483	—	0	40	187	—	1150	—	—	0	0	1	1	0	0
	palliata	N	5.70	17	—	70.0	0.38	720	483	—	0	40	187	45	2000	5.00	1030.00	0	0	1	1	0	0
	villosa	I	6.70	6	7	5.0	—	270	—	—	0	40	187	—	400	5.00	5.00	0	0	1	1	0	1
	seniculus	N	7.20	5	8	20.0	0.10	5570	483	3	0	40	187	—	3200	1.60	120.00	0	0	1	1	0	1
Aotus	*trivirgatus*	N	1.00	3	6	25.0	3.10	6840	358	2	1	40	133	—	3200	1.50	150.00	0	0	1	0	0	0
Ateles	*paniscus*	V	6.20	5	—	15.0	0.30	6570	503	3	0	40	—	—	2500	1.00	45.00	0	0	1	1	0	1
	belzebuth	V	5.80	20	6	13.5	—	—	500	—	0	40	—	—	—	—	—	—	—	—	—	—	—
	geoffroyi	V	7.80	12	6	—	0.60	—	453	—	0	40	229	48	—	—	—	—	—	—	—	—	—
	fusciceps	I	5.80	—	—	—	—	—	530	—	0	40	226	51	—	—	—	—	—	—	—	—	—

(continued)

APPENDIX. Biological and Ecological Features of Primate Species (Continued)

																		19					
1	2	3	4	5	6	7	8	9	10	11	12	13	14	15	16	17	18	a	b	c	d	e	f
Brachyteles	*arachnoides*	E	9.50	8	—	28.0	5.00	500	518	—	0	40	—	—	1500	4.00	52.00	0	0	1	1	0	1
Cacajao	*calvus*	V	3.55	15	—	—	—	430	405	—	0	40	—	60	—	—	—	0	0	1	1	0	1
	melanocephalus	V	3.55	25	—	—	—	640	405	—	0	40	—	36	—	—	—	0	0	1	1	0	1
Callicebus	*moloch*	N	1.00	3	8	13.0	0.01	4940	335	—	0	40	—	—	850	1.80	31.00	0	1	1	1	0	1
	personatus	V	—	—	—	—	—	620	400	—	0	40	—	—	900	—	—	0	0	1	0	0	0
	torquatus	N	1.10	3	6	9.0	0.20	1790	342	—	0	40	—	—	500	2.00	15.20	0	0	1	0	0	0
Cebus	*albifrons*	N	2.20	8	—	15.0	0.20	3660	400	—	0	40	180	43	2000	3.80	34.00	0	0	1	1	0	1
	apella	N	3.10	9	4	8.0	0.10	11730	400	2	0	40	160	42	2700	2.00	111.00	0	0	1	1	0	1
	capucinus	N	2.70	10	6	17.0	0.86	450	402	—	0	40	180	—	2100	50.00	50.00	0	0	1	1	0	1
	nigrivittatus	N	3.10	10	—	4.0	—	2020	400	2	0	40	180	—	2000	1.20	73.80	0	0	1	1	0	1
Chiropotes	*albinasus*	V	2.95	20	—	—	—	600	460	—	0	40	—	—	—	—	—	0	0	1	1	0	1
	satanas	E	2.95	15	—	7.5	0.20	2090	460	—	0	40	—	—	161	3.90	82.40	0	0	1	1	0	1
Lagothrix	*flavicauda*	E	—	7	—	—	—	40	525	—	0	40	225	48	3000	—	—	—	—	—	—	—	—
	lagotricha	V	5.80	10	6	7.0	10.00	3580	485	—	0	40	225	48	3000	6.00	45.00	0	0	1	1	0	1
Pithecia	*monachus*	N	1.41	5	—	7.0	—	2350	400	—	0	40	—	—	762	0.62	30.00	0	0	1	1	0	1
	pithecia	N	1.41	3	6	4.0	7.00	1680	363	2	0	40	163	—	350	2.50	12.00	0	0	1	1	0	1
	hirsuta	N	—	44	—	—	—	—	400	—	0	40	—	—	—	—	—	—	—	—	—	—	—
	albicans	N	—	—	—	—	—	—	380	—	0	40	—	—	—	—	—	—	—	—	—	—	—
Saimiri	*sciureus*	N	0.63	30	6	13.0	0.80	5440	260	2	0	40	170	—	2000	7.50	585.00	0	0	1	1	0	1
	oerstedii	E	0.93	30	6	66.0	—	—	270	2	0	40	—	—	—	—	—	—	—	—	—	—	—
Allenopithecus	*nigroviridis*	N	—	4	6	—	—	440	410	—	0	17	193	—	—	—	—	0	0	1	1	0	1
Cercocebus	*albigena*	N	6.40	15	6	27.0	4.10	1590	508	2	0	17	177	48	1700	8.50	77.00	0	0	1	1	0	1
	aterrimus	N	—	—	—	—	—	1080	508	2	0	17	—	—	—	7.70	500.00	0	0	1	0	0	0
	galeritus	E	5.50	10	—	10.0	2.00	890	480	1	0	17	171	—	—	—	—	0	0	1	1	0	1
	torquatus	N	7.50	20	6	133.0	—	1400	480	1	0	17	171	32	350	0.50	133.00	0	0	1	1	0	1

Cercopithecus	*aethiops*	N	3.50	29	6	135.0	0.30	12300	425	1	0	17	163	30	4500	0.87	154.00	1	1	0	1	0	0
	albogularis	N	4.54	—	—	—	—	—	460	—	—	—	—	—	—	—	—	—	—	—	—	—	—
	ascanius	N	2.90	35	6	140.0	0.28	2710	408	—	0	17	—	—	2000	65.00	190.00	0	1	1	1	0	0
	campbelli	N	3.50	8	6	0.7	—	—	450	—	—	—	—	40	—	—	—	—	—	—	—	—	—
	cephus	N	4.50	10	6	22.0	0.35	710	461	2	0	17	—	—	—	12.00	39.00	0	0	1	1	0	1
	diana	N	4.00	25	6	—	0.25	330	422	3	0	17	—	—	—	—	—	0	0	1	0	0	0
	erythrogaster	N	—	—	—	—	—	40	410	—	0	17	—	—	—	—	—	0	0	1	0	0	0
	erythrotis	N	—	30	—	—	—	90	420	—	0	17	—	—	—	—	—	0	0	1	1	0	1
	hamlyni	N	—	10	—	—	—	420	418	3	0	17	—	—	4600	—	—	0	0	1	0	0	0
	l'hoesti	N	7.50	10	—	6.0	8.50	430	490	1	0	17	—	—	2500	5.00	8.00	0	0	1	1	0	1
	mitis	N	6.50	25	6	42.0	0.73	3530	500	2	0	17	140	62	3300	42.00	183.00	0	0	1	1	0	1
	mona	N	4.00	9	—	—	0.03	2810	418	3	0	17	—	—	2000	—	—	0	0	1	1	0	1
	neglectus	N	4.00	5	6	33.0	0.10	2570	465	1	0	17	182	48	2100	28.00	50.00	0	0	1	1	0	1
	nictitans	N	4.20	20	6	30.0	0.67	1640	505	3	0	17	—	—	—	22.50	30.00	0	0	1	1	0	1
	petaurista	N	1.80	18	—	—	—	630	391	—	0	17	—	—	—	—	—	0	0	1	0	0	0
	pogonias	N	3.00	15	6	22.5	0.78	890	457	3	0	17	—	—	—	22.50	22.50	0	0	1	1	0	1
	wolfi	N	—	—	—	—	—	—	—	—	—	—	—	—	—	—	—	—	—	—	—	—	—
Colobus	*angolensis*	N	—	5	—	190.0	—	1910	—	3	0	18	—	—	3000	180.00	200.00	0	0	1	1	0	1
	badius	R	5.80	50	9	298.0	0.75	1890	553	3	0	18	—	—	—	45.00	880.00	0	0	1	1	0	1
	guereza	N	9.30	11	9	104.0	0.15	2340	563	3	0	18	—	—	4500	0.10	300.00	0	0	1	0	0	0
	polykomos	N	7.10	10	8	40.0	0.20	940	623	3	0	18	170	—	—	23.50	40.00	0	0	1	0	0	0
	satanas	V	9.50	13	5	30.0	0.60	300	—	3	0	18	195	—	3000	—	—	0	0	1	1	0	1
	verus	R	3.50	13	8	—	—	510	463	2	0	18	—	—	—	—	—	0	0	1	1	0	1

(continued)

APPENDIX. Biological and Ecological Features of Primate Species (Continued)

																		19					
1	2	3	4	5	6	7	8	9	10	11	12	13	14	15	16	17	18	a	b	c	d	e	f
Erythrocebus	*patas*	N	7.00	20	—	0.3	52.00	6180	663	0	0	17	170	33	—	0.08	0.08	1	1	0	0	0	0
Macaca	*arctoides*	N	12.00	27	6	—	—	2950	585	1	0	17	—	—	2400	—	—	0	0	1	0	0	0
	assamensis	N	7.00	23	—	—	—	1660	605	1	0	17	—	—	3500	—	—	0	0	1	0	0	0
	cyclopis	N	—	—	—	—	—	30	—	1	0	17	—	—	3048	—	—	—	—	—	—	—	—
	fascicularis	N	4.10	13	6	39.0	0.37	2670	445	3	0	17	162	52	2000	5.00	400.00	0	1	1	1	0	0
	fuscata	N	10.00	75	—	20.0	4.00	180	535	1	0	17	173	42	1500	2.00	104.00	0	0	1	0	0	0
	mulatta	N	6.50	21	—	44.0	2.30	6220	500	1	0	17	167	34	4300	0.01	1000.00	0	1	1	0	1	0
	nemestrina	I	7.80	20	6	20.0	2.00	3060	515	1	0	17	167	35	1700	1.50	51.00	0	0	1	1	0	1
	nigra	N	6.40	8	—	—	—	90	555	1	0	17	—	—	2500	—	—	0	0	1	0	0	0
	pagensis	I	—	—	—	—	—	—	—	—	—	—	—	—	—	—	—	—	—	—	—	—	—
	radiata	N	6.50	30	6	—	2.00	660	539	1	0	17	162	42	2134	—	—	0	1	1	0	1	0
	silenus	E	6.00	20	6	3.8	3.50	40	460	1	0	17	—	—	1500	3.80	3.80	0	0	1	1	0	1
	sinica	N	3.60	35	6	100.0	0.40	50	440	1	0	17	—	—	1524	0.30	100.00	0	1	1	1	0	1
	sylvana	V	—	20	—	41.0	—	20	600	1	0	17	—	46	2160	12.00	70.00	0	0	1	0	0	0
Miopithecus	*talapoin*	N	1.00	85	6	65.0	1.20	1110	355	1	0	17	196	48	700	26.00	92.00	0	0	1	1	0	1
Theropithecus	*gelada*	N	13.60	127	8	82.0	1.00	210	575	0	0	17	170	50	4500	15.00	100.00	1	0	0	0	0	0
Papio	*anubis*	N	12.00	20	8	26.0	0.94	7570	613	0	0	17	180	45	4500	1.10	58.00	1	1	1	0	0	0
	cynocephalus	N	19.00	50	8	10.0	10.00	3170	826	0	0	17	175	51	1000	9.70	60.00	1	1	1	0	0	0
	hamadryas	N	9.40	32	—	1.8	—	1300	800	0	0	17	172	45	2300	1.80	3.40	1	0	0	0	0	0
	leucophaeus	E	10.00	15	8	—	—	150	416	0	0	17	176	42	—	—	—	0	0	1	1	0	1
	papio	N	—	69	—	14.0	12.00	380	600	0	0	17	184	45	—	12.00	15.00	1	1	1	0	0	0
	sphinx	N	15.00	52	—	—	40.00	410	600	0	0	17	173	45	—	—	—	0	0	1	0	0	0
	ursinus	N	18.00	40	8	7.0	10.00	2890	700	0	0	17	187	38	3000	2.30	43.00	1	1	0	0	0	0

Presbytis	*aygula*	N	6.25	11	8	20.0	0.20	1220	—	2	0	18	168	42	2225	3.80	64.00	0	0	1	1	0	1
	melalophos	N	6.60	9	—	74.0	0.21	—	564	3	0	18	168	42	—	74.00	74.00	—	—	—	—	—	—
	cristata	N	6.20	25	9	40.0	0.10	2000	564	3	0	18	168	42	1737	6.40	120.00	0	0	1	1	0	1
	entellus	N	11.40	19	8	57.0	3.40	3700	564	1	0	18	168	42	4267	2.80	135.00	0	0	1	1	1	0
	francoisi	N	8.10	—	—	—	—	100	564	3	0	18	168	42	—	—	—	—	—	—	—	—	—
	frontata	N	6.30	—	—	5.7	—	400	564	3	0	18	168	42	300	5.70	5.70	0	0	1	0	0	0
	geei	R	8.10	10	8	—	4.00	6	564	3	0	18	168	42	2400	—	—	0	0	1	0	0	0
	hosei	N	6.30	8	—	10.0	—	—	—	—	0	18	—	—	—	6.40	12.00	—	—	—	—	—	—
	johnii	V	9.50	13	—	90.0	1.50	20	564	2	0	18	168	42	2400	21.00	60.00	0	0	1	1	0	1
	obscura	N	6.60	10	8	31.0	0.29	220	564	4	0	18	150	42	1800	10.00	10.00	0	0	1	1	0	1
	phayrei	N	8.10	14	—	4.5	—	1200	564	3	0	18	168	42	—	2.10	6.80	0	0	1	0	0	0
	pileata	N	8.10	9	—	16.0	—	320	564	3	0	18	168	42	—	1.50	31.00	0	0	1	1	0	1
	potenziani	I	8.10	4	—	17.0	0.18	8	564	3	0	18	168	42	—	15.00	20.00	0	0	1	1	0	0
	rubicunda	N	6.30	6	—	12.0	—	680	564	3	0	18	168	42	2000	12.00	12.00	0	1	1	0	0	0
	senex	N	7.80	8	9	154.0	0.12	40	564	3	0	18	168	42	2195	92.00	327.00	0	0	1	1	0	0
Pygathrix	*nemaeus*	E	—	8	—	—	—	540	686	—	0	18	165	48	2000	—	—	0	0	1	0	0	0
Rhinopithecus	*avunculus*	N	—	—	—	—	—	100	620	—	0	18	—	—	—	—	—	0	0	1	0	0	0
	bieti	N	7.50	30	7	—	15.00	—	790	—	—	—	195	48	3000	—	—	—	—	—	—	—	—
	brelichi	N	7.50	—	—	—	—	—	750	—	—	—	—	—	—	—	—	—	—	—	—	—	—
	roxellanae	R	7.50	100	7	—	—	1420	620	—	0	18	195	48	3300	—	—	0	0	1	0	0	0
Simias	*concolor*	E	7.20	4	9	9.0	0.28	8	505	—	0	18	—	—	—	5.00	12.00	0	0	1	0	0	0
Nasalis	*larvatus*	V	10.00	20	6	8.9	0.10	740	573	3	0	18	166	—	—	8.90	8.90	0	0	1	1	0	1

(continued)

APPENDIX. Biological and Ecological Features of Primate Species (Continued)

																		19					
1	2	3	4	5	6	7	8	9	10	11	12	13	14	15	16	17	18	a	b	c	d	e	f
Hylobates	*agilis*	N	5.90	5	—	4.3	0.21	—	—	—	0	26	210	—	1212	—	—	—	—	—	—	—	—
	concolor	I	5.70	4	—	4.0	—	660	515	3	0	26	210	72	—	—	—	0	0	1	0	0	0
	hoolock	N	6.70	4	—	4.0	0.23	820	515	3	0	26	210	84	1370	1.60	14.00	0	0	1	0	0	0
	klossii	V	5.80	4	—	2.8	0.20	8	515	3	0	26	210	84	—	26.00	30.00	0	0	1	0	0	0
	lar	N	5.50	3	6	4.0	0.54	1930	515	3	0	26	205	108	2400	0.68	21.00	0	0	1	1	0	0
	moloch	E	5.90	4	6	2.7	0.20	—	—	—	0	—	—	—	—	—	—	—	—	—	—	—	—
	muelleri	N	5.50	4	—	3.2	0.36	—	—	—	0	—	—	—	—	—	—	—	—	—	—	—	—
	pileatus	E	5.50	4	—	5.0	0.25	180	—	3	0	26	225	84	—	2.00	4.00	0	0	1	1	0	0
Symphalangus	*syndactylus*	N	10.60	4	6	0.8	0.40	180	534	3	0	26	231	—	3800	0.50	26.00	0	0	1	1	0	1
Gorilla	*gorilla*	E	93.00	10	7	1.8	6.20	700	1625	0	0	26	256	78	3500	0.30	7.00	0	1	1	1	0	0
Pan	*paniscus*	V	35.00	5	6	0.5	55.00	390	—	1	0	26	238	106	—	0.30	0.90	0	0	1	1	0	1
	troglodytes	V	35.00	2	6	1.3	7.00	2500	—	1	0	26	228	118	1000	0.05	26.00	1	1	1	1	0	0
Pongo	*pygmaeus*	E	42.00	2	6	2.0	5.00	340	768	3	0	26	260	84	2000	0.10	5.00	0	0	1	1	0	1

Column heads and abbreviations: 1, genus name; 2, species name; 3, IUCN Red Data Book status (N, not listed; I, indeterminate; R, rare; V, vulnerable; E, endangered); 4, female body weight (kg); 5, average group size; 6, diet quality; 7, average population density per km^2; 8, home range size in km^2; 9, geographic distribution in 1,000 Km^2; 10, female head-body length (mm); 11, arboreality measure (0, terrestrial; 1, lower strata and ground (0–10 m); 2, middle and lower strata (5–20 m); 3, upper and middle strata (10–30 m); 4, upper strata only); 12, diurnal (0)/nocturnal (1); 13, geological age of taxonomic family (millions of years); 14, gestation length (days); 15, age of female sexual maturation (months); 16, maximum elevation at which species can live (m); 17, minimum recorded population density per km^2; 18, maximum recorded population density per km^2; 19, habitat variable; 0, species cannot occupy habitat; 1, species can occupy habitat: a, open country; b, woodland/bush/shrub; c, canopy forest; d, wet environments (swamps, riparian areas); e, towns, settled areas; f, only lives in wet habitat.

Primate Conservation in the Tropical Rain Forest, pages 83–107

4

Effects of Habitat Disturbance on Rain Forest Primates

C.W. Marsh, A.D. Johns, and J.M. Ayres

Sabah Foundation, Kota Kinabalu, Sabah, Malaysia (C.W.M.), and Department of Anatomy, University of Cambridge, Cambridge CB2 1QS, England (A.D.J., J.M.A.)

INTRODUCTION

Although loss of undisturbed habitat is widely recognized as the most serious threat to the survival of rain forest primates, the different forms this can take, and their effects on primates, have received only limited study to date. Not all disturbance amounts to the total removal of natural forest and its replacement with pasture, crops, or construction. Two of the most widespread forms of exploitation are commercial logging and shifting cultivation, neither of which normally involves the total destruction of forest over large areas but rather its modification. Because the total area of forest disturbed in these ways is increasing rapidly, there is an urgent need to determine the effects on particular primate species and the extent of variation between species and habitats. While the two main practical goals of such research are to predict population declines and, where appropriate, to suggest management action, there may also be opportunities to explore questions of more academic interest, especially ones concerning feeding ecology and mechanisms of population regulation. For purposes of study, therefore, habitat disturbance can be viewed as a form of ecological experiment, albeit usually an unplanned and unwanted one. Our aim in this paper is thus to stimulate research in a field that is both important practically and potentially fertile academically.

For the present purposes most kinds of disturbance to forest can be placed in one of six categories according to their nature and scale and the level of technology employed: 1) removal of selected plant products by traditional means, 2) clearance or damage of the forest understorey, 3) changes in water regime, 4) shifting cultivation, 5) commercial selective logging, and 6) large-scale clearance, usually for agriculture or ranching.

We are aware that these categories are not exclusive and that different kinds of disturbance often follow one another. For example, clear-felling in many areas is preceded by either selective logging or a phase of shifting cultivation. Moreover, logging and road construction open up forest to hunters and trappers, so that the effects of habitat disturbance per se are hard to distinguish from other forms of exploitation. Nevertheless, some discrimination between these problems is the first step towards understanding them more fully.

REMOVAL OF SELECTED PLANT PRODUCTS BY TRADITIONAL MEANS

The first category of disturbance includes a wide range of activities whose only common denominator is that they do not involve gross destruction of the habitat. Indeed, the desirability of continuing many of these practices provides an important reason for conserving the forest against more drastic forms of exploitation [Myers, 1979]. Probably all traditional cultures in rain forest collect wild fruits, and from this practice have come cultivated orchards and improved domestic varieties. Nevertheless, extensive harvesting of, for example, fruits of the leguminous tree, *Parkia speciosa*, in a Malaysian forest could reduce the supply of an important source of food for langurs, *Presbytis* spp. [Curtin, 1980]. In the Shola forests of south India, harvesting of scarce *Artocarpus* fruits similarly deprives the lion-tailed macaque, *Macaca silenus*, of one of its most important year-round foods [Green and Minkowski, 1977].

More commonly, exploitation of forest products requires whole trees to be cut. This may be done for incidental reasons, such as access to a bees' nest that is otherwise inaccessible, or for some intrinsic property of the wood or foliage. For example, destructive exploitation of *Castilloa ulei* (used to produce latex), and *Couma macrocarpa* and *Ragala sanguinolenta* (used in the production of nonelastic gums) is common in parts of the Amazon basin. These trees are an important food source for bearded sakis, *Chiropotes albinasus* and *C. satanas*, and woolly monkeys, *Lagothrix lagotricha* [Ayres, 1983].

Another widespread practice of this kind is the cutting of selected trees for fashioning into dug-out canoes. Since the finished product must be hauled to water, this is mainly a problem of riverine forests. In one study, in Kenya, a survey of canoes along 25 km of the Tana River revealed 75 canoes, made from six species of tree, each with a characteristic cutting time and useful life on the river [Marsh, 1980]. The most commonly used tree was a species of free-standing fig, *Ficus sycomorus*, which could be carved in 2 weeks, but usually rotted in little over a year. Cutting requirements to maintain the existing number of canoes were estimated at 18 trees per annum of this species. Unfortunately, as this species is also the principal food source of the endangered Tana subspecies of red colobus, *Colobus badius rufomitratus*, and crested mangabey, *Cercocebus galeritus galeritus*, which occur in forest patches totalling only about 25 km^2 [Marsh, 1978], the removal of even a few large trees each year might significantly affect the prospects for survival of these populations. In this case, the study suggested a management alternative; it was found that one of the most common forest trees, *Diospyros mespiliformis*, provides almost no food for red colobus (although somewhat more for mangabeys [Homewood, 1978]), and is also suitable for making canoes. Indeed, because *Diospyros* trees take almost 3 weeks to cut and last for 3 years, as compared with a 2-week cutting time and a 1-year life for figs, they seem to offer a better investment.

In general, then, the short-term consequences for wildlife of selectively removing a plant species or part depend principally on its food value to particular animals. On this basis, three classes of effect can be distinguished: harmful, neutral, and beneficial. Although harmful effects are often all too clear, those that may be neutral or beneficial are more problematic. In principle, the removal of an uncommon tree species not used for food by a particular primate may be of little or no consequence for it. Examples—for leaf-eating species at least—may include certain trees from which oils can be extracted for perfume manufacture (e.g., *Aquillaria malaccensis* in Thailand and *Aniba duckei* in the middle Amazon) and others yielding poisons (e.g., *Antiaris toxicaria* in Malaysia). The problems with such categorization are, first, that it can only be upheld from the parochial viewpoint of one kind of animal (what about others?) and, second, that it is very hard to be sure that the removal of one species has no indirect consequences for another. Gilbert [1980] gives some examples of animals that are significant factors in the persistence of several plant species, which in turn support otherwise separate food webs. He calls these *mobile links* in the ecosystem.

The third possible effect of selective tree removals is that they benefit primates by removing nonfood species and encouraging the growth of food species. Proud and Hutchinson [1980] have even recommended that such "liberation thinning" around fruit trees be considered as a management technique to increase the density of frugivorous primates. However, since the appropriate experiments have yet to be carried out (and would require long-term monitoring in which hunting was excluded as a population determinant), this suggestion is at present of mainly academic interest [see also comments by P. Ashton in Proud and Hutchinson, 1980].

In the long-term, an alternative means of satisfying demand for a specific forest product is to cultivate it, a classic case being provided by the history of rubber production. Tapping of wild *Hevea brasiliensis* and other species began as a rather inefficient and destructive forest practice in nineteenth century Amazonia, but, after the successful introduction of *H. brasiliensis* to Malaya in 1877, it has since developed into a large and sophisticated plantation industry. Unfortunately, in neither host nor parent country has this benefitted natural forests. In Malaysia and elsewhere in Southeast Asia, large areas were cut down to make way for plantations; in Brazil the collapse of the rubber economy removed a prime reason for maintaining natural forest, thus inviting more destructive later forms of exploitation such as cattle ranching [Santos, 1980]. Undoubtedly, there are many other species that are at present harvested on a modest scale but which could be cultivated to increase production. Rattans (stemless palms), for example, are on the verge of becoming a commercial crop in parts of Southeast Asia. The effects of minor forest product removal on wildlife thus present a dilemma for conservation. On the one hand, the exploitation of some species need not be detrimental to other wildlife and constitutes a good reason for maintaining tropical forests. On the other hand, if demand for a "minor" forest product grows, either overexploitation of wild stocks or clearance for its cultivation will follow.

CLEARANCE OR DAMAGE OF THE UNDERSTOREY

A second class of habitat disturbance is the clearance of forest undergrowth, leaving the middle and upper canopies intact. This might not be immediately harmful to primate populations where these do not rely on low vegetation for food resources. In parts of the Western Ghats of South India both the lion-tailed macaque, *Macaca silenus*, and the Nilgiri langur, *Presbytis johnii*, are widespread in cardamum plantations, which require shade and are planted beneath a natural forest canopy following removal of the understorey [Green and Minkowski, 1977]. Similarly, at Polonnaruwa, in Sri Lanka, where the undergrowth has been cleared around some of the archaeological ruins, primates still occur at high densities in the canopy [Hladik and Hladik, 1972; Dittus, 1977]. In parts of Amazonia, clearing of small areas of the forest understorey is followed by the establishment of nurseries for certain tree crops, such as *Paulinia cupana* and *Hevea brasiliensis*. Cutting of understorey trees for firewood or production of charcoal is also common around small towns and villages. The consequences of such activities are unknown, but they are likely to affect most those primates that spend a large proportion of their time in the lower forest levels and feed partially or largely on insects (e.g., *Saimiri sciurus*, *Cebus nigrivitatus*, *C. apella*, *Saguinus midas* [Mittermeier and Van Roosmalen, 1981] and *Cebus albifrons*, *Callicebus moloch*, *Callithrix humeralifer* [Ayres, 1983]). The more general problem associated with removal of the understorey is that natural regeneration is prevented when large canopy trees die and they are not replaced. Thus the forest is doomed in the long term.

Whereas under-shade plantations, archeological sites, and firewood collection are overt forms of human disturbance, essentially the same result can be caused by burning of the understorey. Some forest formations in the monsoonal tropics, such as the miombo woodlands (*Brachystegia-Julbernardia-Isoberlina*) of southern Africa [Malaisse, 1978] and the sal (*Shorea robusta*) and teak (*Tectona grandis*) forests in parts of India and Southeast Asia [Seth and Kaul, 1978], are adapted to regular burning of the understorey and indeed are maintained by fire. In many other forest types, however, fire can severely damage saplings, thus reducing the eventual carrying capacity of the forest for primates. This problem is particularly acute in areas of relict forest, such as those along the Tana River, in Kenya, that abut onto grasslands where high-temperature fires can be generated in the dry season. This has undoubtedly contributed to compression of groups' ranging patterns and possibly to local extinctions [Marsh, 1978].

Forest regeneration also can be reduced or prevented in less direct ways. For instance, elephants can flatten the undergrowth over wide areas and show selective preferences for certain species of sapling. Increases in elephant densities are themselves often a result of traditional ranging patterns becoming compressed by human pressures surrounding protected areas. Laws et al. [1975] showed this in the Budongo Forest, Uganda, and suggested that dominance in some areas by ironwood (*Cynometra alexandri*) is a deflected climax induced by elephants. This could benefit chimpanzees, *Pan troglodytes*, which feed heavily on *Cynometra* fruits, at the expense of certain monkeys that do not [Albrecht, 1976].

CHANGES IN WATER REGIME

Alteration of drainage patterns and river regimes by roads, dams, and irrigation schemes can critically alter the vegetation both above and below the primary site of disturbance. Roads, and particularly logging roads, in the humid tropics are often built with inadequate culverts for the streams they cross, with the result that trees upstream become inundated and die. Because these areas often form a district riverine vegetation community that may be preferentially used by certain species (e.g., *Cebus apella* in South America or *Macaca fascicularis* in Southeast Asia), such disturbance can cause a reduction in their populations.

Dams achieve the same effect only on a larger scale, and a big dam, like the Tucurui dam on the Tocantins River, in Brazil, has flooded as much as 200,000 ha of forest. Less drastic but still important effects can also be created downstream of a dam through changes in the flow regime. The Tana River forests, in Kenya, for instance, are supported by ground water and seasonal inundation of a flood plain. Homewood [1975] has argued that the crested mangabey, *Cercocebus galeritus*, is particularly adapted to this unstable environment and might be eliminated by competition with other species following construction of upstream dams that will decrease the frequency of flooding and thus stabilize the river course and the vegetation dynamics dependent on it.

SHIFTING CULTIVATION

It has been estimated that shifting cultivation is practiced by some 250 million people in the tropics [UNESCO, 1978] and accounts for the loss of some 200,000 km^2 of forested land each year [Myers, 1980]. It is the dominant form of land use over large areas, especially in hilly regions. Forest farmers are thus by far the most important factor in the conversion of tropical moist forest.

Because shifting cultivation increases the proportion of secondary forest at the expense of primary forest, we can expect a change in the composition of primate communities in favor of species adapted to secondary forests. On the whole, in the Old World tropics, this change favors cercopithecines at the expense of most apes and colobines. For example, in Peninsular Malaysia, and possibly in much of Southeast Asia, the primate biomass in primary forest is dominated by langurs, *Presbytis* spp., with gibbons, *Hylobates* spp., also occurring widely but at lower densities [Rodman, 1978; Wilson and Wilson, 1976; Chivers and Davies, 1979; MacKinnon and MacKinnon, 1980; Marsh and Wilson, 1981a,b]. By contrast, the two most widely distributed macaques, *Macaca fascicularis* and *M. nemestrina*, are often less common, groups of the former species concentrated mainly in swamp and riverine habitats and the latter more evenly distributed but at very low densities. However, in areas of secondary forest interspersed with domestic orchards, rubber small-holdings, and padi fields, both macaques are common—especially *M. fascicularis* [Southwick and Cadigan, 1972; Wilson and Wilson, 1976]. Langurs occasionally venture into sites of old cultivation, but gibbons almost never do so [Wilson and Johns, 1982]. Thus low-to-moderate-intensity agriculture causes not only a reduction in total biomass but also a major change in species composition in favor of macaques.

In Africa, the role of survivor in secondary forest and commensal with man is shared broadly between the baboons, *Papio* spp., in more open country, and the green monkey, *Cercopithecus aethiops*, in taller woodland. A lesser role in forested areas has been reported for *Cercopithecus mitis* in the east and south of the continent and *C. mona* in the west and center [Gartlan and Brain, 1968; De Vos and Omar, 1971; Kavanagh, 1980].

Among New World primates, differences are again apparent in ability to survive in either isolated pockets of primary forest or in secondary forests that develop following shifting agriculture. In northern Colombia, Bernstein et al. [1976] surveyed forest pockets isolated by agricultural development and found woolly monkeys, *Lagothrix lagotricha*, to be most adversely affected, whereas spider monkeys, *Ateles belzebuth*, survived in areas with only isolated tall trees, and white-fronted capuchins, *Cebus albifrons*, occupied areas of secondary growth. White-faced tamarins, *Saguinus leucopus*, apparently thrive in these conditions. Unfortunately, little is as yet known about the rates at which these changes occur or their demographic mechanisms. In a 20-ha forest pocket isolated within the northeastern part of Manaus city area about 10 years ago only two primate species now remain: the brown-faced saki, *Pithecia pithecia*, and the pied tamarin, *Saguinus b. bicolor*. The latter species appears able to use the secondary vegetation growing within this isolated forest pocket, but the saki is restricted to the primary forest patches [Oliveria and Lima, 1981]. There is a need for long-term monitoring of populations at selected sites of this kind [Lovejoy, 1980].

Most species that are successful in shifting cultivation mosaic habitat show a number of features in common. First, most have an opportunistic diet and are thus able to cope with alterations in plant community composition and in the relative availability of different food types. Thus more specialized feeders (e.g., *Pan* spp., *Pongo pygmaeus*, *Ateles* spp., *Chiropotes* spp.) tend to survive less well in mosaic habitats than opportunists such as *Macaca* spp., *Papio* spp., *Saguinus* spp., and *Presbytis entellus* [Johns, 1983]. Among folivores, species that preferentially feed on regenerating or colonizing vegetation, even in undisturbed forest, can survive well alongside low-intensity cultivation (e.g., *Gorilla gorilla* [Goodall and Groves, 1977; Murnyak, 1981] and *Colobus guereza* [Oates, 1977].

A second feature that can influence a primate's adaptability is its degree of terrestriality. Most Old World species capable of colonizing secondary forest are at least semiterrestrial in habits, which must facilitate survival in small forest patches and also predisposes them to crop raiding. Although no New World species are naturally semiterrestrial, marmosets and tamarins are frequently observed on the ground in disturbed habitats and travel easily across roads or between forest patches (J.M. Ayres, personal observations).

Third, there is evidence for rapidly acquired changes in behavior associated with an ability to survive in cultivated areas. Kavanagh [1980] has discussed the invasion of cultivated forest in southwestern Cameroon by green monkeys, *Cercopithecus aethiops*, that has occurred during the last 70 years. As a result, the animals have now become important agriculture pests and are often hunted or harassed. Compared with conspecifics in the savannah, they have adopted less predictable ranging patterns, a less detectable call repertoire, stronger patterns of male vigilance behavior, and a habit of hiding from dogs rather than confronting them with loud alarm calls as they do where canids are not associated with humans. All these differences can be interpreted as adaptations to increased hunting pressure. Conversion of forest land to shifting cultivation thus creates opportunities for colonization and by presenting strong new selection pressures can lead to rapid changes in behavior.

COMMERCIAL SELECTIVE LOGGING

Commercial logging in tropical forests can take several forms, most of which involve the removal of selected trees rather than the clear-cutting of whole stands. In this respect, they differ from most forestry practices in temperate zones where clear-cutting is the rule. Nevertheless, the number of species of tropical timber that are considered commercial is increasing steadily, and extraction rates are rising accordingly. Moreover, certain products based on wood chips can now be made using all species and sizes of tree present in a forest, thus paving the way for much more intensive operations in the future. The long periods required for natural regeneration of selectively logged forests combined with rising demand for timber have also encouraged the replacement of natural forest in some areas with plantations of fast-growing exotic species (e.g., in eastern Brazil and the Philippines). Hence commercial logging is both a major disturbance of forest in its own right and, in practice, often a prelude to its replacement by tree plantations or other agricultural crops.

Timber production is basically an exercise in transport, at minimum cost, of logs from source to processing mill and then of lumber or plywood to the consumer. In most parts of the tropics, log transport is nowadays a capital-intensive operation using bulldozers and heavy trucks in place of hand-sledges, bullock-teams, or elephants. More traditional rafting of logs is still practiced where large rivers permit (e.g., in Burma, Borneo, and Amazonia), but, as accessible lowland forests have become depleted, attention has shifted to upland forests where big rivers are fewer and the costs of extraction—both economic and environmental—are much higher. In a few cases, cable logging techniques are in use, and even "sky-crane" helicopters have been tried, but these methods are unlikely to become widespread on account of their higher basic costs.

At a global level, differences in haulage and shipping distances largely explain why the major sources of tropical hardwood entering world trade have been from West Africa and Southeast Asia while forests in Amazonia and central Africa have been little exploited. However, this pattern is changing as traditional sources of supply decline and government policies in producer countries favor more processing (sawmilling and veneer and plywood production) at source by domestic companies. In the short term, this is leading multinational firms to seek concessions in areas previously considered uneconomic (e.g., Papua New Guinea and Gabon) and, in the longer term, towards more plantation forestry and probably less reliance on native hardwoods [Gammie, 1981].

The components of a typical large-scale logging operation in the field can be summarized as follows: 1) mapping the concession area and inventory of timber trees, 2) building of access roads, 3) felling, 4) removal of logs, and 5) possible replanting or "timber stand improvement" operations.

In this sequence, the first major disturbance is the construction of roads. In a study at Sungai Tekam, in hill forest in West Malaysia [Johns, 1982], construction of major access roads caused a loss of 4.8% of the total stand. Clearing of ridgetops for log-loading areas and setting up a cable-logging system destroyed a further 3.6%. Loss of 8.4% of the stand thus occurred even before the start of logging. Felling subsequently occurred at a density of approximately 18 stems/ha (equivalent to 3.3% of the stand), but due to associated damage only 49.1% of the original stand remained on completion of logging (Figs. 1 and 2). A great deal of the stand was thus destroyed incidentally to the removal of saleable timber [see also Burgess, 1971]. Other adverse physical effects of logging are well known and include soil compaction by heavy machinery, higher rates of water run-off, and accelerated soil erosion [Myers, 1979]. These problems are much more severe on steep land, where an increasing proportion of logging operations now takes place.

Effects of logging on primates are of at least three kinds, operating on somewhat different timescales. First, the noise of logging and the shock of habitat destruction can produce immediate, if temporary, changes in behav-

Fig. 1. Ridgetop logging access road at Sungai Tekam, Malaysia.

Fig. 2. Logging damage at Sungai Tekam.

ior. Second, the destruction of many food trees and aerial pathways in the canopy causes changes in diet and ranging behavior, which, in turn, can influence social behavior and the dynamics of the population. Third, because the postlogging stand does not regenerate quickly and might never return to its former condition, there can be long-term changes in species' relative abundance and total biomass.

Considering first the immediate effects of logging, species reactions can differ markedly. For example, at Sungai Tekam, groups of lar gibbons, *Hylobates lar*, ceased calling during logging and became very agitated but remained within their territories. Directly afterwards, and for several months, there was a dramatic increase in calling and confrontations between groups but no noticeable shifts in territories. By contrast, groups of banded langurs, *Presbytis melalophos*, escaped from the advancing logging front by moving several hundred meters away from their former ranges into areas not yet logged, or undisturbed patches left behind after logging. This option may be closed to gibbon groups because they are highly territorial, so that if they moved they would be challenged by their immediate neighbors [Johns, 1983]. Migration away from logging areas has also been reported in orang-utans, *Pongo pygmaeus* [MacKinnon, 1974], colobus and diana monkeys, *Colobus* spp. and *Cercopithecus diana* [Martin and Asibey, 1979], and indris, *Indri*

indri [Petter and Peyrieras, 1976]. Following logging, however, most species appear to reoccupy their former ranges as far as possible, with minor variation to conform to alterations in habitat topography (e.g., major access roads frequently form the new boundaries to ranges because most species are reluctant to descend to the ground to cross them).

Animals that remain within their former ranges after logging are faced with a loss of food trees and disruption of their canopy pathways. In principle, two options are open to them in obtaining their food: 1) increasing searching time, thus accepting increased energetic costs, and 2) reducing selectivity of food items, thus minimizing cost at the expense of quality (though the costs of digestion will increase accordingly and can become important).

After logging of Johns' study site at Sungei Tekam, groups of both gibbons and banded langurs spent more time resting, moved less far each day, and showed a dietary shift to eating less fruit and more young leaves, which suggests a strategy of greater energy conservation. Comparable data for other primates are scant, but a cross-species analysis of relative abundance in primary and disturbed habitats showed a negative correlation between species' abundance in disturbed forest and the proportion of fruit and flowers in their diets when in primary forest. Johns suggests that frugivores generally fare less well than more folivorous or insectivorous species but, more especially, that opportunistic feeders do better than those with more fixed daily requirements whether for a few or many kinds of food [Johns, 1983].

A notable comparison is provided by Struhsaker and Oates [1975] in their studies of red colobus (*Colobus badius*) and black-and-white colobus (*C. guereza*) in the Kibale Forest, Uganda. Both species are substantially folivorous, and *C. badius* eats from a wider variety of species each day and is the more abundant animal in primary forest. On the other hand, in old logged forest, *C. guereza* is relatively much more abundant, because it can subsist on the leaves of just one or two common colonizing tree species when other foods are scarce. Lacking this adaptability in diet, *C. badius* is much less common.

The consequences of logging for an animal's feeding depend not only on its digestive capabilities but also on the choice of commercial tree species. Marsh and Wilson [1981a] have suggested that one reason why langur, *Presbytis* spp., densities decline proportionally more than those of lar gibbons, *Hylobates lar*, following logging is because langurs feed more heavily from several species of legume that are also important as timber trees (e.g., *Intsia palembanica*, *Sindora* spp., *Dialium* spp.). More generally in Southeast Asia, the main timber trees are species of the Dipterocarpaceae, which are not normally used as sources of food by primates [see studies in Chivers, 1980]. In Amazonia, the most valuable timber species (e.g., mahogany, *Swietenia macrophylla*, and cedar, *Cedrela odorata*) are again not widely used by primates for food. However, these are very heavy woods, and other trees are cut to act as floats while rafting logs downstream. The most widely used float tree is *Virola* spp., which is an important food species of black-capped capuchins, *Cebus apella*, and red-handed howlers, *Alouatta belzebul*. In Africa, the main commercial trees can themselves form a larger part of the diet of primates, resulting in greater population declines following logging (e.g., red colobus, *Colobus badius*, and diana monkeys, *Cercopithecus diana*, in Ghana [Asibey, 1978]).

A factor related to the choice of commercial species is the extraction intensity: Where more commercial species are present more trees will be cut and incidental damage to the stand will be higher. This can be an important determinant of effects on primate populations. Wilson and Wilson [1975] suggest relatively slight changes in primate populations following logging at a level of 8–10 trees/ha in East Kalimantan. This is much lower than on most concessions in Malaysia. In an area logged at an intensity of 18 trees/ha, Johns [1983] noted some decrease of primate populations, and in an area logged at an intensity of 25 trees/ha Stevens [Burgess, 1971] points to severe effects on the resident fauna. Clearly, comparative survey data should be qualified by information on extraction intensity as well as the species of tree cut and their use as food by particular primates.

Very little is known of the demographic mechanisms causing population declines following logging. Densities can drop through the disappearance of social groups [e.g., MacKinnon, 1974; Petter and Peyrieras, 1976] or by reduction in their size [e.g., Rudran, 1978]. Most probably, reductions in group size are effected principally by increased mortality among infants and juveniles related to food shortages and dominance rank [Dittus, 1979]. However, the opening up of the canopy by logging can also increase susceptibility to natural predators (especially raptors), diseases (especially mosquito-borne pathogens), and falls [Skorupa, in preparation]. Johns [1981] has reported a high level of infant mortality in all species following logging, although the causes of death were not known. At Ulu Segama, in Sabah, MacKinnon [1974] found that, following logging operations 15–30 km away, orang-utan densities in his study area increased by immigration, but the number of infants per female was lower, indicating less breeding. This finding suggests that migration in response to logging does not prevent population declines because the invaded areas are presumably already occupied at close to carrying capacity. In the longer term, populations can either stabilize at lower levels or recover their former densities as secondary forest matures. Marsh and Wilson [1981a,b] have proposed a model of this process to account for density estimates of gibbons, *Hylobates lar*, and langurs, *Presbytis* spp., surveyed at sites in West Malaysia that had been logged up to 27 years previously (Fig. 3). The model depicts temporary emigration and a rapid decline in the langur group density, followed by a recovery within 10–15 years to levels close to those in the original forest. Gibbon group densities, by contrast, remain constant at first, then decline slightly and recover more slowly than in langurs. Although some of the details of this model might need revision [Johns, 1983], the approach has the merit of emphasizing the long-term consequences of logging and the consequent need for long-term monitoring of sample populations with good supporting data on behavior and habitat quality.

Evidence that some species survive quite well in logged forest needs to be qualified to prevent its misinterpretation in relation to conservation. Even in Southeast Asian dipterocarp forests, where deleterious effects of logging may be less than in other regions, some threatened species are known to be very badly affected at even low cutting intensities. These include the orang-utan,

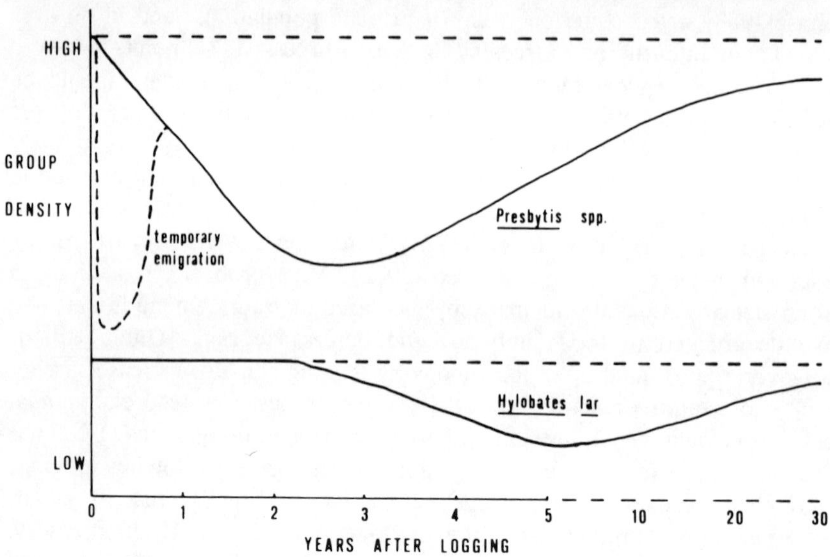

Fig. 3. Hypothesized effects of logging on populations of langurs and gibbons in West Malaysia. The model depicts temporary emigration and a rapid decline in langur population density followed by recovery within 10–15 years to levels close to those in the original forest. Gibbon densities, in contrast, remain constant at first, then decline slightly and recover more slowly than in langurs. After Marsh and Wilson [1981a].

Pongo pygmaeus [MacKinnon, 1974; Rijksen, 1978], the proboscis monkey, *Nasalis larvatus* [Wilson and Wilson, 1975], and all four Mentawaian species, *Hylobates klossii*, *Presbytis potenziani*, *Simias concolor*, and *Macaca pagensis* [Whitten, 1982]. Furthermore, Peninsular Malaysia is unusual among countries in the humid tropics in supporting generally only a low level of hunting for primates [Marsh and Wilson, 1981a]. Elsewhere, logging is usually followed by heavy hunting pressure [See Mittermeier, chapter 5, this volume].

Although the practical conservation argument against logging to conserve stocks of primates and most other wildlife is overwhelming, logging has sometimes been permitted within reserves under the official sanction of "habitat improvement." This has happened, for example, in at least two important lowland reserves in Indonesia (the Sekundar Reserve, North Sumatra, and the Kutai Reserve, East Kalimantan) and illustrates a dilemma for the biologist. Although it is essential that research findings be published objectively and in full, there is an unavoidable risk that evidence of the survival of some species following logging will be used to justify the violation of reserves. It might therefore be unwise to generalize between species or even between areas.

A final point to be considered is the long-term effect of forestry management operations. Although only a small proportion of logged forests receive any form of treatment at present, this proportion is rising as remaining areas of virgin forest decline and governments are obliged to increase the priority they give to forestry. Basically, forests can be managed in two ways. Either they are replanted with fast-growing species, which are usually exotics to the area concerned, or else they are left for natural regeneration to occur assisted by stand-improvement operations designed to accelerate the growth of commercial species and reduce the proportion of "weed" species.

Replanting with exotics, such as *Acacia mangium*, *Gmelina arborea*, *Pinus* spp., and *Eucalyptus* spp., is occasionally done as an adjunct to natural regeneration by replanting bare ground around spar trees and log-loading points. More commonly, though, the remnant stand is clear-felled and replanted to produce an even-aged stand, which differs little from an agricultural tree crop, like rubber, except that the undergrowth is usually allowed to grow up and the area is less disturbed until cutting or thinning time. With a species such as *Albizia falcataria*, growing on good soils, trees over 30 m in height can be cut after as little as 8 years.

Where natural regeneration is practiced, management operations include the cutting of lianes (including strangling figs, *Ficus* spp.) and the poisoning or ring-barking of noncommercial species. Because many of these noncommercial species are food sources for primates, their removal is likely to reduce the carrying capacity of the forest further. There is an urgent need for research in this field to formulate guidelines for maintaining primate and other wildlife populations in managed forest areas.

LARGE-SCALE CLEARANCE, USUALLY FOR AGRICULTURE OR RANCHING

Although all of the preceding forms of habitat disturbance pose serious threats to rain forest fauna, the most drastic form of disturbance is outright clearance of forest and its permanent conversion to agriculture, cattle ranches, and other uses. In some regions, this results from an intensification of traditional shifting agriculture to a point at which forest is never permitted to regenerate. Elsewhere, conversion is being carried out more systematically by private firms or government agencies.

In Peninsular Malaysia, for example, conversion of forest is now almost always a large-scale mechanized operation in which the first stage is the extraction of commercial timber under concession. Next, a contractor takes over and cuts all the remaining trees leaving them to dry for 6–8 weeks before burning. Finally, the remaining stumps and debris are piled up for a second burning and the area is then graded and terraced ready for planting with rubber, oil palm, or cocoa. In Brazilian Amazonia, by contrast, forest is normally burned in its entirety with no attempt to harvest timber, and the most common use of the land is for cattle ranching. To date, some 7.7 million ha have been cleared, representing 3% of the total area. Although this is still a small fraction of the total, the rate is increasing at 169% per year and is localized in certain areas in Pará, Mato Grosso, and Rondonia states and the vicinity of the city of Manaus [Valverde, 1981].

Whereas the means, scale, and end uses of conversion vary widely, the consequences for most primates are probably similar. With few exceptions (e.g., the slow loris, *Nycticebus coucang* [Davies and Payne, 1982], and Sykes' monkey, *Cercopithecus mitis* [De Vos and Omar, 1971], primates do not survive in monoculture conditions, although some species feed at the forest margin [e.g., Hernandez-Camacho and Cooper, 1976; Izawa, 1977; Wilson and Johns, 1982]. During the conversion process, therefore, most animals must either die in situ or else shift to remaining uncut areas. Marsh and Wilson [1981a] conducted surveys in primary forest at two West Malaysian sites, Ulu Sebol and Sungai Selai, that were immediately adjacent to areas recently cleared for agriculture. Comparison of group densities indicated a much higher proportion of *Presbytis melalophos* groups at these recent forest-edge sites than at five control sites, whereas *Hylobates lar* group densities were similar in both cases. As with responses to selective logging, gibbons seem unable to move away from an area even if it has been cleared completely, whereas at least a proportion of langurs can escape and increase their density nearby.

Notwithstanding this difference, there is evidently a large mortality among all species in areas being clear-felled (see Figs. 4 and 5). Moreover, mortality rates among animals that manage to escape from cleared land are likely to be much higher than would be the case in undisturbed forest, because densities presumably exceed the carrying capacity of the habitat. Thus the end result is probably the same for all species—a population reduction proportional to the loss of habitat.

As forest clearance proceeds in a country, not only does the area of forest decline but its distribution becomes fragmented. At first, highways linking major urban centers divide the country's forest into blocks, although these may still be thousands of km^2 in area (Fig. 6). Later, as encroachment proceeds further, blocks are fragmented into archipelagoes of forest "islands" that may completely isolate sections of what was previously one population and thus impede or prevent gene flow. Similar fragmentation of

Fig. 4. Large-scale clearance for agriculture in Malaysia.

Fig. 5. Remains of a banded langur, *Presbytis melalophos*, found close to the vehicle shown in Figure 4.

habitat by river systems, coupled with the former withdrawal of species to Pleistocene refugia, is likely to have been important in creating the present diversity of primate species, especially in the Amazon region [Haffer, 1969; Vanzolini, 1973]. One hypothetical consequence of modern forest fragmentation, therefore, is that it could lead to more speciation. Unfortunately, this is a process measurable in thousands of years, whereas other consequences of insularization portend a drastic collapse of faunal diversity on a much shorter timescale [Wilcox, 1980].

Fragmentation of a population increases the probability of local extinctions in several ways [Terborgh and Winter, 1980]. First, insofar as habitat is rarely uniform in quality, some population fragments may have become isolated in areas of poor habitat from which they cannot move. Second, the area of the island might be insufficient to provide food at all times of the year. This is particularly likely for frugivorous species that have wide-ranging patterns. Third, there may be imbalances in the food web created by extinctions of other species from these first two causes. As an example, frugivorous primates depend on fruit trees, which may rely on frugivorous birds for the dispersal of their seeds. Thus extinction of disperser and

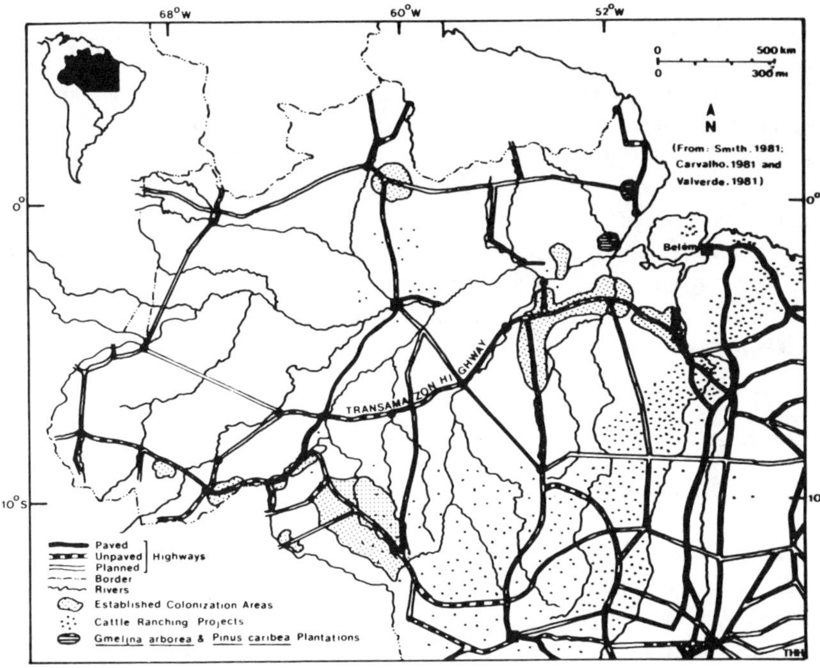

Fig. 6. Present and planned systems of highways and agricultural development in Brazilian Amazonia. Adapted from Ayres [1983], with permission.

pollinator species can have multiplier effects on other animals. Fourth, with small population sizes, levels of inbreeding rise, leading to progressive loss of phenotypic fitness and hence a decline in the population. Although there are marked differences in species resistance to inbreeding depression, Franklin [1980] suggests that an effective population of 50 animals is a minimum for survival and that for long-term maintenance of genetic variance a figure of 500 is more appropriate. Clearly, species that naturally occur at low densities will be particularly at risk from habitat fragmentation. In that there is an inverse relationship between population density and body size in primates [Clutton-Brock and Harvey, 1979], many of the largest species are among the most threatened (e.g., indris, *Indri indri*; spider monkeys, *Ateles* spp.; great apes, Pongidae).

We know of no systematic study of primate populations from the perspective of island biogeography, but evidence of local extinctions on isolated islands abounds. Rodgers [1981] has described the distribution of three species of colobus monkey among scattered forest relicts in Tanzania and cites a number of localities where species were present within the last century but are now absent. At a finer level, Marsh [1978] described absences of red colobus (*Colobus badius*) and crested mangabeys (*Cercocebus galeritus*) from a number of forest patches on the Tana River that lie within the animals' geographical ranges and appear to be of suitable habitat. As the animals are not hunted and many of these patches are now separated from one another by a kilometer or more of grassland and the animals rarely cross open ground, the simplest explanation is that local extinctions have occurred among isolated subpopulations. Further work comparing population structures in occupied patches of differing size and habitat could be revealing for the way in which extinctions occur and would have obvious implications for species management.

SUMMARY

Habitat disturbance can take a variety of forms, which are commonly interlinked and thus cumulative in their effects on primate populations. In this review, six classes of disturbance are distinguished according to their nature, scale, and the level of technology employed.

1. The lightest form of disturbance is the removal of selected plant products by traditional means. This includes activities such as fruit and rattan collecting and the cutting of trees to make canoes or to collect honey. The effect of such removals depends mainly on the importance of the items concerned in the diets of particular primate species.

2. A more insidious threat to forests arises when the undergrowth is cleared for agriculture or otherwise destroyed by fire or elephants. Although primates can continue to occupy the canopy, the forest is doomed in the long term because regeneration is prevented.

3. Building of dams or bad siting of roads can cause extensive flooding, particularly of riverine forest. This can be expected to cause a reduction in the populations of primates proportional to the flooded area and can also affect downstream populations through changes in vegetation patterns that are dependent on the river.

4. Shifting cultivation creates a mosaic of small-holder farmland and secondary forest in which some primates thrive but to which other species from the original forest are unable to adapt. Successful colonizer species in the Old World, such as baboons, vervets, and various macaques, tend to be opportunistic frugivores, semiterrestrial in habit, and adept at crop-raiding and avoiding predation by man.

5. Commercial selective logging operations in tropical forests usually destroy at least half the timber stand. In Southeast Asia, some species, such

as langurs and orang-utans, tend to escape, at least temporarily, from the immediate vicinity of the operation, whereas more territorial species, such as gibbons, are apparently unable to do so. In the longer term, less selective feeders may generally survive better than those with more specialized diets. However, there can be considerable regional differences according to the species of tree cut and their importance in primate diets. Changes in population demography following logging probably persist for several decades, and there is a need for long-term studies. Unfortunately, it is seldom possible to distinguish the effects of logging per se from those of hunting and shifting agriculture that often follow.

6. During large-scale conversion of forest to agriculture, a few primates may escape, causing temporarily increased densities in any remaining forest nearby, but these are likely to be countered by higher rates of mortality. As forest areas decline, fragmentation of the remnant areas compounds the threat to wildlife by isolating small populations, which are then more vulnerable to local extinctions.

ACKNOWLEDGMENTS

We wish to acknowledge support from the Sabah Foundation, Malaysia, and the Department of Anatomy, University of Cambridge, United Kingdom. Field work in Peninsular Malaysia (1978–81) by C.W.M. and A.D.J. was supported by the National Cancer Institute of the U.S. Department of Health, Education and Welfare through a contract with the University of Cambridge, U.K., as part of a collaborative "Malaysian Primates Research Programme", involving also Universiti Kebangsaan Malaysia and Universiti Pertanian Malaysia. Comments on drafts of the paper were given by E. Barrett, E.L. Bennett, D.J. Chivers, J.O. Caldecott, A.G. Davies, M. Kavanagh, J. Skorupa, and T.T. Struhsaker. Figure 4 was drawn by Mrs. S. Caldecott. To all these we are grateful.

POSTSCRIPT

Since this paper was first written, information has appeared on widespread fires that occurred in Borneo, in 1983, mostly in logged-over forests, but also in primary forest [Mackie, 1984; Beaman et al., 1986]. Although burning of the understorey is common in savannahs and monsoonal forest types, Whitmore [1984] has stated that "primary rain forest within the principal ever-wet parts of the Far East appears to be mainly noninflammable." The 1983 events in Borneo force some revision of this assumption and prompt the inclusion of burning as a category of habitat disturbance for rain forest primates. Carbon-14 dated charcoals in soils of mature rain forests in Venezuela likewise demonstrate that numerous fires have occurred over the past 6000 years in that region [Sanford et al., 1985].

In late 1983, Duff et al. [in press] surveyed a tract of logged-over forest in Sabah, Malaysia, that had been intensively burned four months earlier leaving almost no living trees. Although mammal densities were lower than in unburned logged forest, tree shrews, and four species of primates were seen. Leighton and Wirawan [in press] also reported remarkable resilience among primates following burning in the Kutai National Park, East Kalimantan, but noted an almost complete absence of large frugivorous hornbills and a decline in diurnal squirrels. Neither of these studies, however, has assessed long term changes in the primate community.

REFERENCES

Albrecht H (1976): Chimpanzees in Uganda. Oryx 13:357–361.

Asibey EOA (1978): Primate conservation in Ghana. In Chivers DJ, Lane-Petter W (eds): "Recent Advances in Primatology, Vol. 2. Conservation." London: Academic Press, pp 55–59.

Ayres JM (1983): Conservation of primates in Brazilian Amazonia: Problems and strategies. In Harper D (ed): "Proc Symp Conservation of Primates and Their Habitats, Vol. I. Primate Conservation in the Wild." Leicester: University of Leicester (Vaughan Paper No. 31), pp 2–33.

Beaman RS, Beaman JH, Marsh CW, Woods PV (1986): Drought and forest fires in Sabah in 1983. J Sabah Society.

Bernstein IS, Balcen P, Dresdale L, Gouzoules H, Kavanagh M, Patterson T, Neyman-Warner P (1976): Differential effects of forest degradation on primate populations. Primates 17:401–411.

Burgess PF (1971): Effects of logging on hill dipterocarp forest. Malay Nat J 24:231–237.

Chivers DJ (ed) (1980): "Malayan Forest Primates." New York: Plenum Press.

Chivers DJ, Davies G (1979): Abundance of primates in the Krau Game Reserve, Peninsular Malaysia. In Marshall AG (ed): "The Abundance of Animals in Malesian Rain Forests: Trans. VIth Aberdeen-Hull Symp on Malesian Ecology." Aberdeen: Aberdeen University, pp 9–36.

Clutton-Brock TH, Harvey PH (1979): Home range size, population density and phylogeny in primates. In Bernstein IS, Smith EO (eds): "Primate Ecology and Human Origins." New York: Gartland, pp 201–214.

Curtin SH (1980): Dusky and banded leaf monkeys. In Chivers DJ (ed): "Malayan Forest Primates." New York and London: Plenum, pp 107–145.

Davies AG, Payne JB (1982): "A Faunal Survey of Sabah." Kuala Lumpur: World Wildlife Fund (Malaysia).

De Vos A, Omar A (1971): Territories and movements of Sykes monkeys (*Cercopithecus mitis kolbi* Neuman) in Kenya. Folia Primatol 16:196–205.
Dittus WPJ (1977): The socioecological basis for the conservation of the toque monkey (*Macaca sinica*) of Sri Lanka (Ceylon). In Prince Rainier III of Monaco, Bourne GH (eds): "Primate Conservation." New York: Academic, pp 234–265.
Dittus WPJ (1979): The social regulation of primate populations: a synthesis. In Lindberg DG (ed): "The Macaques: Studies in Ecology, Behaviour and Evolution." New York, Van Nostrand-Reinhold, pp 263–286.
Duff AB, Hall RA, Marsh CW (in press): A survey of wildlife in and around a commercial tree plantation in Sabah. Malaysian Forester.
Franklin IR (1980): Evolutionary change in small populations. In Soulé ME, Wilcox BA (eds): "Conservation Biology." Sunderland, MA: Sinauer, pp 135–149.
Gammie JI (1981): "World Timber to the Year 2000." London: Economist Intelligence Unit Special Report No. 98.
Gartlan JS, Brain CK (1968): Ecology and social variability in *Cercopithecus aethiops* and *C. mitis*. In Jay PC (ed): "Primates: Studies in Adaptation and Variability." New York, Holt, Reinhart and Winston, pp 253–292.
Gilbert LE (1980): Food web organization and conservation of neotropical diversity. In Soulé ME, Wilcox BA (eds): "Conservation Biology." Sunderland, MA: Sinauer, pp 11–33.
Goodall AG, Groves CP (1977): The conservation of Eastern gorillas. In Prince Rainer III, Bourne GH (eds): "Primate Conservation." New York: Academic, pp 599–637.
Green S, Minkowski K (1977): The lion-tailed monkey and its South Indian rain forest habitat. In Prince Rainier III, Bourne GH (eds): "Primate Conservation." New York: Academic, pp 290–338.
Haffer J (1969): Speciation in Amazonian forest birds. Science 165:131–137.
Hernandez-Camacho J, Cooper R (1976): The nonhuman primates of Colombia. In Thorington RW, Heltne PC (eds): "Neotropical Primates: Field Studies and Conservation." Washington DC: National Academy of Sciences, pp 35–69.
Hladik CM, Hladik A (1972): Disponsibilities alimentaires et domaines vitaux des primates à Ceylon. Terre Vie 2:149–215.
Homewood KH (1975): Can the Tana Mangabey survive? Oryx 13:53–59.
Homewood KH (1978): Feeding ecology of the Tana mangabey (*Cercocebus galeritus galeritus*) (Mammalia: Primates). J Zool (London) 186:375–391.
Izawa K (1979): Studies on peculiar distribution pattern of *Callimico*. Kyoto University Overseas Research Reports of New World Monkeys 1979:1–19.
Johns AD (1981): The effects of selective logging on the social structure of resident primates. Malay Appl Biol 10:221–226.
Johns AD (1983): "Ecological Effects of Selective Logging in a West Malaysian Rainforest." PhD thesis, University of Cambridge.
Kavanagh M (1980): Invasion of the forest by an African savannah monkey: Behavioural adaptations. Behavior 73: 238:260.
Laws RM, Parker ISC, Johnstone RCB (1975): "Elephants and their Habitat: The Ecology of Elephants in North Bunyoro, Uganda." Oxford: Oxford University Press.
Leighton M, Wirawan N (in press): Catastrophic drought and fire in Bornean tropical rain forest associated with the 1982–83 ENSO event. In Trance GT (ed): "Tropical Rain Forest and World Atmosphere." New York: Proc. Symp. AAAS, May 1984.
Lovejoy TE (1980): Discontinuous wilderness: Minimum areas for conservation. Parks 5.
Mackie C (1984): The lessons behind East Kalimantan's forest fires. Borneo Research Bulletin 16:63–74.

MacKinnon J (1974): The behaviour and ecology of wild orang-utans (*Pongo pygmaeus*). Animal Behavior 22:3–74.

MacKinnon JR, MacKinnon KS (1980): Niche differentiation in a primate community. In Chivers DJ (ed): "Malayan Forest Primates." New York and London: Plenum, pp 167–190.

Malaisse F (1978): The Miombo ecosystem. In UNESCO (ed): "Tropical Forest Ecosystems." Paris: UNESCO-UNEP.

Marsh CW (1978): Problems of primate conservation in a patchy environment along the lower Tana River, Kenya. In Chivers DJ, Lane-Petter W (eds): "Recent Advances in Primatology, Vol. 2. Conservation." London: Academic, pp 85–86.

Marsh CW (1980): Primates and economic development on the Tana River, Kenya: The monkey in the works. In Furtado JI (ed): "Tropical Ecology and Development: Proc Vth Int Symp Trop Ecol." Kuala Lumpur: Int Soc Trop Ecol, pp 373–376.

Marsh CW, Wilson WL (1981a): "A Survey of Primates in Peninsular Malaysian Forests." Kuala Lumpur: Universiti Kebangsaan Malaysia.

Marsh CW, Wilson WL (1981b): Effects of natural habitat differences and disturbance on the abundance of Malaysian primates. Malay Appl Biol 10:227–249.

Martin C, Asibey EOA (1979): Effects of timber exploitation on primate populations and distribution in the Bia rain forest area of Ghana. Bangalore: Paper presented at Vth Congress of the Int Primatol Soc.

Mittermeier RA, Van Roosmalen MGM (1981): Preliminary observations on habitat utilization and diet in eight Surinam Monkeys. Folia Primatol 36:1–39.

Murnyak DF (1981): Censusing the gorillas in Kahuzi-Beiga National Park. Biol Conservation 21:163–176.

Myers N (1979): "The Sinking Ark." Oxford: Pergamon.

Myers N (1980): The present status and future prospects of moist tropical forests. Environ Conservation 7:101–114.

Oates JF (1977): The guereza and Man: How Man has affected the distribution and abundance of *Colobus guereza* and other black colobus monkeys. In Prince Rainier III, Bourne GH (eds): "Primate conservation." New York: Academic, pp 419–467.

Oliveria JM, Lima MG (1981): "Observacoes Sobre a Ecologia e o Compartmento dos Parauacus (*Pithecia pithecia*, Cebidae: Primates)." Manaus: INPA (unpublished report).

Petter JJ, Peyrieras A (1976): A study of the population density and home range of *Indri indri* in Madagascar. In Martin RA, Doyle GA, Walker AC (eds): "Prosimian Biology." London: Duckworth, pp 39–48.

Proud KRS, Hutchinson ID (1980): Management of natural reserves to maintain faunal diversity: A potential use for forest silviculture. In Furtado JI (ed): "Tropical Ecology and Development. Proc Vth Int Symp Trop Eco." Kuala Lumpur: Int Soc Trop Ecol, pp 247–255.

Rijksen HD (1978): "A Field Study on Sumatran orang-utans (*Pongo pygmaeus abelii* Lesson 1827)." Wageningen: Veenam H, Zonen BV.

Rodgers WA (1981): The distribution and conservation status of colobus monkeys in Tanzania. Primates 22:33–45.

Rodman PS (1978): Diets, densities and distribution of Bornean primates. In Montgomery GG (ed): "The Ecology of Arboreal Folivores." Washington, DC: Smithsonian Institution Press, pp 465–478.

Rudran R (1978): "Socio-Ecology of the Blue Monkeys (*Cercopithecus mitis stuhlmannii*) of the Kibale Forest, Uganda." (Smithsonian Contrib Zool No 249). Washington, DC: Smithsonian Institution Press.

Sanford RL, Jr et al. (1985): Amazon rain-forest fires. Science 227:53–55.
Santos R (1980): In: "Historia Economica da Amazonia (1800–1920)" T.A. Lulinoz TA (ed). Sao Paulo: Ltda.
Seth SK, Kaul ON (1978): Tropical forest ecosystems of India: The teak forests. In UNESCO (ed): "Topical Forest Ecosystems." Paris: UNESCO-UNEP.
Southwick CH, Cadigan FC (1972): Population studies of Malaysian primates. Primates 13:1–19.
Struhsaker TT, Oates JF (1975): Comparison of the behaviour and ecology of red colobus and black-and-white colobus monkeys in Uganda: A summary. In Tuttle RH (ed): "Socio-ecology and Psychology of Primates." The Hague: Moulton, pp 103–123.
Terborgh J, Winter B (1980): Some causes of extinction. In Soulé ME, Wilcox BA (eds): "Conservation Biology." Sunderland, MA: Sinauer, pp 119–134.
UNESCO (ed) (1978): "Tropical Forest Ecosystems: A State-of-Knowledge Report." Paris: UNESCO-UNEP.
Valverde O (1981): Ecologia e desenvolvimento da Amazonia. Rev Bras Tecnol 12(4):3–16.
Vanzolini PE (1973): Paleoclimates, relief and species multiplication in equatorial forest. In Meggers BJ, Ayensu ES, Duckworth WD (eds): "Tropical Forest Ecosystems in Africa and South America: Comparative Review." Washington DC: Smithsonian Institution Press, pp 255–258.
Whitmore TC (1984): "Tropical Rainforests of the Far East." 2nd Ed. Oxford: Clarendon.
Whitten AJ (1982): "The Gibbons of Siberut." London: Dent.
Wilcox BA (1980): Insular ecology and conservation. In Soulé ME, Wilcox BA (eds): "Conservation Biology." Sunderland, MA: Sinauer, pp 95–118.
Wilson CC, Wilson WL (1975): The influence of selective logging on primates and some other animals in East Kalimantan. Folia Primatol 23:245–274.
Wilson CC, Wilson WL (1976): Behavioural and morphological variation among primate populations in Sumatra. Yearbook Phys Anthropol 20:207–233.
Wilson WL, Johns AD (1982): Diversity and abundance of selected animal species in undisturbed forest, selectively logged forest and plantations in East Kalimantan, Indonesia. Biol Conservation 24:205–218.

Primate Conservation in the Tropical Rain Forest, pages 109–146

5
Effects of Hunting on Rain Forest Primates

Russell A. Mittermeier

World Wildlife Fund, Washington, D.C. 20037 and Department of Anatomical Sciences, HSC, State University of New York, Stony Brook, New York 11794

INTRODUCTION

Habitat destruction is, without a doubt, the major threat to wild populations of nonhuman primates around the world. However, hunting of primates for a variety of different purposes can be a very significant factor as well depending on the region and the species under consideration. Indeed, hunting can cause more damage than forest destruction in certain regions and has already resulted in the loss of local populations even in areas of otherwise suitable habitat. The significance of hunting as a threat to nonhuman primates has long been recognized, but the topic has never been reviewed on a global basis and its role remains poorly understood.

The purpose of this chapter is to provide a review of what is known of primate hunting worldwide and to assess the effects of threats in the major regions in which these animals occur. I begin with a discussion of the different reasons for which primates are hunted, followed by a discussion of the reasons why primates are *not* hunted in certain areas. Next, I review what is known of primate hunting in the Neotropical region, Africa, Madagascar, and Asia, where almost all living primates occur. Finally, I finish with some generalizations on primate hunting and its effect on the conservation of wild primate populations around the world.

This chapter is based on a preliminary review of the relevant literature on this topic (of which there is very little), on the author's own field observations in some 25 countries, and on the results of a questionnaire to which 46 field primatologists responded. The questionnaire provided much new and useful information, most of which is presented here for the first time. It should be emphasized that the topic of primate hunting is a complex one deserving of a much more detailed analysis than can be provided in a review chapter of this kind. A more thorough investigation will be available in a book on this topic that is currently in preparation.

Although the international trade in live primates can be considered a form of primate hunting, this topic is not covered here, since it is reviewed elsewhere in this book (see Kavanagh et al., Chapter 6, this volume). However, I do include discussion of local use of primates as pets, since this is often a byproduct of the other kinds of hunting covered in this chapter.

REASONS FOR HUNTING PRIMATES

The reasons for hunting primates can be divided into several major categories, each of which is described here.

Hunting of Primates as a Food Source

The main reason for hunting of primates appears to be to eat them, i.e., as a food source for our own species. This, of course, it not a new development, since the genus *Homo* has been hunting its nonhuman primate relatives ever since it first appeared on the scene. The earliest evidence that exists for primate hunting dates back to between 400,000 and 700,000 years ago and is from Olorgasaille in Kenya. Shipman et al. [1981] have shown that an unidentified hominid, probably *H. erectus*, was systematically butchering giant gelada baboons (*Theropithecus oswaldi*), and it is inferred that these animals were actually hunted by the early hominids and not just scavenged.

Although hunting and eating of primates is alien to most western cultures, many people in tropical countries consider primates perfectly appropriate sources of meat, and some even prefer primates to larger, more traditional (to westerners, at least) game animals [e.g., Sabater Pi and Groves, 1972; Mittermeier, 1977]. This is especially true in tropical forest areas, where primates are among the larger and more conspicuous mammals and where animal protein may be in short supply (Fig. 1).

Use of primates as a source of food varies greatly from area to area and, within a given area, from species to species. Whether or not primates are eaten depends on a variety of factors, which are discussed in more detail in the next section. Nonetheless, it is clear that hunting of primates as a source of food is a serious factor in at least two major regions of the world, the Guineo-Congolian forest region of West and Central Africa and the Amazonian region of northern South America, and is a less important but nonetheless noteworthy factor in a number of other areas.

It is also important to distinguish between hunting for local consumption and market hunting to supply a demand for meat in larger towns and cities in a given region. Hunting for local consumption does not necessarily present a major problem if the human population is low and if large tracts of undis-

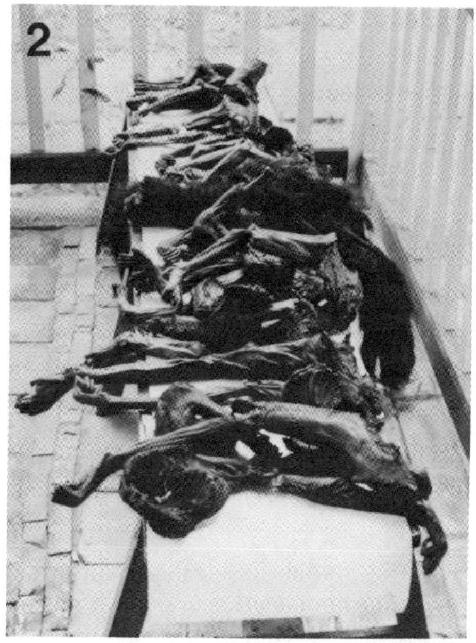

Fig. 1. The main reason for hunting primates is as a source of food for man. Here, a Trio Indian woman from Suriname prepares to cook a black spider monkey (*Ateles p. paniscus*). Photo by M.J. Plotkin.

Fig. 2. Market hunting of primates is a serious problem in some parts of the world. These smoked spider monkeys were confiscated 1978 from the market in Suriname, where this species is protected. Photo by J.G. Fleagle.

turbed habitat are available (e.g., parts of Brazilian Amazonia, the Guianas). Market hunting (Fig. 2), on the other hand, is highly destructive and has already led to the extermination of local populations of certain species in areas of otherwise suitable habitat (e.g., *Ateles* and *Lagothrix* over much of Peruvian Amazonia [Freese, 1975; Soini, 1982; J. Terborgh, personal communication]).

Hunting of Primates for Medicinal Purposes

Another reason for hunting primates, which is closely connected with meat hunting, is the use of primates or certain primate body parts for their supposed medicinal value. In most cases, this is a minor factor in primate conservation, but if it involves a particular endangered species that happens to be in heavy demand it can be far more serious.

Examples of this threat to primates include use of the slender loris (*Loris tardigradus*) in black magic ceremonies in Sri Lanka (R. Rudran, personal communication), use of slow loris (*Nycticebus coucang*) bones as charms and killing of *Presbytis hosei* to obtain bezoar stones in Sarawak (M. Kavanagh, personal communication; K. Proud, personal communication), and use of monkey fat, applied either externally to the back or chest or taken internally with hot water, as a cure for respiratory ailments in Costa Rica (C. Freese, personal communication). In the Philippines, blood taken directly from the heart of the crab-eating macaque (*Macaca fascicularis*) is thought to be a cure for asthma [M. Kavanagh, personal communication]; in Thailand, langur blood is thought to be a tonic for vitality or vigor (A. Eudey, personal communication); in south India the meat of the Nilgiri langur (*Presbytis johnii*) and sometimes the lion-tailed macaque (*Macaca silenus*) is thought to have value as an aphrodisiac and to have other medicinal properties.

In South America, the cuplike hyoid apparatus of male howler monkeys (*Alouatta* spp.; Fig. 3) is used as a drinking vessel because it is thought to have many medicinal properties. In Colombia, where the red howler monkey (*Alouatta seniculus*) is known as *cotomono* ("goiter monkey"), it is believed that drinking from a howler hyoid will cure goiters [Hernandez-Camacho and Cooper, 1976]. In Suriname, drinking from the hyoid is said to cure stuttering, and in some parts of Brazilian Amazonia it is thought to ease labor pains during birth. Another unusual howler medication, this one from Peruvian Amazonia, is to take a pile of howler hair, grind it up, put into boiling water, cool, and then drink to cure a cough (C. Freese, personal communication).

The above-mentioned examples involving an endangered species, the lion-tailed macaque, and a vulnerable species, the Nilgiri langur, are situations in which primate hunting for medicinal purposes can be a serious conservation problem. However, in most cases, it is not a major factor and frequently is nothing more than a byproduct of meat hunting.

Hunting of Primates as Evil Omens

Related to the hunting of primates for medicinal purposes is killing of these animals as evil omens. Only one example of this has come to our attention, but it is worthy of note because it involves a highly endangered, unique species. The animal in question is the aye-aye (*Daubentonia madagascariensis*) from Madagascar, which is greatly feared by people living in the eastern forest region where this species occurs. It is thought to bring bad luck to any village in which it appears, the only escape being to kill the animal and possibly burn down and relocate the entire village. It is even thought to have

Fig. 3. The howler monkey hyoid apparatus is thought to have many medicinal properties. In this case, it is being used as part of a necklace. Photo by R.A. Mittermeier.

Fig. 4. Primates are shot to bait spotted cat traps in some parts of Amazonia. This hunter along Brazil's Transamazonian Highway has shot an endangered white-browed spider monkey (*Ateles belzebuth marginatus*) for this purpose. Photo by N.J.H. Smith.

occult power when dead, and a sorcerer who keeps its elongated middle finger is believed to have special powers [Jolly, 1980]. This unfortunate belief has almost certainly contributed to bringing this animal to its current precarious state.

Hunting of Primates for Bait

Primates are sometimes hunted to use their meat as bait to capture other animals. Although never really highlighted before, this factor, surprisingly, can be quite significant in certain areas. It is an important factor in parts of Amazonia, where monkeys, and especially larger species (e.g., *Lagothrix*, *Ateles*), are shot to bait traps for spotted cats (Fig. 4). The technique used is to drag the dead animal along the forest floor for several hundred meters, and then place it in a crude wooden live trap. The cat, usually an ocelot (*Felis pardalis*) and sometimes a jaguar (*Panthera onca*), follows the scent trail into the trap and is caught alive. It is then easily dispatched with a head shot or strangled by a nylon cord, providing the hunter with an undamaged pelt. In addition to monkeys, cat hunters also use sloths and other forest mammals and larger birds such as curassows (*Crax* spp.), guans (*Penelope* spp., *Pipile* spp.), and trumpeters (*Psophia* spp.) to bait such traps.

Primates are also sometimes used as fish bait or turtle bait in Amazonia, and rare species like black uakaris (*Cacajao melanocephalus*) are sometimes shot for this purpose (personal observation, 1973; Fig. 5).

In Mexico, *Ateles* and *Alouatta* are both shot to bait basket traps for shrimp, which are caught for commercial purposes (E. Rodriguez Luna, personal communication).

Outside the Neotropics, there are only a few reports of primates being used for bait. In Sri Lanka, all three monkey species (*Macaca sinica*, *Presbytis senex*, *P. entellus*) native to that island are sometimes used as crocodile bait by skin hunters [R. Rudran, personal communication], and in the Tana River region of Kenya baboons are occasionally used for the same purpose [J. Allaway, quoted in Marsh, 1986].

Hunting of Primates to Obtain Skins or Other Body Parts for Ornamentation

Primates are occasionally hunted for their skins or other body parts to produce various kinds of human ornamentation. The best known case is that of the African colobus monkeys (*Colobus polykomos*, *C. guereza*). Tradition-

Fig. 5. Another example of a rare primate being shot as bait. Here, two black uakaris (*Cacajao melanocephalus*) have been shot as fish bait in the Rio Cuiuni, a tributary of the Rio Negro in Brazilian Amazonia, 1973. Photo by R.A. Mittermeier.

ally, the pelt of these monkeys was used by a number of African tribes to make headdresses, caps, and cloaks, and this custom still exists in a number of African countries. Colobus skins were also popular at the end of the last century and the early part of this century in Europe and, to a lesser extent, the U.S. The species whose skins were most used in this trade was the West African *C. polykomos*, with those from the area that is now Ghana and the Ivory Coast (*C. p. vellerosus*) being most in demand because of the length and glossiness of the hair [Poland, 1892, Brass, 1925; Bachrach, 1946; Oates, 1977]. According to available statistics, 1,750,000 skins of "Afrikanische Affen" (almost entirely *Colobus*) were auctioned in London between 1871 and 1891, with yearly figures ranging from 19,814 in 1871 to 223,599 in 1889. Germany was the destination for most of these skins, but many went to the U.S., Canada, and Italy as well [Poland, 1892; Brass, 1925; Oates, 1977].

In recent years, colobus monkey rugs were popular in tourist shops in Kenya and Ethiopia, the species used in this case being *Colobus guereza* (Fig. 6). This trade accounted for the loss of thousands of colobus monkeys

Fig. 6. Colobus monkey *(Colobus guereza)* rugs in a tourist shop in Nairobi, Kenya, in 1972. Photo by R.A. Mittermeier. The sale of such products is now illegal in Kenya.

Fig. 7. Skin of the Peruvian yellow-tailed woolly monkey (*Lagothrix flavicauda*) being used as a wall decoration in the northern Peruvian Andes, 1983. Photo by R.A. Mittermeier.

in the 1960s and early 1970s [Mittermeier, 1973; Oates, 1977], with Dunbar and Dunbar [1975] estimating some 200,000 skins in Ethiopian shops in the early 1970s. These were sold mainly within the countries in question, but some export occurred as well [Mittermeier, 1973; Oates, 1977]. In 1972–1973, for instance, there seems to have been a renewed interest in these skins for clothing, with large sales going to Italy and some to France and Germany [Oates, 1977]. There may also be continuing demand for guereza skins in the long established markets of the Middle East [Koford, 1972; Oates, 1977]. Kenya and Ethiopia have now prohibited the skin trade, but it is unclear how many of these monkeys are still killed illegally.

It remains to be determined what effect the trade in colobus monkey skins has had (and continues to have?) on these animals, since an in-depth study of the topic has never been conducted. *C. guereza* is a widespread and adaptable species, but several of its subspecies are very restricted in range and could be endangered by hunting pressure of this kind. A detailed analysis of the history of the colobus skin trade, together with a survey of the practice of using colobus monkey skins, should be carried out in several key African countries.

Skins of other African monkeys (e.g., *Cercopithecus mitis stuhlmanni*) appeared in Kenyan tourist shops in the early 1970s [Mittermeier, 1973], and can occasionally be seen in tourist shops in many African countries. However, this trade is usually quite small and probably not significant in most cases.

In the case of rare or endangered species, however, the demand for skins or other body parts can be a source of concern. The best example of this is the slaughter of mountain gorillas (*Gorilla gorilla beringei*) in the Virunga Volcanoes to obtain skulls and hands for sale to tourists [Fossey, 1983]. Another example, from the same region, is the golden monkey (*Cercopithecus mitis kandti*) from the Virungas and possibly some neighboring forest areas. This attractive monkey is shot for its skin, which can sometimes be seen for sale in the town of Goma in Zaire (D. McMeekin, personal communication, 1985). The extent of this trade remains to be determined, but it could be a serious factor in the survival of this rare animal.

In South America, monkey skins are sometimes used for ornamentation as well, although never on the scale of the colobus monkey example in Africa. In Colombia and Peru, the skins of *Alouatta*, *Lagothrix*, and *Aotus* are occasionally employed as brow bands or horse bridles or as saddle covers, and in the northern Peruvian Andes the skins of *Lagothrix flavicauda* and *Ateles belzebuth belzebuth* are popular wall decorations (Fig. 7). Stuffed monkeys are also popular in this region and can often be seen in local homes

(Fig. 8). In Brazilian Amazonia, skins of *Lagothrix lagotricha* are also used to fabricate a device called the *cuica*, which is used by jaguar hunters to imitate the vocalizations of these cats [Ayres and Ayres, 1979].

Various monkey ornaments are also offered for sale in tourist shops in larger Amazonian towns (e.g., Iquitos, Peru; Leticia, Colombia). These include monkey skulls, monkey skins, monkey skin hats, stuffed monkeys, and necklaces made of monkey teeth, monkey bones, monkey hands and feet, and the cup-shaped hyoid bone of male *Alouatta* (Figs. 9–11).

Many of the above-mentioned items are imitations of ornaments made by various Amerindian tribes or modifications on several basic Indian themes. The more traditional of the Amerindian tribes continue to make monkey ornaments as well, an example of which is given in Figure 12.

Another example from Amazonia is use of the bushy tails of *Pithecia* and *Chiropotes* as dusters. The tails are cut off at the base, the vertebrae are removed and a stick is inserted, making a reasonable equivalent of a feather duster (Fig. 13). These tail dusters are used locally throughout Amazonia and sometimes appear in tourist shops as well.

Fig. 8. Stuffed infant Peruvian yellow-tailed woolly monkey (*Lagothrix flavicauda*) in a home in Rioja, northern Peru, 1983. Photo by R.A. Mittermeier.

Fig. 9. Hat made of *Pithecia* fur for sale in a tourist shop in Iquitos, Peru, 1980. Photo by R.A. Mittermeier.

Fig. 10. Stuffed animals for sale in a tourist shop in Cayenne, French Guiana, 1979. Included on the shelves are kinkajous (*Potos flavus*), golden-handed tamarins (*Saguinus m. midas*), iguanas (*Iguana iguana*), and caimans (*Caiman crocodilus*). Snake skins are on the table at the bottom of the picture. Photo by R.A. Mittermeier.

Fig. 11. Monkey tooth necklace from Brazilian Amazonia. Photo by R.A. Mittermeier.

Fig. 12. Trio Indian comb made from the radius of a spider monkey Suriname, 1976. Photo by R.A. Mittermeier.

Fig. 13. *Pithecia* tail duster being used by a Suriname Bushnegro, 1978. Photo by R.A. Mittermeier.

In the Neotropical region, the use of primate body parts as ornamentation need not be considered a serious threat. It is usually small-scale, and in most cases nothing more than a byproduct of meat hunting.

Hunting of Primates for Pets

Another reason for hunting primates is to obtain infants as pets (Figs. 14–16). In areas where primates are regularly hunted as a source of food, capture of pets is usually a byproduct of meat hunting, and it is rare that a hunter will go out only in search of primates for pets. However, it frequently happens that a hunter will seek out a female with infant and shoot at her in preference to other group members. The usual procedure is to attempt to kill the mother with a carefully placed shot and then retrieve the live infant from her body after the two have fallen to the ground. However, since shotguns are frequently the preferred weapon in primate hunting, especially in places like Amazonia, it often happens that the infant is killed by stray shot or by the fall to the ground.

This kind of selective hunting for females with infants is especially serious for animals like the South American woolly monkeys (*Lagothrix* spp.), which are both the favorite primate food species and the most sought-after monkey pet in Amazonia (Fig. 17). By shooting a mother woolly monkey, a hunter gets both a source of highly desirable meat and an infant that can be sold at considerable profit in a nearby town. As a result, he will often make a special effort to find woolly monkeys.

Fig. 14. White-fronted capuchin monkey (*Cebus albifrons*) being kept as a pet in northern Peru, 1978. Photo by R.A. Mittermeier.

Fig. 15. Juvenile night monkey (*Aotus* sp.) being kept as a pet by a Yagua Indian woman in Peruvian Amazonia, 1972. Photo by R.A. Mittermeier.

Although capture of primate pets is usually a byproduct of meat hunting with those species that are regularly eaten, some species are sought after only for the purpose of keeping them as pets. This is especially true of callitrichids, which most people consider too small to eat. There is a small-scale trade in live *Cebuella*, *Saguinus*, and *Callithrix* locally in Amazonia and a quite significant internal pet trade in *Callithrix jacchus*, *C. penicillata*, and, to a much lesser extent, several other eastern Brazilian species as well (personal observation). This is especially serious for highly endangered and very attractive species like the lion tamarins (*Leontopithecus* spp.), which are endemic to Brazil's Atlantic forest region and are down to only a few hundred individuals each. *L. rosalia* still occasionally appears in the trade (L. Harper, personal communication; J. Dietz, personal communication),

Fig. 16. Black spider monkey being kept as a pet by a Trio Indian boy, Suriname. Photo by M.J. Plotkin.

and a recent internal and international trade has developed in *L. chrysomelas* [Mallinson, 1984]. This trade involved at least 50 individuals exported illegally from Brazil and an unknown number (but at least several dozen individuals) sold within Brazil.

In those parts of the world where primates are *not* regularly hunted as a source of meat, there may occasionally be hunting of primates *only* to obtain infants as pets. For instance, this sometimes happens with south Indian species such as the bonnet macaque (*Macaca radiata*, Fig. 18) and with all three monkey species of Sri Lanka (*Macaca sinica*, *Presbytis senex*, *P.*

Fig. 17. Infant woolly monkeys are the most popular primate pets in Amazonia. This baby woolly is from Brazilian Amazonia, 1973.

entellus). Orang-utans and gibbons also used to be shot (and probably still are) just to obtain infants, and this certainly continues with gorillas and chimpanzees in Africa as well, witness the recent case of seven infant gorillas from Cameroun and five from the Congo Republic being offered on the international market.

Hunting of Primates for "Sport"

Hunting of primates for "sport" is another threat that has to be considered. It appears to be most prevalent in areas like logging camps and military zones in the developing countries, where arms are plentiful and where people consider themselves above the law or are remote from it, and can even involve the killing of animals like orang-utans (B. Galdikas, personal communication).

In Brazil and in Mexico, there are also cases of local people hunting *Brachyteles* and *Ateles* simply for target practice. The *Brachyteles* example is especially grotesque, since the report indicated that hunters would try to shoot off only the animal's tail to demonstrate their marksmanship. In Kenya, young men of some tribes use baboons for spear-throwing practice (C. Marsh, personal communication).

Children are also offenders in many areas. In South America, children in the interior are commonly armed with slingshots and, to a much lesser extent, air rifles, and will sometimes shoot at smaller primates like callitrichids, which are common in secondary forests around towns and villages. There is

Fig. 18. Bonnet macaques (*Macaca radiata*) being kept as pets and to attract tourists in Mahabalipuram, Madras State, south India, 1979. Photo by R.A. Mittermeier.

no data whatsoever on their effect, but they probably account for a good many small primates each year.

Even organized "big-game hunting" of primates is still of interest to some people, as indicated by the case of a 1978 application to the U.S. Federal Wildlife Permit Office to shoot five gorillas (*Gorilla gorilla*), five orangutans (*Pongo pygmaeus*), five Tana River red colobus monkeys (*Colobus badius rufomitratus*), five black colobus monkeys (*C. satanas*), and an unspecified number of Zanzibar red colobus monkeys (*C. b. kirkii*). This application was submitted by the Safari Club International, an organization of wealthy big-game hunters, and was rejected [Federal Register, 1978].

Another example was an advertisement for a gorilla hunting safari offered by a French tour company in 1981. On the whole, "sport" hunting of primates has to be considered a minor factor in primate conservation unless a particular, highly depleted, endangered species happens to be involved.

Hunting of Primates as Agricultural Pests

Finally, it is necessary to consider hunting of primates as agricultural pests, which really belongs in a major category by itself and, in some cases, is probably acting as a serious drain on primate populations. This is especially true in parts of Asia and Africa, where primates like baboons, vervets, and macaques can in fact be quite significant crop raiders and are persecuted as a result.

Probably the most striking example of killing of primates as crop pests, and the only case for which substantial documentation is available, is from Sierra Leone. From 1947 to 1962, the Sierra Leone government sponsored regular "monkey drives" to eliminate primates from agricultural areas. According to government records, some 245,000 primates were destroyed in this period, at a yearly average of 18,885 (range 4,500–39,021), and these were only the animals that were actually recorded. Bounties were paid either for primate heads or for tails, and there was no control over the species killed. In any case, only three species of the 14 known to occur in Sierra Leone were considered harmless, all the others being fair game in the drives. Monkeys were either shot or were driven into nets and clubbed to death [Tappen, 1964; Jones, 1950]. Of course, monkey drives of this kind are not the rule, but this example from Sierra Leone does serve to indicate that primates are considered a menace to agriculture in certain areas and that large numbers of them may be killed as a result.

The major crop raiders appear to be the baboons and vervets in Africa and some of the more abundant macaques in Asia, but dozens of other species have been implicated as well. Some examples from Africa include mona monkeys, green monkeys, and mangabeys in Sierra Leone (G. Teleki, personal communication), *Cercopithecus mitis* and *Cercocebus galeritus* in Kenya (C. Marsh, personal communication), *Cercopithecus mona*, *Cercopithecus erythrotis*, *Papio leucophaeus*, *Pan troglodytes*, *Gorilla gorilla*, and *Erythrocebus patas* in Cameroun (J.S. Gartlan, personal communication), *Cercopithecus ascanius* and *Colobus guereza* in Uganda (T.T. Struhsaker, personal communication), and introduced *Macaca fascicularis* in Mauritius (R.W. Sussman, personal communication). In Asia, the most serious pests seem to be the pig-tailed macaque (*Macaca nemestrina*), the crab-eating macaque (*Macaca fascicularis*), and the rhesus macaque (*Macaca mulatta*), but other species mentioned as crop raiders include orang-utans, esp. in fruit orchards in Sarawak and Kalimantan (K. Proud, personal communication; B. Galdikas, personal communication), *Presbytis rubicunda* in Kalimantan (B. Galdikas, personal communication), and *Presbytis entellus* and *Macaca radiata* in India (M.A. Reza Khan, personal communication; J.M. Johnson, personal communication). In Madagascar, black lemurs (*Lemur macaco*) are known to raid coffee plantations, and even the very rare aye-aye (*Daubentonia madagascariensis*) enters coconut and lychee plantations (R. Albignac and J.J. Petter, personal communication).

Crop raiding by primates appears to be less of a problem in Central and South America, perhaps because all Neotropical species are primarily or exclusively arboreal. The capuchin monkeys (*Cebus* spp.) seem to be the major offenders (F. Medem, personal communication; C. Freese, personal communication; P. Soini, personal communication; personal observation), and the vernacular names for them in several countries reflect this and also the fact that they are especially fond of corn. For instance, the Colombian name for *Cebus apella* is *maicero*, and the Suriname name for *Cebus nigrivittatus* is *nyangkaru mongi*, both of which translate as "corn eater." Squirrel monkeys (*Saimiri* spp.) also raid fruit orchards in the Amazon region, howlers may do the same in Belize (R. Horwich, personal communication), and even an endangered species like *Lagothrix flavicauda* occasionally raids corn fields (A.B. Luscombe, personal communication). On the whole, though, crop raiding is a minor factor in the Neotropics, and primates are rarely heavily persecuted for this reason alone.

In general, it appears that the more abundant and adaptable species are the ones that usually raid crops and that hunting of primates as crop pests does not yet seem to have driven any primate species into the endangered category. Nonetheless, this important subject is very poorly understood and would certainly warrant much further investigation.

The human response to crop raiding by primates varies from simple chasing, throwing sticks and stones, or shooting with slingshots, to hunting with dogs and guns, to major killing operations like the one in Sierra Leone. Dead primates are used as "scarecrows" in a number of countries from Mauritius (R.W. Sussman, personal communication) to Thailand (A. Eudey, personal communication) to Venezuela (T. Blohm, personal communication), bounties have been paid in Barbados for introduced *Cercopithecus aethiops*, and electric fences have been erected (with some effect) to keep away introduced *Macaca fascicularis* in Mauritius (R.W. Sussman, personal communication). Recently, several studies have been conducted in Kenya to investigate several different methods of preventing baboons from raiding crops.

Included are playback of baboon alarm vocalizations, use of lion and leopard scent, taste aversions using *Lithium chloride* and several other substances, and planting of *Euphorbia* barrier fences [Quick, 1984 ; S. Strum, personal communication]. Findings from this research could have widespread application in Africa and Asia if inexpensive and effective techniques can be developed.

It is also important to note that, in many areas, primates, especially the more terrestrial species, are probably forced into crop raiding because of loss of natural habitat to agriculture. In some cases, they have clearly learned to coexist with man by using his crops as a significant part of their diet, meaning that a man-animal conflict situation has become part of their ecology. Conflicts of this kind are likely to increase in the future as the human population continues to grow in the developing world, and they may hamper other primate conservation efforts by creating negative feelings towards primates in general.

REASONS FOR NOT HUNTING PRIMATES

Having taken a look at the reasons why primates are hunted, it is also informative to discuss reasons why they are *not* hunted in many areas.

Legal restrictions

Primates are legally protected in many of the countries in which they occur, and this occasionally serves as a deterrent against hunting. However, it is clear that legal restrictions are not effective in most of the areas occupied by primates, especially in the more remote regions (e.g., tributaries of the Amazon, the Congo Basin), where enforcement of protective measures is impossible. In general, legal restrictions are more or less effective in towns and cities and can cut down on market hunting of primates if strictly enforced; in the bush, however, it is difficult to prevent hunting if it has long been part of the local culture.

Religion

By far the most important reason for not hunting primates is religion. Certain religions forbid the killing or eating of primates; others pay no attention to this matter. Islam, for example, prohibits the eating of most wild mammals, primates included, because they are considered unclean and not fit for human consumption. As a result, one can often see primates, even large ones, in close proximity to towns and villages in Muslim areas. Primates also occupy a special position in Hinduism because of the monkey god Hanuman, a heroic figure in Hindu mythology (Figs. 19,20). Consequently, Hindus do not eat primates and usually don't kill them either (except in areas where they are perceived to be really serious crop pests). Devout Buddhists apparently do not eat primates either, although this restriction seems to become less important as devoutness declines.

These three religions provide the best protection against hunting of primates, and some of the other smaller religions (e.g., Jains, Parsis) of the Indian subcontinent are similarly effective. However, most other religions do not have restrictions on eating of primates and other wildlife, and primate hunting can be quite common in Christian areas, in areas where there is some mix of Christianity and pagan religions, and, in general, in areas where religious strictures are not closely followed. Primate hunting is especially common among pagan tribal groups that still live more or less traditional life-styles even in areas where the majority of the population does not hunt these animals (e.g., many small Indian tribes hunt primates in largely Hindu areas; *orang asli* tribesmen hunt primates in Muslim Peninsular Malaysia).

Taboos

Closely connected with tribal religions are local taboos against eating primates. These taboos can vary from region to region, from tribe to tribe, and even from individual to individual. In Suriname, for instance, one person in a particular Bushnegro village may have a personal taboo (*trefu*) against eating the black spider monkey (*Ateles p. paniscus*), whereas his next door neighbor may happily eat spider monkeys but wouldn't consider eating a golden-handed tamarin (*Saguinus m. midas*). Another individual may have a taboo against all monkeys; yet another in the same village will happily eat all primate species without any restrictions [Mittermeier, 1977].

In Madagascar, many lemur species were traditionally protected by regional taboos (*fadys*), among them the indri (*Indri indri*) in the eastern rain forests and Verreaux's sifaka (*Propithecus v. verreauxi*) in the southern spiny desert region. On the other hand, the aye-aye (*Daubentonia madagascariensis*) is a species that has suffered from its taboo, since it is considered an evil omen to be killed on sight.

In southeast Asia, orang-utans were considered taboo in Sarawak, and the crab-eating macaque (*Macaca fascicularis*) is considered taboo both in Sarawak and in Kalimantan, one reason being the belief that these monkeys scavenge human corpses and lick fluids from burial jars and coffins (F.W. King, personal communication). A similar taboo exists in Brazilian Amazonia against the giant armadillo (*Priodontes giganteus*), which is said to dig tunnels in cemeteries.

Figs. 19,20. Hanuman, the Monkey God, occupies an important position in Hindu mythology. Consequently, primates are still considered sacred over much of India and are not molested. These photos are from a series of paintings of the Ramayana in a temple in Bangkok, Thailand. Photos by R.A. Mittermeier.

Other species protected by taboos include Kloss's gibbon (*Hylobates klossi*) and the light phase of *Simias concolor* on the Mentawai Islands off Sumatra (WWF, 1980; A. Mitchell, personal communication; A. Whitten, personal communication), *Saguinus m. midas* in Suriname, and gorillas and chimpanzees in some parts of Gabon. Kloss's gibbon has various taboos associated with it. It may be eaten only by young men, it may not be hunted while it is calling, and it may never be taken into a clan house in most parts of Siberut [WWF, 1980]. *Saguinus m. midas* is not eaten by many people in Suriname because it is considered too pretty by some and crazy by others because it falls out of trees; it is also taboo for twins or people with twin offspring because it produces twins itself [Mittermeier, 1977]. Gorilla and chimpanzee meat is taboo for pregnant women in some parts of Gabon because it is thought that, by eating such meat, the woman will give birth to a baby ape (C. Tutin, personal communication).

Resemblance to Man

In some areas, people won't eat any primates or won't eat certain species because they look or act too much like people. In Brazilian Amazonia, for instance, uakaris (*Cacajao calvus* spp.) are sometimes protected because their naked red faces look too human, and the black spider monkey (*Ateles p. paniscus*) is not eaten by some Surinamers because it appears too human or because "it cradles its baby like a woman" [Mittermeier and Coimbra-Filho, 1977; Mittermeier, 1977; Johns, 1986].

In West Africa, there are similar restrictions. In Gabon, chimps are often not eaten because of their resemblance to man, and the Fang people in Equatorial Guinea won't eat chimpanzees for the same reason, although they happily eat gorillas and other primates that occur in their region (Sabater Pi and Groves, 1972; C. Tutin, personal communication). On the Mentawai Islands off Sumatra, Kloss's gibbon (*Hylobates klossi*) is the least persecuted of the four species endemic to these islands in part because of its human appearance (A. Mitchell, personal communication; A. Whitten, personal communication).

Educational Level

Another factor preventing hunting of primates is educational level. In some parts of South America, for example, it is considered "primitive" to eat primates, and many people, especially those living in towns, won't engage in such behavior because they feel that they have reached a level at which it is below them to eat an animal like a primate. To do so would, in effect, make them equivalent to an "indio" or "wild Indian" still living in the forest, which is still considered an insult in much of Amazonia.

Size

The size of a particular primate species is another consideration for hunters, especially in South and Central America. Although most Indian tribes will eat even the smallest species, other hunters usually won't shoot marmosets, tamarins, or squirrel monkeys simply because they are too small. This trend is less evident in Africa and Asia, where the monkey species that make up the bulk of the primates hunted are more or less similar in size, but it is certainly true in the Neotropics. It is especially the case where the shotgun is the main hunting weapon, since a small primate will barely recompense the hunter for the cost of his often quite expensive shotgun shell.

Availability of Other Sources of Protein

Primate hunting is most intense in those areas in which primates are one of the main sources of protein, as is the case in some of the more remote tropical forest regions. When fish, wild ungulates, other wild game animals (e.g., paca, agouti, deer, large birds in South America; duikers, canecutter rats in West Africa, etc.), or domestic animals are available, primate hunting is usually much less of a problem. Nonetheless, it may still occur because primates are often a less expensive form of bushmeat, and domestic animals may simply be too valuable to their owners to slaughter on a regular basis.

REGIONAL REVIEW OF PRIMATE HUNTING

In this section, I review some of the patterns in primate hunting that can be identified for the major regions of the world in which nonhuman primates occur: the Neotropics (including South America, Central America, and Mexico), Africa, Madagascar, and Asia.

The Neotropical Region

Primates are hunted in all of the Neotropical countries in which they occur, and the main reason for hunting them is as a source of food. Indeed, in many areas, monkey are a preferred food source [e.g., Mittermeier, 1977], and have been described as "among the most delicious of all Brazilian mammals" [Holt, 1923].

Primates have been hunted, at least to some extent, by Indian tribes almost everywhere [e.g., Thompson, 1968], and they are an important element in the diet of most of the Amazonian Indian tribes for which data are available [e.g., Vickers, 1979; Hames, 1979; Yost and Kelley, 1983] (Fig. 21). Indeed, for the Waorani of Ecuador, monkeys account for a total of more than 35% of all animals taken by gross weight, with the woolly monkey (*Lagothrix*

Fig. 21. Primates are an important element in the diet of many Amazonian Indian tribes. Here a Trio Indian hunter from Suriname has killed a tufted capuchin monkey (*Cebus a. apella*) with his bow and arrow. Photo by M.J. Plotkin.

Fig. 22. Suriname Bushnegro with three bearded sakis (*Chiropotes satanas chiropotes*) shot for food in the interior of Suriname, 1976. Photo by R.A. Mittermeier.

lagotricha) at 21% and the howler monkey (*Alouatta seniculus*) at 12% leading the list [Yost and Kelly, 1983]. This percentage is the highest that I have encountered in the literature, but it is approached by several other South American groups for which data are available [e.g., Mittermeier, 1977].

The practice of eating monkeys has also been picked up by other people living in Amazonia, including the Brazilian *caboclos* or neo-Amazonians (which are of mixed European and Indian origin), the Suriname Bushnegroes, the French Guianan Creoles, and various waves of Spanish-speaking *colonos* or colonists that have moved and are moving into the Amazonian portions of Venezuela, Colombia, Peru, Bolivia, and Ecuador in increasing numbers (Figs. 22,23).

The percentage of primates in the diet of these groups is usually somewhat lower than that of the Indian tribes studied, but the figures overlap. Smith

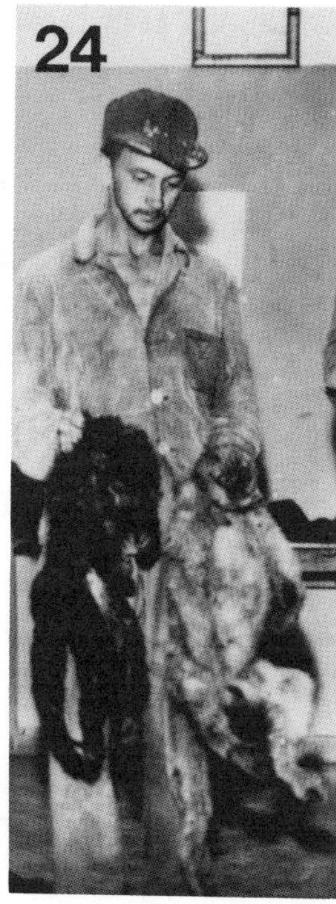

Fig. 23. White-bellied spider monkey (*Ateles b. belzebuth*) shot by Spanish-speaking *colonos* in Colombia. Photo by F. Medem.

Fig. 24. Capuchin monkey (*Cebus apella nigritus*) and muriqui (*Brachyteles arachnoides*) poached in the Carlos Botelho State Reserve, São Paulo, Brazil, 1973. Photo by Bento Vieira de Moura Neto.

[1976], for instance, found that monkeys accounted for 1.3% of the total game weight taken in three *agrovilas* that he studied along the Transamazonian Highway in Brazil, whereas Pierret and Dourojeanni [1967] estimated that monkeys made up 9.3% of the game weight consumed in Peruvian Amazonia. Percentages from studies in other areas investigated are between these two figures [e.g., Castro et al., 1976; Ayres and Ayres, 1979].

Hunting in Amazonia is now largely carried out with shotguns, although Indian tribes living traditional life-styles or in the process of acculturation continue to use the bow and arrow and the blowgun. Shotguns are quite common in Amazonia, and virtually every family of *caboclo* or *colono* owns one, and increasingly Indians are acquiring them as well. A study by Yost and Kelley [1983] indicated that use of the shotgun increased the amount of game taken by one-third, meaning increasing pressure on local game populations.

Every monkey species, including the pygmy marmoset (*Cebuella pygmaea*), is a potential candidate for the cooking pot, but there is no doubt that a preference exists for the larger species in the Neotropics, especially *Lagothrix*, *Ateles*, *Alouatta*, *Cebus apella*, and *C. albifrons* [Smith, 1976; Mittermeier, 1977; Mittermeier and Coimbra-Filho, 1977; Ayres and Ayres, 1979; Hames, 1979; Vickers, 1979; Soini, 1982; Yost and Kelley, 1983; Johns, 1986]. *Lagothrix* appears to be the most sought-after monkey food species wherever it occurs, and hunting pressure can sometimes be quite heavy. As far back as the middle of the last century, Bates [1863] noted that one small tribe of Indians living on the Colombian-Brazilian border accounted for some 200 *Lagothrix* per year, and Yost and Kelley [1983] found that this species was the most commonly taken animal among the Waorani of Ecuador (562 killed out of a total of 3,165 animals of 37+ species in 867 recorded hunts). Castro et al. [1976] further indicated that *Lagothrix* accounted for 937 of 1,507 monkeys found in the city market of Iquitos, Peru, during a 7 month period in 1973. *Lagothrix* and *Ateles* have slow reproductive rates and consequently have suffered the most from heavy hunting pressure. Indeed, they have already been exterminated from large areas of Amazonia with otherwise suitable forest habitat [Freese, 1975; Hernandez-Camacho and Cooper, 1976; Mittermeier and Coimbra-Filho, 1977; Soini, 1982]. *Lagothrix* is especially vulnerable because infant woollies are also the most prized pet monkeys, meaning that a hunter will often shoot preferentially at a female with infant.

Alouatta and *Cebus* are much more adapatable than the two ateline genera and usually persist for longer periods of time in areas where primate hunting is prevalent. Nonetheless, extreme pressure through market hunting can result in local extinctions of these species as well, as has been the case in parts of Peruvian Amazonia [Freese, 1975; J. Terborgh, personal communication].

Other genera that are heavily hunted include *Chiropotes* and *Cacajao*, and their situation in some areas may be as serious as that of *Lagothrix* and *Ateles* [Mittermeier and Coimbra-Filho, 1977] (P. Soini, personal communication). *Pithecia* and *Callicebus* are also hunted in many areas, but *Callithrix*, *Saguinus*, *Callimico*, *Saimiri*, and *Aotus* are usually under little pressure from hunting. The first four are too small to recompense the hunter for the cost of his shotgun shell, and *Aotus*, being nocturnal, is more difficult to find. If nothing else is available, these animals may go into the pot as well, but they are more often captured alive as pets for local people or for export than shot as a source of food.

In other parts of South America outside Amazonia, primates are also hunted from time to time. In the Atlantic forest region of eastern Brazil, which is the highest conservation priority in the Neotropics, only about 1–5% of the original forest cover remains, and much of that is in isolated parks and reserves. Nonetheless, poaching of primates still takes place (Fig. 24), and is especially serious for the larger species like *Alouatta fusca* and *Brachyteles*, which are sometimes represented by only a few dozen individuals in these protected areas. The situation is especially serious for *Brachyteles*. It is the largest monkey in the New World and has always been subjected to heavy hunting pressure, to the point that expeditions of some of the early naturalists in the Atlantic forest region would at times live almost entirely from its meat [Wied-Neuwied, 1821]. Now, with so little habitat remaining, any hunting at all could be disastrous. Several other highly endangered species in the Atlantic forest are also at great risk from hunting pressure on remaining populations. The northern brown howler (*Alouatta fusca fusca*) from northern Espirito Santo, southern Bahia, and adjacent parts of Minas Gerais and the buff-headed tufted capuchin (*Cebus apella xanthosternos*) from southern Bahia are two monkeys that are now at more risk than *Brachyteles* and could be quickly hunted to extinction in the few forests in which they still occur.

In other parts of the New World, such as the forests of Central America and southern Mexico, the *llanos* or plains of Venezuela and Colombia, the *chaco* region of Paraguay and Bolivia, the *cerrado* of central Brazil, and the *Pantanal* swampland of the Brazilian Mato Grosso, there are far fewer species of primates, and other larger, more abundant animals, either domestic or wild, usually serve as food for man. Consequently, primate hunting, though it occurs, is less common than in greater Amazonia. Nonetheless, in combination with habitat destruction, it can be a serious threat for certain species, such as the nine subspecies of *Ateles geoffroyi* from Central America and southern Mexico, most of which have quite restricted ranges [Konstant et al., 1985].

Africa

Primate hunting is widespread over much of the Guineo-Congolian rain forest zone of Africa and especially in certain West and Central African countries like Liberia, Sierra Leone, Nigeria, Cameroun, Equatorial Guinea, Congo Republic, and Zaire. Both subsistence hunting for local use and large-scale market hunting are important factors in these countries and are undoubtedly a major reason for the decline of many primate species. Indeed, in general terms, hunting is probably a more serious threat to primates in the Guineo-Congolian rain forest zone of West and Central Africa than anywhere else in the world.

Unfortunately, very little is available in the way of quantitative data on primate hunting in Africa. Sabater-Pi and Groves [1972] interviewed 100 Fang people in Equatorial Guinea (then Rio Muni) and found that 33% preferred higher primates above all other sources of food. Oates et al. [1982] conducted a 3 month market and roadside survey in the Bendel state of Nigeria and noted that monkeys (plus two pottos) accounted for 16.7% of 1,084 animals sold as bush meat. Teleki [1982] studied hunting and trapping practices in Sierra Leone and estimated that Basa tribesmen from Liberia (who apparently hunt monkeys by preference) cross the border into Sierra Leone and take out as many as 2,000 smoked monkeys per truck trip, with several trucks crossing the border each month.

With heavy hunting pressure of this kind, it is almost certain that primates have been greatly reduced or even exterminated from many areas, but solid data correlating primate decline and hunting unfortunately are not yet available. What is clear, however, is that certain rare and endangered species with very restricted ranges are under serious and immediate threat because of hunting, good examples being the white-throated guenon (*Cercopithecus erythrogaster*) from Nigeria [Oates and Anadu, 1982] and the drill (*Mandrillus leucophaeus*) from Equatorial Guinea, Cameroun, and Nigeria [J.S. Gartlan, personal communication].

In general, the preferred species seems to be the guenons (*Cercopithecus* spp.), the mangabeys (*Cercocebus* spp.), and the mandrill and drill (*Mandrillus* spp.). Colobus monkeys vary in popularity. In some areas, they are not eaten at all because their flesh is considered too dry (e.g., in Equatorial Guinea [Sabater-Pi and Groves, 1972], in Gabon [C. Tutin, personal communication]). On the other hand, they are often the most common species in the markets of Liberia and also account for 65% of the prey taken by the Mbuti Pygmy bow hunters in Zaire [Harako, 1982], indicating that significant regional preferences for different species exist.

Gorillas are also killed for food in some countries (e.g., Gabon [C. Tutin, personal communication], Equatorial Guinea [Sabater-Pi and Groves, 1972]), and chimps may be as well, although taboos against the eating of chimps seem to be more common than for other species. For instance, Sabater-Pi and Groves report that gorillas are one of the favorite sources of meat for the Fang people of Equatorial Guinea (although they are getting difficult to obtain), but that the Fang will not eat chimpanzees because these animals are "almost human" and eating them would be a sin. A similar belief apparently exists in Gabon (C. Tutin, personal communication). In the Lomako region of Zaire, eating pygmy chimps (*Pan paniscus*) is considered tantamount to cannibalism (N. Badrian, personal communication) and in the Bossou area of southeastern Guinea, chimps are not killed because of villagers' "sympathy" for them (Y. Sugiyama, personal communication).

In general, the smallest species like the talapoin monkey (*Miopithecus talapoin*) and the nocturnal galagos, pottos, and angwantibos are not preferred food items, but even they are taken in certain areas (e.g., Nigeria [Oates, 1982], Zaire [Harako, 1982; N. Badrian, personal communication]). However, the correlation between size and hunting pressure, which is so obvious in the Neotropical region, is less striking in Africa, with taste, local customs, and general abundance playing more important rules.

African primates are hunted mainly with shotguns, bows and arrows, and vine or wire snares, with a variety of other traps being employed as well [Marks, 1976] and even nets and dogs in some areas [Harako, 1982]. Frequency of use of different weapons depends on local traditions and on the general availability of firearms. For instance, guns are common in Liberia but more strictly controlled in Sierra Leone.

In general, primate hunting is less frequent outside the major Guineo-Congolian forest zone of West and Central Africa. It does occur in other countries, but seems to be restricted to certain tribal groups and is not the serious factor in primate decline that it is in the major forest zones (T.T. Struhsaker, personal communication; J. Skorupa, personal communication; R. Rudran, personal communication; J.F. Oates, personal communication).

Hunting appears to be restricted to Christian and pagan peoples. As in Asia, it is not practiced by Muslims, who consider primates unclean and not fit for human consumption. Consequently, primate hunting for food is apparently nonexistent in predominantly Muslim countries like Somalia (E. Visalberghi, personal communication).

Madagascar

Although primate hunting is not nearly as great a threat to Madagascar's endemic lemur fauna as is forest destruction, hunting can be quite important as well [Richard and Sussman, 1975]. In the eastern rain forest region, for example, many lemurs are trapped for food, from the smaller species (e.g., *Lepilemur, Cheirogaleus*) to some some of the largest (e.g., *Varecia, Propithecus diadema*), although the largest of all, the indri (*Indri indri*) seems to be still protected by taboos in at least part of its range. For some of these animals (e.g., *Varecia, Propithecus diadema*), hunting is in fact a very significant threat to survival, locally at least surpassing forest destruction as an immediate problem. The very restricted subspecies of gentle lemur, *Hapalemur griseus alaotrensis*, of the reed beds of Lac Alaotra is also threatened more by hunting than by habitat destruction [Petter et al., 1977].

Guns are not common in Madagascar, and it appears that most hunting using guns is carried out by Chinese and Europeans living in Madagascar, and by Malagasy store keepers and others who have the means to purchase firearms and who have begun to eschew traditional beliefs [Richard and Sussman, 1975; F. Emmanuel, personal communication]. Nonetheless, lemur hunting also appears to be part of the traditional culture in some parts of the country, particularly in the eastern forest region. Lemurs traps are very common in some parts of this region (e.g., the Masoala Peninsula [Fig. 25]; the region of Ranomafana to the east of Fianarantsoa [Fig. 26]), and they are quite specialized for capturing lemurs of different sizes, indicating that they have been used and perfected for a long time. Blowguns are used to hunt lemurs as well. Taboos do exist in the eastern forest region (e.g., the Tanala do not hunt *Avahi* or *Indri*, and the Betsimasaraka avoid hunting *Indri* as well [Richard and Sussman, 1975], but others (e.g., *Varecia, Propithecus diadema*) are not protected and may be severely threatened by hunting.

In the western dry deciduous forest, lemur hunting occurs as well, with the prime targets being members of the genera *Lemur* and *Lepilemur*. These are captured either in traps or, in the case of *Lepilemur*, even caught by hand in their daytime sleeping sites. *Propithecus verreauxi*, on the other hand, is protected by a taboo among the Sakalava people, who occupy a large portion of western Madagascar [Richard and Sussman, 1975].

In the southern spiny desert region, there is also some hunting of lemurs, with the widespread *Lemur catta* apparently sometimes being hunted with the aid of dogs [Richard and Sussman, 1975], but the most conspicuous species, *Propithecus v. verreauxi*, is again protected over part of its range by a taboo of the Antandroy people [Richard and Sussman, 1975]. Nonetheless, as traditions break down and as groups of people move from area to area in search of a living, the old taboos on hunting are losing their importance, and even species like *P. v. verreauxi* are being hunted in places where they were never previously disturbed (J. DeHeaulme, personal communication).

Madagascar is also the only place where a primate species is endangered because it is killed as a bad omen. As mentioned above, this is the case for the aye-aye, which is killed on sight in the eastern rain forest region because it is considered a harbinger of evil (although part of the reason for killing it might also stem from the fact that it occasionally raids coconut and lychee plantations [R. Albignac and J.-J. Petter, personal communication]).

Fig. 25. A snare trap for larger lemurs (e.g., *Lemur fulvus albifrons, Varecia variegata rubra*) on the Masoala Peninsula in northeastern Madagascar, 1984. Photo by R.A. Mittermeier.

Fig. 26. Trap used to capture small nocturnal lemurs in the Ranomafana region of southeastern Madagascar, 1985. Photo by R.A. Mittermeier.

The relatively recent loss of a major portion of Madagascar's primate fauna may also be at least partly attributable to hunting. Six genera of giant lemurs and a total of at least 14 species have disappeared in Madagascar since the arrival of man some 1,500–2,000 years ago [Walker, 1967; Tattersall, 1982; Sussman et al., 1985], almost all of them diurnal and larger than the surviving species. Although there is considerable controversy over the actual role of the human primate in these extinctions (R. MacPhee, personal communication), at least some authorities [e.g., Walker, 1967] believe that overhunting was a factor.

Asia

In general terms, primate hunting in Asia appears to be less of a factor in primate decline than in the other two major continental areas. Primates are not hunted at all over a large portion of Asia because of religious restrictions. As in Africa, they are not hunted by Muslims because they are considered unclean, and they are not hunted over much of India either because monkeys play a prominent role in Hinduism and are still considered sacred in many areas. Primate hunting is not prevalent in Buddhist areas either, although the restrictions in this religion are less clearly defined than in the other two.

Most of the primate hunting that does take place in Asia is done by aboriginal groups such as the orang asli of Peninsular Malaysia, the Dayaks and Ibans of Borneo, the hill tribes of Thailand and Burma, the Veddahs of Sri Lanka, and the many small tribal groups of India. The Chinese are also quite fond of eating monkeys, both in China itself [Tan Bangjie, 1985] and in the other countries with Chinese populations.

In areas where aboriginal people occur, primate hunting pressure can approach that in other tropical forest regions; however, with few exceptions (e.g., Borneo), these groups make up only a small percentage of the population in Asian countries.

Examples of areas where primate hunting is quite common are the Mentawai Islands off Sumatra, where the aboriginal inhabitants still hunt all four endemic primate species to varying extents [Tilson, 1979; WWF, 1980; A. Mitchell, personal communication; A. Whitten, personal communication], and the interior of Kalimantan, where the Dayak people eat almost all primates, including the orang-utan (B. Galdikas, personal communication). Surprisingly, in Kalimantan Tengah, at least, the Dayaks have a taboo against killing the crab-eating macaque (*Macaca fascicularis*), one of the commonest species (B. Galdikas, personal communication).

Hunting of primates in China was apparently common, although it is now coming under more and more legal restrictions [Tan Bangjie, 1985]. Indeed, for some Chinese, monkeys are even considered a delicacy, the best example being the illegal and oft-cited custom of eating the brains of a live macaque strapped into a special restraining table. Various monkey body parts are also used for medicinal purposes in China, and in some cases endangered species are involved (e.g., *Presbytis leucocephala, P. francoisi* [Tan Bangjie, 1985]).

Chinese living in other countries will also hunt primates on occasion, even in Muslim areas where primates are otherwise not eaten (e.g, Sarawak, Brunei). Nonetheless, this is a minor factor in such countries, since the Chinese populations are usually relatively small, do not live in the forest, and only occasionally hunt or purchase primates for food.

Even fewer quantitative data exist for primate hunting in Asia than for South America and Africa. Preferences for different species vary from country to country, but macaques, leaf monkeys, and gibbons are all possible sources of food in areas where hunting takes place. In general, there seems to be little correlation between size and preference, since these three groups of Asian primates are all similar in size (C. Marsh, personal communication; M. Kavanagh, personal communication; B. Galdikas, personal communication). Lorises and tarsiers are rarely if ever eaten, and only in the interior of Kalimantan does there appear to be a preference for the largest Asian primate, the orang-utan (B. Galdikas, personal communication).

Over much of Asia, primates may be more persecuted as crop pests than as a source of food, although again there is little data on the effects of either hunting or destruction of primates as agricultural pests.

Primates are hunted in Asia mainly with shotguns and blowguns, and sometimes bow and arrow, traps, poisoned bait, and dogs. Where available, automatic weapons purchased on the black market may be used as well (A. Eudey, personal communication).

CONCLUSION

It should be clear from the information presented in this chapter that primate hunting can be a significant factor in the decline of wild primate populations in many parts of the world. Although no primate species or subspecies has thus far gone extinct in this century, there is unquestionably a risk of losing certain species before the year 2000 and human predation could play an important role in several cases. Based on the review presented in this chapter, several generalizations on primate hunting can be made.

1. Habitat destruction is by far the major threat to wild primate populations around the world, but primate hunting can play a major role for certain species in certain areas and is even more important than habitat destruction in some cases.

2. Of all the reasons for hunting primates, hunting them as a source of food is by far the most important.

3. Killing of primates as crop pests is definitely the next most significant factor, and in some parts of Africa and Asia may surpass food hunting as a problem.

4. Other forms of primate hunting are usually very minor compared to food hunting and destruction of primates as crop pests, although they can occasionally be a major problem for particular endangered species (e.g., *Macaca silenus*) or species that are in very high demand (e.g., *Colobus guereza, Colobus polykomos*).

5. Market hunting, especially in Amazonia and the Guineo-Congolian forest region of West and Central Africa, is a far more serious threat than subsistence hunting.

6. Taboos protect some primate species, but their effects are limited and they are breaking down in many areas.

7. In the Neotropical region, hunting pressure is heaviest in the Amazonian region. It is part of the tradition of most Amazonian Indian tribes, some of which count on primates for as much as 35% of their meat diet. The practice of hunting primates has also been picked up by many other more recent human groups in Amazonia, among them the Brazilian *caboclos*, the Suriname Bushnegroes, the French Guianan Creoles, and the Spanish-speaking *colonos* of Hispanic Latin America.

The primate species most affected by hunting in the Neotropics are clearly the largest ones, and particularly the Atelinae (*Lagothrix, Ateles, Brachyteles*). *Ateles* and *Lagothrix* are subjected to hunting pressure over much of their respective ranges in Amazonia, and they reproduce slowly and are less adaptable than most of the other Neotropical genera. Consequently they are usually the first species to disappear in areas of heavy hunting. The species and subspecies with the smallest ranges (e.g., *Lagothrix flavicauda* from northern Peru; the Central American and Mexican subspecies of *Ateles*) are most at risk. Even the highly endangered ateline genus, *Brachyteles*, now down to a few hundred individuals in the Atlantic forest region of eastern Brazil, is still threatened by illegal hunting. *Alouatta* and *Cebus* are also heavily hunted in many areas, and may be eliminated by especially heavy hunting pressure, but in general are more adaptable than the Atelinae. *Cacajao* and *Chiropotes* are subjected to heavy hunting pressure as well and are disappearing in some parts of their ranges because of it. Once again, those species and subspecies with the smallest ranges (e.g., *Chiropotes satanas satanas* from lower Amazonian Brazil) are most at risk.

Primate hunting does occur elsewhere in South and Central America as well, and it may be a serious factor for certain species like the *Ateles* mentioned above. By and large, however, the bulk of primate hunting in the Neotropics takes place in Amazonia. Killing of primates as crop pests in South and Central America is a minor problem and not comparable to the situation in Africa and Asia.

8. In Africa, hunting primates is a major problem in the Guineo-Congolian forest zone of West and Central Africa, with tens of thousands being killed for local consumption and for sale as bush meat in towns and cities. All monkey species are affected by this, even the great apes in some areas. Hunting of primates as a source of food looms as a greater threat to primate survival in this region than anywhere else in the world, and it may in fact already be threatening entire primate faunas in some of the smaller West African countries. It is a source of major concern in Upper Guinean countries

such as Sierra Leone, Liberia, and Nigeria, where forest destruction is also a serious problem. In other countries, such as Zaire, Gabon, Cameroun, and the Congo Republic, it may be somewhat less serious at the moment simply because of the large areas of forest that still remain. An in-depth study (or series of studies) of primate hunting and its effects on primate populations in the entire Guineo-Congolian forest region is a top priority in international primate conservation.

Primate hunting in Africa is carried out mainly by Christian and pagan groups and does not occur in Muslim areas. Size is much less of a factor in determining which species are hunted in Africa than in the Neotropics.

9. In Madagascar, the importance of primate hunting has probably been underemphasized in the past. Several highly endangered species such as the diademed sifaka (*Propithecus diadema* sspp.), the ruffed lemur (*Varecia variegata* sspp.), and the Lac Alaotra gentle lemur (*Hapalemur griseus alaotrensis*) are hunted for food, and this may be the major threat to them in some areas. The aye-aye (*Daubentonia madagascariensis*) is killed as an evil omen, the only primate species for which this is known to be true.

10. In Asia, primate hunting appears to be less of a problem than in the other areas in which primates occur, in large part because three of the major religions prohibit or restrict killing and eating of primates. Most of the primate hunting in Asia is done by aboriginal groups and by the Chinese, both in China and in other countries in which they live. Hunting may be a threat to several Asian species with very restricted ranges, such as the simakobu (*Simias concolor*) of the Mentawai Islands, which is the most heavily persecuted of the Mentawai species [Tilson, 1979; WWF, 1980] or some of the endangered *Presbytis* and *Rhinopithecus* of China [Tan Bangjie, 1985].

11. Primate hunting is difficult to control because, in many cases, it is part of a traditional way of life and also because much of it occurs in remote areas. Introduction of alternative sources of protein may be a solution in some areas, but, ultimately, improved education of local people and control of human population growth provide the best solutions.

ACKNOWLEDGMENTS

I would like to express my gratitude to the many researchers who responded to my hunting questionnaire and who provided new and interesting information from their own field experiences. In particular, I would like to thank the following: P. Andau, C. and R. Aveling, N. Badrian, T. Blohm, M. Borner, F. Emmanuel, A. Eudey, J.G. Fleagle, C. Freese, B. Galdikas,

J.S. Gartlan, K. Green, R. Horwich, J.M. Johnson M. Kavanagh, F.W. King, Kunlong Ben, A.B. Luscombe, J. McNeely, C. Marsh, F. Medem, A. Mitchell, J.F. Oates, M.A. Parthasarathy, M.J. Plotkin, K.R.S. Proud, M.A. Reza Khan, M.G.M. van Roosmalen, R. Rudran, A.B. Rylands, J. Skorupa, N.J.H. Smith, P. Soini, C. Southwick, T.T. Struhsaker, Y. Sugiyma, R. Susman, R.W. Sussman, G. Teleki, C. Tutin, A. Vedder, E. Visalberghi, A. Whitten, and R. Wrangham. The Suriname portion of this work was partly supported by a grant from the New York Zoological Society; the remainder was supported by World Wildlife Fund—U.S.

REFERENCES

Ayres JM, Ayres C (1979): Aspectos da caça no alto Rio Aripuaná. Acta Amazonica 9:287–298.

Bachrach M (1946): "Fur: A Practical Treatise." New York: Prentice-Hall.

Bates HW (1863): "The Naturalist on the River Amazons." London: John Murray.

Brass E (1925): "Aus dem Reiche der Pleze." Berlin: Neuen Pelzwaren-Zeitung und Kürschner-Zeitung.

Castro N, Revilla J, Neville M (1976): Carne de monte como una fuente de proteinas en Iquitos, con referencia especial a monos. Rev. Forestal Peru 6(1-2):19–32.

Dunbar RIM, Dunbar EP (1975): Guereza monkeys: Will they become extinct in Ethiopia? Walia 6:14–15.

Federal Register (1978): Endangered species permit: receipt of application. Vol. 43(239):5812.

Fossey D (1983): "Gorillas in the Mist." Boston: Houghton Mifflin Co.

Freese C (1975): A census of nonhuman primates in Peru. In: "Primate Censusing Studies in Peru and Colombia" (Report to the National Academy of Sciences on the activities of Project AMRO-0719, Pan American Health Organization, WHO).

Hames RB (1979): A comparison of the efficiencies of the shotgun and the bow in Neotropical forest hunting. Hum Ecol 7:219–252.

Harako R (1982): The cultural ecology of hunting behavior among Mbuti Pygmies in the Ituri Forest, Zaire. In Harding, RSO, Teleki, G (eds) "Omnivorous Primates." New York: Columbia University Press, pp 499–555.

Hernandez-Camacho J, Cooper RW (1976): The nonhuman primates of Colombia. In Thorington RW, Heltne PG (eds): "Neotropical Primates: Field Studies and Conservation." Washington, DC: National Academy of Sciences.

Holt EG (1923): Monkeys as human food. J Mammal 4:193–194.

Johns A (1986): Effects of habitat disturbance on rainforest wildlife in Brazilian Amazonia. Unpublished report to World Wildlife Fund.

Jolly A (1980): "A World Like Our Own." New Haven: Yale University Press.

Jones TS (1950): Notes on the monkeys of Sierra Leone. Sierra Leone Agric. Notes 22:1–10.

Koford C (1972): Guereza monkeys of Kibale Forest, Uganda. Unpublished manuscript.

Konstant WR, Mittermeier RA, Nash SD (1985): Spider monkeys in captivity and in the wild. Primate Conserv. 5:82–109.

Mallinson JJC (1984): Golden-headed lion tamarin contraband a major conservation problem. IUCN/SSC Primate Specialist Group Newsletter 4:23–25.

Marks SA (1976): "Large Mammals and a Brave People." Seattle: University of Washington Press.
Marsh C (1986): A resurvey of the Tana River primates and their habitat. Primate Conserv. 7:72–82.
Mittermeier RA (1973): Colobus monkeys and the tourist trade. Oryx 12:113–117.
Mittermeier, RA (1977): "Distribution, Synecology and Conservation of Surinam Monkeys." PhD dissertation, Harvard University.
Mittermeier RA, Coimbra-Filho AF (1977): Primate conservation in Brazilian Amazonia. In Prince Rainier of Monaco, Bourne GH (eds): "Primate Conservation." New York: Academic Press, pp 117–166.
Oates JF (1977): The guereza and man. In Prince Rainier of Monaco, Bourne GH (eds): "Primate Conservation." New York: Academic Press, pp 419–467.
Oates JF et al. (1982): A study of the bushmeat trade in Bendel State, Nigeria. Unpublished report to the New York Zoological Society and World Wildlife Fund.
Oates JF, Anadu P (1982): Report on a survey of rainforest primates in southwest Nigeria. IUCN/SSC Primate Specialist Group Newsletter 2:17.
Petter J -J, Albignac R, Rumpler Y (1977): "Faune de Madagascar 44: Mammiferes Lemuriens (Primates Prosimiens)." Paris: ORSTOM/CNRS.
Pierret P, Dourojeanni M (1967): Importancia de la caza para alimentación humana en el curso inferior del Río Ucayali, Peru. Rev. Forestal Peru 1(2):10–21.
Poland H (1982): "Fur-Bearing Animals in Nature and in Commerce." London: Gurney and Jackson.
Quick DF (1984): Reduction of crop damage by olive baboons (*Papio anubis*): the feasilibity of conditioned taste aversion. PhD dissertation, University of California, Los Angeles.
Richard AF, Sussman RW (1975): Future of Malagasy lemurs: Conservation or extinction? In Tattersall I, Sussman RW (eds): "Lemur Biology." New York: Plenum Press, pp 335–350.
Sabater Pi J, Groves C (1972): The importance of higher primates in the diet of the Fang of Rio Muni. Man 7:239–243.
Shipman P, Bosler W, Davis KL (1981): Butchering of giant geladas at an Acheulian site. Curr. Anthropol. 22(3):257–268.
Smith NJH (1976): Utilization of game along Brazil's Transamazon Highway. Acta Amazonica 6:455–466.
Soini P (1982): Primate conservation in Peruvian Amazonia. Int. Zoo Yearbook 22:37–47.
Sussman RW, Richard AF, Ravelojaona G (1985): Madagascar: Current projects and problems in conservation. Primate Conserv. 5:53–59.
Tan Bangjie (1985): The status of primates in China. Primate Conserv 5:63–81.
Tappen N (1964): Primate studies in Sierra Leone. Curr Anthropol 5:339–340.
Tattersall I (1982): "The Primates of Madagascar." New York: Columbia University Press.
Teleki G (1982): Hunting and trapping wildlife in Sierra Leone: Aspects of exploitation and exportation. Unpublished report to World Wildlife Fund.
Thompson JE (1968): Ethnology of the Mayas of southern and central British Honduras. Publ. Field Mus. Nat. Hist. Chicago, Anthropol. Ser. 17(1–4):23–214.
Tilson R (1979): Der unbekannte Affe. Die ersten Bilder de Pageh Stumpfnase. Tier 5:20–23.
Vickers WT (1979): Native Amazonian subsistence in diverse habitats: The Siona-Secoya of Ecuador. In Moran EF (ed): "Changing Agricultural Systems in Latin America." Studies in Third World Societies, Publ. No. 7:6–36.

Walker A (1967): Patterns of extinction among the subfossil Madagascan lemuroids. In Martin PS, Wright HE Jr (eds): "Pleistocene Extinctions: The Search for a Cause." New Haven: Yale University Press, pp 426–432.

Wied-Neuwied, Maximilian Prinz zu (1821): Reise nach Brasilien in den Jahren 1815 bis 1817.

World Wildlife Fund (1980): "Saving Siberut: A Conservation Master Plan." Bogor, Indonesia: World Wildlife Fund Indonesia Programme.

Yost JA, Kelley PM (1983): Shotguns, blowguns, and spears: The analysis of technological efficiency. In Hames RB, Vickers WT (ed): "Adaptive Responses of Native Amazonians." New York: Academic Press, pp 189–224.

Primate Conservation in the Tropical Rain Forest, pages 147–177

6

The Effects of Live Trapping and Trade on Primate Populations

Michael Kavanagh, Ardith A. Eudey, and David Mack

Wildlife Trade Monitoring Unit, IUCN Conservation Monitoring Centre, Cambridge CB3 ODL, England (M.K.), Department of Anthropology, University of Nevada, Reno, Nevada 89557-0006 (A.A.E.), and TRAFFIC (U.S.A.), World Wildlife Fund — United States, Washington, D.C. 20009 (D.M.)

INTRODUCTION

Tens of thousands of wild primates are trapped in habitat countries every year in numbers that are far in excess of those that actually reach their destinations in user countries. In this chapter, we provide an overview of the international trade in live primates and consider it in the context of other pressures on wild populatioins, such as habitat destruction and subsistence hunting.

The topic is both vast and complex, ideally requiring, for example, the consideration of trade statistics from diverse, worldwide sources and a knowledge of national legislation and management practices in scores of countries, as well as a grasp of international agreements and their effects on the trade. While some traders have been helpful to this research, others have been secretive and unhelpful. A detailed monograph on the subject has recently been published [Mack & Mittermeier, 1984]; and we have produced the present chapter in the course of researching our respective contributions to that review, which should be consulted for further details, especially sources of data.

HISTORICAL PERSPECTIVE

Humans have been trading in other primates for 5,000 years or more: the earliest known examples of this were part of an ancient Egyptian pet trade.

Since then, primates have appeared frequently in historical accounts as pets, zoo specimens, agricultural laborers, circus exhibits, objects of religious veneration or cultural denigration, and laboratory research subjects [Morris & Morris, 1966; Schultz, 1969; Kavanagh, 1983a].

During the twentieth century, the biggest stimulus to the primate trade has been the demand for laboratory animals. The initial impetus for the use of primates in large numbers was the search for polio vaccines during the 1950s and early 1960s [LeCornu and Rowan, 1978, 1979]. The pet trade appears to have been the second most important stimulus to the trapping of primates, especially New World species. Trade in Neotropical monkeys began in the 1940s. It both stimulated and was dependent on the growth of commercial air freight transportation between Miami, Florida and the major Amazonian collection centers of Iquitos, Peru and Leticia, Colombia [Cooper, 1968]. Zoos and circuses also have contributed to the demand for primates, stimulating a relatively small sector of the trade but one that has been important in relation to the capture of endangered species such as the great apes and lemurs [Morris and Morris, 1966; Kavanagh, 1984].

No reliable figures have ever been available for the total worldwide trade volume, but see LeCornu and Rown [1978, 1979] and Mack and Mittermeier [1984] for estimates. The principal importer and user for primates during the twentieth century has been the United States [Kavanagh, 1984]. U.S. primate imports reached their peak of about 200,000 per year during the latter half of the 1950s when polio vaccine development and production was at its height. More recently, U.S. imports have declined from approximately 127,000 in 1968 to less than 20,000 in 1982 [Mack, 1983; Mack and Eudey, 1984]. There has been a corresponding decline in the range of taxa imported by the United States from approximately 107 species during the period 1968–1972 to approximately 79 species during 1976–1980 [Mack and Eudey, 1984].

Figures comparable to those compiled for the United States do not exist as far back for the other major importing countries, but there appears to be a similar pattern of declining numbers. The second largest importer, the United Kingdom, experienced a drop in totals between 1965 and 1975 from over 30,000 per year to about 8,000 per year [Burton,1978]. Currently, British imports are running at just over 6,000 per year [Kavanagh, 1983a]. Japan, the third largest importer of primates, appears to have had its peak at a slightly later date: more than 22,000 were imported in 1972 but the numbers have steadily declined to approximately 4,700 in 1981 [Anonymous, 1979; Kavanagh, 1984].

No single reason can account for the overall decline in the primate trade, but two factors appear to be most important; 1) technical advances and higher costs have reduced the demand for primates in the production and testing of vaccines, and 2) laws and regulations have been promulgated governing the export and import of primates for health and/or conservation reasons. Technical developments associated with the reduced use of primates in the production and testing of polio vaccines are reviewed by LeCornu and Rowan [1979]. The natural decline in the numbers of people seeking vaccination against polio after the massive immunization program of the early 1960s also has contributed the the reduced demand.

In the mid-1970s, both the United Kingdom and the United States implemented health-related controls on primate imports. The United Kingdom enacted the Rabies (Importation of Dogs, Cats, and Other Mammals) Order 1974, under which all imported primates must be quarantined for 6 months in goverment-approved premises. The effect of this regulation was to reduce all forms of demand, especially that for pets, by increasing the cost of imported primates [Burton, 1978]. In 1975, the U.S. Public Health Service amended federal quarantine regulations to prohibit primate imports except for scientific or educational purposes or for exhibition. This rule eliminated primates from the pet trade. It was implemented, however, after Brazil, Colombia, and Peru had already banned the export of primates and after India had significantly lowered its quota on worldwide rhesus macaque (*Macaca mulatta*) exports [Held, 1982; Mack and Eudey, 1984].

As of 1980, no similar health-oriented regulations had been implemented in Japan or in Taiwan, the second biggest primate importer in Asia. However, several European countries implemented health controls in the 1960s and 1970s: these include Denmark, Finland, West Germany, Italy, and Switzerland [Kavanagh and Bennett, 1984]. In West Germany, the imposition of controls followed an outbreak of Frankfurt-Marburg virus in 1969 [Meister, 1977].

Concern over the conservation of primates has been translated into action in both habitat and user countries. The most important development has been the imposition of export bans by countries that were formerly major suppliers of primates. In the early 1970s, Peru, Colombia, India, and Thailand supplied more than 75% of the primates imported by the United States and large numbers of those imported by other countries. Between 1973 and 1978, all four of these major source countries implemented total bans on the commercial export of primates, bringing about significant reductions in the numbers of primates traded worldwide and shifts in the species composition of the trade [Eudey, 1978; Mack and Mittermeir, 1984]. Since then other countries have taken steps to reduce or to eliminate export pressures on their native primate populations: Malaysia, for example, since 1980 has limited the export of macaques to those that are certified as being for genuine scientific research purposes and where the exports do not appear to jeopardize the viability of populations within the country [Nordin and Hasnah Samian, 1981]. As of June, 1984, the commercial export of Malaysian primates was prohibited altogether.

In user countries, conservation-oriented regulations with applications to primates were intially very few, although they did include a little-enforced provision of the U.S. Lacey Act of 1900, which makes it a civil offense to obtain wildlife taken in violation of any law of a foreign country [Inskipp and Wells, 1979; Mack and Eudey, 1984]. In recent years, what has been of much more importance is the development of an international agreement regulating trade in wild animals and plants. Throughout the 1960s, there was a growing movement within the International Union for Conservation of Nature and Natural Resources (IUCN) to protect the natural heritages of habitat countries by restricting the importation elsewhere of threatened species and their products. In early 1973, representatives of 87 countries met in Washington, D.C. where, on March 3, 21 of them signed the Convention on International Trade in Endangered Species of Wild Fauna and Flora (CITES, or the "Washington Convention"). This came into force on July 1, 1975, having by then been duly ratified by more than the ten countries required to bring the convention into effect. Eighty-three countries are now Parties to CITES (Table I).

Since 1975, all primate species, in recognition of their potential vulnerability to uncontrolled trade, have been listed in the CITES appendices (Table II). CITES Appendix I specifies taxa that are threatened with extinction. Trade in them requires permits from the countries of both origin and import, which should only be granted when the export is not detrimental to the survival of the species and is not primarily for commercial reasons. CITES Appendix II lists species that may become threatened with extinction unless trade is subject to strict regulation. Permits are required only from the country of export for such species, and there is no restriction on trade for commercial purposes. CITES Appendix I species that are bred in captivity are treated as Appendix II species. Permits are issued by national management authorities after the consultation with national scientific authorities, and Parties to CITES must require similar documentation from both member and nonmember countries. For further details, see Inskipp and Wells [1979].

Parties to the convention are required to submit annual reports to the CITES Secretariat in Switzerland. Because all primates are listed in the appendices, any trade in them involving CITES Parties should be on record at the Secretariat. To date, such reporting has not been complete [Kavanagh,

TABLE I. CITES Parties on January 1, 1984 With the Dates on Which Their Memberships of the Convention Entered Into Force

83)[a] Algeria	21-02-84	77) Malawi	06-05-82
66) Argentina	08-04-81	41) Malaysia	18-01-78
30) Australia	27-10-76	12) Mauritius	27-07-75
76) Austria	27-04-82	45) Monaco	18-07-78
53) Bahamas	18-09-79	21) Morocco	14-01-76
75) Bangladesh	18-02-82	69) Mozambique	23-06-81
82) Belgium	01-01-84	13) Nepal	16-09-75
54) Bolivia	04-10-79	39) Nicaragua	04-11-77
43) Botswana	12-02-78	19) Niger	07-12-75
17) Brazil	04-11-75	2) Nigeria	01-07-75
71) Cameroon	03-09-81	29) Norway	25-10-76
11) Canada	09-07-75	25) Pakistan	19-07-76
66) Cent. African Rep.	25-11-80	47) Panama	15-11-78
9) Chile	01-07-75	23) Papua New Guinea	11-03-76
67) China	08-04-81	34) Paraguay	13-02-77
73) Colombia	29-11-81	14) Peru	25-09-75
81) Congo	01-05-83	72) Philippines	16-11-81
15) Costa Rica	28-09-75	65) Portugal	11-03-81
6) Cyprus	01-07-75	62) Rwanda	18-01-81
37) Denmark	24-10-77	79) Saint Lucia	15-03-83
8) Ecuador	01-07-75	38) Senegal	03-11-77
44) Egypt	04-04-78	35) Seychelles	09-05-77
26) Finland	08-08-76	16) South Africa	13-10-75
46) France	09-08-78	52) Sri Lanka	02-08-79
40) Gambia	24-11-77	78) Sudan	24-01-83
20) Germany, East	07-01-76	63) Suriname	15-02-81
24) Germany, West	20-06-76	5) Sweden	01-07-75
22) Ghana	12-02-76	3) Switzerland	01-07-75
56) Guatemala	05-02-80	57) Tanzania	27-02-80
74) Guinea	20-12-81	80) Thailand	21-04-83
36) Guyana	25-08-77	48) Togo	21-01-79
27) India	18-01-76	4) Tunisia	01-07-75
51) Indonesia	28-03-79	33) U.S.S.R.	08-12-76
32) Iran	01-11-76	7) United Arab Emirates	01-07-75
59) Israel	17-03-80	31) United Kingdom	31-10-76
55) Italy	31-12-79	1) United States	01-07-75
60) Japan	04-11-80	10) Uruguay	01-07-75
50) Jordan	14-03-79	42) Venezuela	22-01-78
49) Kenya	13-03-79	28) Zaire	18-10-76
68) Liberia	09-06-81	64) Zambia	22-02-81
58) Liechtenstein	28-02-80	70) Zimbabwe	17-08-81
18) Madagascar	18-11-75		

[a]Sequence of joining.

TABLE II. Primates Listed in CITES Appendix I on January 1, 1984[a]

Cheirogaleidae	*Allocebus* spp.; *Cheirogaleus* spp.; *Microcebus* spp.; *Phaner* spp.
Lemuridae	*Hapalemur* spp.; *Lemur* (including *Varecia* spp.; *Lepilemur* spp.
Indriidae	*Indri* spp.; *Lichanotus* (=*Avahi*) spp.; *Propithecus* spp.
Daubentoniidae	*Daubentonia madagascariensis.*
Callithrichidae	*Callithrix aurita; Callithrix flaviceps; Leontopithecus* (=*Leontideus*) spp.; *Saguinus bicolor; Saguinus leucopus; Saguinus oedipus* (=*S. geoffroyi oedipus*).
Callimiconidae	*Callimico goeldii.*
Cebidae	*Alouatta palliata; Ateles geoffroyi frontatus; Ateles geoffroyi panamensis; Brachyteles arachnoides; Cacajao* spp.; *Chiropotes albinasus; Lagothrix flavicauda; Saimiri oerstedii.*
Cercopithecidae	*Cercocebus galeritus galeritus; Cercopithecus diana (roloway); Colobus pennantii* (=*badius*) *kirkii; Colobus rufomitratus* (=*C. badius rufomitratus*); *Macaca silenus*; *Nasalis* (including *simias*) spp.; *Papio* (=*Mandrillus*) *leucophaeus; Papio* (=*Mandrillus*) *sphinx; Presbytis entellus; Presbytis geei; Presbytis pileatus; Presbytis potenziani; Pygathrix nemaeus.*
Hylobatidae	Hylobatidae spp.
Pongidae	Pongidae spp.

[a]All other primate species are listed in Appendix II.
The taxonomy follows that given in the CITES appendices. For common names, see Napier and Napier [1967], Kavanagh [1983b], or Kavanagh and Bennett [1984]. For further information on the appendices, or on any other aspect of the convention, contact the CITES Secretariat, 6 Rue du Maupas, 1000 Lausanne 9, Switzerland.

1984; Mack and Eudey, 1984] and, as a consequence, the present study has not yielded statistics for the worldwide primate trade that may be relied on as comprehensive and definitive.

THE CURRENT VOLUME OF INTERNATIONAL TRADE

The only year for which it has been possible to collate the numbers of all known imports and exports is 1979: the tabulations coincide approximately at just under 65,000 [Kavanagh, 1984]. This figure may be questioned owing to a lack of credible import data from countries such as Taiwan, the U.S.S.R., most eastern European countries and France. As an estimate of total trade volume, however, it gains some credence when the 22,801 well-documented U.S. imports for 1979 [Mack and Eudey, 1984] are added to the 45,443 primates that are estimated to be used in laboratories elsewhere in the world and that are not captive bred [Caldecott and Kavanagh, 1984]. The resulting figure of 68,244 is within 5% of 65,000. These figures may be contrasted with the late 1950s total of approximately 200,000 primates entering the United States alone every year [Conway, 1966]. This gives some indication of the decline in international trade volumes that is well exemplified by the 80% drop in U.S. annual imports between 1964 and 1981 [Mack, 1982]. The most recent American figures show a further 26% decline in U.S. imports from 1981 to 1982 [Mack, 1983]. A reversal of this downward trend seems unlikely although an annual global demand for many thousands of wild-caught primates (mainly monkeys) is likely to continue at least into the 1990s [Caldecott and Kavanagh, 1984].

These figures ignore mortalities between trap-sites and destinations, although these certainly are significant. They also ignore the presence or absence of tree shrews (Tupaiidae) in the data because their numbers are not high enough to be significant at this level of analysis [Mack and Mittermeier, 1984].

The evident broad trends are encouraging and wild primate populations may be expected to benefit from the diminishing volume of trade. Nevertheless, practical conservation measures must go beyond a general attempt to minimize the numbers of animals traded: they must relate to particular species and populations, and they therefore require a more detailed consideration of the effects of the trade in relation to other factors that work on wild populations.

MORTALITY RATES

The number of primates taken from the wild cannot be assessed solely from trade statistics: mortality from capture to final destination takes its toll. In the 1960s, Latin American primate losses from capture to final recipient were estimated at between 25% and 80% [Warland, 1972]. In holding compounds alone, one South American dealer gave the following losses: 15% for squirrel monkeys (*Saimiri*) and capuchins (*Cebus*); 20% for night monkeys (*Aotus*); 27% for woolly monkeys (*Lagothrix*); and a range from 3 to 30% for marmosets and tamarins (Callitrichidae) [Thorington, 1972].

Since the 1970s, very little has been written about primate trade mortality rates. According to one American importer, though, trade-related mortality is generally believed to be less than it was a decade ago [J. Porter, personal communication, 1982]. The higher costs of primates today have caused dealers to take better care in capture and handling. While pointing out the difficulties of obtaining reliable data on death rates, Nordin and Hasnah

Samian [1981] report that 12% of trapped Malaysian macaques die before reaching the exporter's compound, and another 8% die before they can be flown out. One Indonesian exporter, however, estimates that 71% of macaques may die between the trap-site in Indonesia and the end-purchaser in the country of import [Darsono, 1979].

Data collected by the Center for Disease Control, U.S. Public Health Service, and analyzed for 1978–1979 by Mack and Eudey [1984] make it possible to quantify shipment-related deaths of primates imported into the United States. The differential vulnerability of species and/or populations to the stresses involved is reflected in these figures. Out of 28,558 primates imported in 1978, 5,204 (18%) were dead-on-arrival or died with 90 days of entering the United States; and in 1979, the comparable figure was 3,818 (17%) out of 22,726 primates. Regional mortality ranged from a low of 15% among African species in 1978 to a high of 25% among New World species during the same year.

When the figures for individual taxa are examined, it transpires that some species, usually those imported in very small numbers such as some spider monkeys (*Ateles*) and forest guenons *(Cercopithecus)*, did not suffer any trade-related deaths. On the other hand, 26 silvered langurs (*Presbytis cristata*) had a mortality rate of 89%. In both 1978 and 1979, the greatest number of deaths were recorded for the long-tailed macaque (*Macaca fascicularis*): 4,787 (17%) out of 28,415 were dead-on-arrival or died with 90 days during the 2-year period. Mortality rates for this species, generally regarded to be fairly hardy, averaged 14% for imports from the Philippines, 18% for Indonesia, and 19% for Malaysia. In contrast, only 221 (4%) of 4,979 rhesus monkeys (*Macaca mulatta*) that were imported from India in 1978 died. The lower mortality of the rhesus monkeys appears to have been partly a consequence of controls implemented by the Indian government over conditions of capture and transport [Muckenhirn, 1975]. The African green monkey (*Cercopithecus aethiops*) is another common trade species that may be especially vulnerable. For the combined years 1978 and 1979, 1,196 (26%) of 4,677 green monkeys imported died, with deaths averaging 47% for imports from Ethiopia, 23% from Kenya, and 18% from Somalia. See Table III for species with above average mortality rates.

In 1980, Thailand became one of the few countries to take legal action in response to shipment mortalities when the government banned the export of tree shrews (*Tupaia glis*) because of fatalities in excess of 40% on arrival in the United States [Anonymous, 1980]. The ban was briefly rescinded in June 1981 (S. McGreal, personal communication, 1981), but no tree shrews appear to have been exported from Thailand since then (TRAFFIC [U.S.A.], unpublished data).

TABLE III. Above Average Mortality Rates Among Primates Imported to the United States, 1978–1979

Species	No. imported	No. dead	%
1978			
Presbytis cristata	26	23	88
Tupaia glis	1,320	782	59
Aotus trivirgatus	667	251	38
Callithrix jacchus	248	87	35
Saguinus mystax	173	60	35
Cebus capuchinus	14	4	29
Galago senegalensis	176	42	24
Saimiri sciureus	1,986	458	23
Cercopithecus aethiops	2,073	437	21
Macaca fascicularis	13,893	2,570	18
1979			
Callithrix argentata	91	64	70
Callithrix pencillata	19	11	58
Tupaia glis	505	230	46
Callithrix jacchus	464	138	30
Cercopithecus aethiops	2,604	759	29
Saimiri sciureus	1,071	231	22
Cebuella pygmaea	16	3	19

Annual average mortalities for all importations were 18% (1978) and 17% (1979). Number dead refers to animals that were dead-on-arrival plus those that died within 90 days of arrival. The percentages are number dead as a percentage of number imported.
U.S. Center for Disease Control information from Mack and Eudey [1984].

COUNTRIES OF ORIGIN

Direct comparisons of export figures are, of course, heavily biased in favor of those few countries that keep accurate records. We have therefore combined the data on countries of origin from five major importing countries from which records are available (Table IV). While this does not provide exact numbers of animals exported, it does permit the determination of relative contributions, and it is probably accurate in identifying Indonesia, the Philippines, Bolivia, Kenya, and Malaysia as the major exporting habitat countries during the period 1978–1981. We have also illustrated the regional sources of primates imported into the United States and Japan since 1970 (Fig. 1). Data from these two nations are used because no other major importers provide comparable statistics.

In the mid-1970s Asia replaced Latin America as the major regional source of primates imported into the United States. This was after several South American countries had curtailed exports, and the effect was strengthened by the 1975 U.S. ban on the import of primates as pets, which reduced the

TABLE IV. Approximate Relative Levels of Primate Exports, Based Solely on Records of Imports by the Netherlands, the United Kingdom, Japan, Canada, and the United States

Exporting	1978		1979		1980		1981		
countries	No.	%	No.	%	No.	%	No.	%	Total
Indonesia	11,102	23	14,568	32	10,703	27	9,765	26	46,138
Philippines	3,486	7	4,920	11	8,766	22	7,286	20	24,458
Bolivia	4,617	10	4,595	10	4,307	11	7,912	21	21,431
Kenya	4,879	10	5,942	13	6,176	15	3,976	11	20,973
Malayasia	9,188	19	6,697	15	1,960	5	2,481	7	20,326
United States	3,311	7	3,465	8	2,626	7	2,438	7	11,840
India	5,134	11	22	—	15	—	—	—	5,171
Ethiopia	1,451	3	1,659	4	1,051	3	970	3	5,131
Somalia	1,032	2	1,216	3	1,404	4	2	—	3,654
Guyana	543	1	448	1	1,000	2	542	1	2,533
Thailand[a]	1,403	3	471	1	482	1	100	—	2,456
Paraguay	179	—	513	1	569	1	—	—	1,261
West Indies	171	—	131	—	204	1	470	1	976
Senegal	351	1	179	—	236	1	195	1	961
Bangladesh	486	1	388	1	—	—	45	—	919
Panama	696	1	32	—	—	—	—	—	728
Peru	—	—	40	—	136	—	252	1	428
Belgium	2	—	208	—	124	—	—	—	334
China	—	—	69	—	97	—	131	—	297
Canada	13	—	18	—	151	—	63	—	245
Tanzania	—	—	—	—	—	—	203	1	203
United Kingdom	—	—	—	—	—	—	25	—	25
Nigeria	—	—	—	—	—	—	3	—	3
Total	48,044		45,581		40,007		36,859		170,491

[a]Thailand's 1976 ban on the export of primates did not include tree-shrews until 1980 (see text). Data are from CITES report for Canada [1981], Mack [1982], Kavanagh [1984], and Mack and Eudey [1984].

Note that this table gives approximate indications of relative export levels and should only be used in conjunction with the explanation in the text. U.S. import data include tree-shrews; other data do not.

demand for some Neotropical species. Asian countries supplied the majority of Japanese primate imports until 1980, at which time Asian and Neotropical imports became approximately equivalent. For both the United States and Japan, Africa has long been of much lesser importance than the other tropical regions as a source of primates, at least in terms of numbers. However, many African nations have strong economic links with countries in Western Europe and a growing general trade with Eastern Europe, including the U.S.S.R. There may be significant primate trading along these routes that we have not been able to document.

Following the Indian ban on the export of rhesus macaques in 1978, Indonesia, the Philippines, and Malaysia increased exports of long-tailed macaques. According to data collected by World Wildlife Fund Indonesia [1980], an average of 16,527 monkeys, including 16,484 macaques, was exported annually from Indonesia during 1975–1977, followed by 28,142 live monkeys, including 28,134 macaques, in 1978. More recent government figures [W. Kadri, personal communication, 1982] report that the annual export of macaques averaged 14,125 during 1979–1981. If the two sets of figures are directly comparable, they suggest that the Indonesian export surge of 1978 was a temporary phenomenon. However, Darsono [1979] reports the export of 17,907 monkeys from Indonesia during the first 8.5 months of 1979, versus the official figure of 13,815 macaque exports for all of 1979 [W. Kadri, personal communication, 1982]; and the World Wildlife Fund data suggest that more than 99% of the monkeys exported from Indonesia are macaques. Whatever the real figure, Indonesia, as Table IV indicates, is probably the world's major exporter of primates.

Both official and commercial export figures for the Philippines generally give lower annual numbers than those shown in Table IV, but they also show a marked upturn in 1979, which is maintained through 1981. Official figures record 2,860 macaques exported in 1978, 5,225 in 1979, 6,138 in 1980, and 5,959 in 1981 [Kavanagh, 1984]. Exports from Malaysia started to follow the same trend, jumping from a steady annual average of approximately 10,000 primates in 1974–1977 to nearly 14,000 in 1978. This upward trend continued into 1979, but exports were cut to about 3,000 in 1980 owing to the imposition of strict governmental controls [Nordin, 1981; Nordin and Hasnah Samian, 1981].

The imposition of bans on the export of primates from Peru and Colombia during the early 1970s gave rise to the emergence of Bolivia as the major source of Neotropical monkeys. Bolivia exports both its native primates and those smuggled from neighboring countries in a context of unregulated trade, even though the country became a party to CITES in 1979 [Mack and

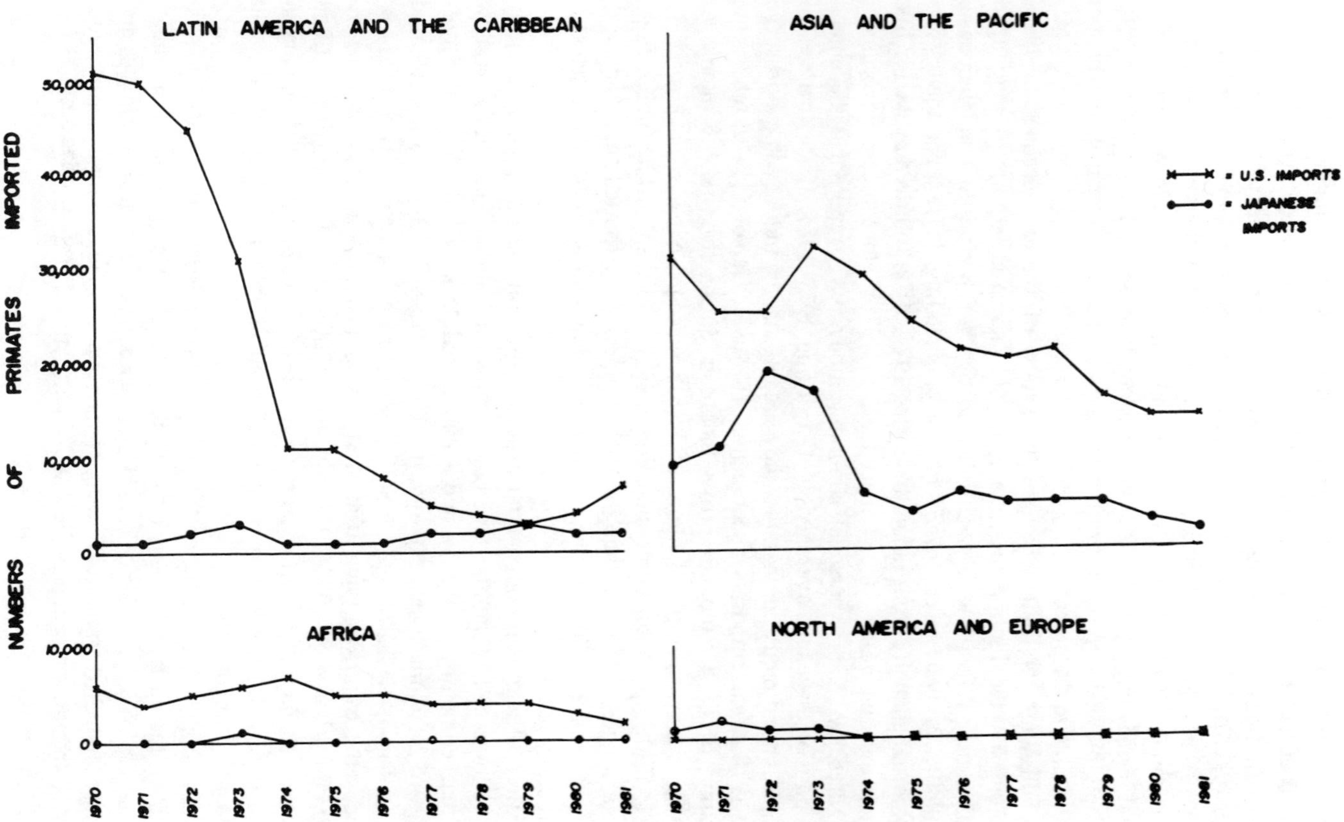
LATIN AMERICA AND THE CARIBBEAN
ASIA AND THE PACIFIC
AFRICA
NORTH AMERICA AND EUROPE
= U.S. IMPORTS
= JAPANESE IMPORTS
NUMBERS OF PRIMATES IMPORTED
0
10,000
20,000
30,000
40,000
50,000
1970
1971
1972
1973
1974
1975
1976
1977
1978
1979
1980
1981

Mittermeier, 1984]. A 1-year ban on all primate exports from Bolivia was put into effect in May 1984, and it remains to be seen what the consequences of this measure will be.

Table IV is probably accurate in attributing most African exports to Kenya and Ethiopia, with Somalia in third place, but in December 1981 Kenya banned further primate exports except for those animals that had already been stockpiled [Anonymous, 1982a]. Tanzania recently began to export a few hundred monkeys per year to the United States [Mack, 1983], but it appears that few other African countries export large numbers of primates and the small numbers (annually less than 20) that official figures suggest have been sent out since 1979 from such countries as Gabon and Uganda are probably typical [Kavanagh, 1984].

EFFECTS OF THE TRADE ON WILD POPULATIONS

In this section, the likely effects of the trade are considered in relation to species and particular populations that may be threatened by current levels of live-trapping.

Species Traded in Large Numbers

Cercopithecidae. The majority of primates currently traded are Old World Monkeys. Their relative importance has risen since New World Monkeys became harder to obtain in the mid-1970s, and in 1980 they made up approximately 77% of U.S. imports and 90% of British imports [Mack, 1982; Kavanagh, 1984]. Macaques (*Macaca*), guenons (*Cercopithecus*), baboons (*Papio*), and patas monkeys (*Erythrocebus*) comprised almost all the cercopithecids traded, with macaques being by far the most important. Rhesus monkeys (*M. mulatta*) have declined greatly in the number traded in recent years [Held, 1982] and long-tailed macaques or cynomolgus monkeys (*M. fascicularis*) now account for about half the total worldwide primate trade. Indonesia, the Philippines, and Malaysia export mainly *M. fascicularis* and 53% of the exports collated in Table IV originate from these three countries. Rhesus and long-tailed macaques possess many physiological, morphological, and behavioral characteristics in common and this has per-

---

Fig. 1. Regional origin of Japanese and U.S. primate imports, 1970–1981. Japanese data from Anonymous [1979] and Y. Kawanishi (personal communication). U.S. data are from U.S. Imports (IM-146), U.S. Customs Service statistics, compiled by the Bureau of Census, U.S. Department of Commerce. See Mack and Mittermeier [1984] for a more detailed statistical breakdown.

mitted the substitution of the latter for the former species in many projects, since rhesus monkeys are no longer available in large numbers from the wild. Recently, the United States for the first time became self-sufficient in the production of rhesus monkeys [Mack, 1982] and the conservation status of this fairly common South Asian Species is not currently being undermined by trapping.

Nevertheless, there can be little doubt that the extensive trapping between the early 1950s and the imposition of the 1978 Indian primate export ban greatly affected the rhesus population of that country, albeit working in conjunction with habitat loss and an increasing tendency for rural Indians to view monkeys as agricultural pests rather than revered creatures to be protected. Current estimates, including one based on a recently completed 5-year survey by the Zoological Survey of India, suggest that there are between 140,000 and 200,000 rhesus monkeys in India, the population having declined by about 90% during the past 20 years (C. Southwick, personal communication, 1982). There is, however, some disagreement about these figures among Indian primatologists (R. Ali, personal communication, 1983). Since the imposition of the export ban, the age structure of sample groups at four sites has shown improvement, strongly supporting the hypothesis that the trade was an influential factor in the observed population decline (C. Southwick, personal communication, 1982). The species is now well protected by trade bans in habitat countries, of which India and Bangladesh are the most important, but there was a controversial attempt in 1977 to restart exports from Bangladesh on a regular, long-term basis. This project was cancelled in 1979 amid much public acrimony [Long, 1982]. The rhesus monkey is the most obvious example, but by no means the only one, saved from further trapping by the export bans of habitat countries.

Reprieve for the rhesus has brought additional pressure to bear on the long-tailed macaque [Nordin and Hasnah Samian, 1981]. In general, this is a widespread and adaptable species that does well in the mixed secondary forest and farmland that covers large areas of Southeast Asia, as well as in undisturbed riverine forest [Marsh and Wilson, 1981; MacKinnon, 1983]. There are, however, no quantitative data with which to assess the effects of trapping, to which must be added intense hunting pressures in some localities [Mittermeier, this volume]. Furthermore, subspecific variations remain to be established for the species and it is quite possible that particular populations may be threatened, leading to local extinctions. Although Malaysia banned the commercial export of all primates in 1984, both Indonesia and the Philippines are likely to continue to export long-tailed macaques in the foreseeable future [K.R. Hobbs, personal communication, 1982; Mac-

Kinnon, 1983]. Detailed reviews on the status of macaques in relation to the trade are provided by MacKinnon [1983] for Indonesia and Eudey [1984] for all of Asia.

The guenons that are traded in large numbers are exclusively green monkeys (including vervets) of the very variable species, or superspecies, *Cercopithecus aethiops*. Once again, management information is sadly lacking, but there is no reason to suppose that live-trapping currently threatens this monkey. Predominantly a savannah and woodland animal, it is highly adaptable and widespread in sub-Saharan Africa [Gartlan and Brain, 1968; Kavanagh, 1980]. As most of the traded green monkeys are from East Africa, they are presumably vervets (*C. a. pygerythrus*, including *johnstoni*) and grivets (*C. a. aethiops*) [Dorst and Dandelot, 1972]. The West Indian green monkeys that are exported from Barbados [J. Baulu, personal communication, 1981; Mack, 1982] are all feral callitrix monkeys (*C. a. sabaeus*) of West African origin [Napier and Napier, 1967].

The majority of the baboons that are traded go to countries other than the United States, which makes them difficult to trace because of inadequate and diverse sources of data. The four savannah species (olive, *P. anubis*; yellow, *P. cynocephalus*; Guinea, *P. papio*; and chacma, *P. ursinus*) are seldom distinguished from each other in the records. In recent years, Kenya has been a major source, presumably of olives and yellows, although the recent export ban must now be altering this. Ethiopia may well be the biggest single source of savannah baboons for research (olives), although records of its trade are largely unavailable, expecially those that relate to exports to Eastern Europe. Senegal is suspected to be a growing source of Guinea and olive baboons and must be a prime candidate for the hundreds that France appears to need annually for terminal projects, although the evidence for the latter is circumstantial. South Africa utilizes 2,000 or more chacma baboons annually for laboratory research, drawing them from its own wild populations [see Caldecott and Kavanagh, 1984; Kavanagh, 1984]. Nevertheless, savannah baboons in general are abundant and they probably suffer far greater losses from pest eradication programs than they do from live-trapping [Kavanagh and Bennett, 1984].

The utilization of hamadryas baboons (*P. hamadryas*) must be of greater concern because of their more limited distribution and numbers in Ethiopia and Somalia, although it is not known how many are exported annually. Kenya used to be a regular conduit for Ethiopian hamadryas baboons (Dutch Ministry of Culture data, 1982). The U.S.S.R. has a long tradition of *hamadryas* use [Asanov and Mirvis, 1972], and it is believed to be importing them on a regular basis, probably in large numbers if a single shipment of 750 in 1981 is representative (confidential source, 1981).

The patas monkey (*Erythrocebus patas*) is insignificant among current U.S. imports and appears to enter the United Kingdom and Japan only in very small numbers. Chad and Nigeria are both supplier countries [Kavanagh, 1984; Mack and Eudey, 1984]. There is, however, considerable circumstantial evidence that France uses well over 1,000 patas monkeys annually in terminal studies [Caldecott and Kavanagh, 1984]. The possible conservation implications of this cannot be assessed without more factual information, including trapping areas and actual numbers involved.

Cebidae and Callitrichidae. Neotropical monkeys are far less numerous in the trade now than they they were a decade ago (Fig. 1) and, of them, only squirrel monkeys (*Saimiri*) and tamarins (*Saguinus*) are still being trapped for biomedical research in large numbers. Although the common marmoset (*Callithrix jacchus*) is widely traded, most of these animals are now captive bred, the export of wild-born individuals from their native Brazil having been banned since the passage of the 1967 Brazilian Fauna Protection Law. They are, however, still threatened by an internal Brazilian pet trade [Caldecott and Kavanagh, 1984; Eudey and Mack, 1984; Mittermeir & Coimbra-Filho, 1982].

Squirrel monkey populations appear to have been remarkably resistant to depletion by trapping even on the large scale that pertained in the Amazon region during the 1960s. However, the genus has just been revised by Hershkovitz [1984], who recognizes four species and nine subspecies. Whether or not one or more of these newly defined taxa were hurt by past or present trading remains to be determined. In any case, the distinct Central American species, *S. oerstedi*, was certainly reduced in numbers by live-trapping in the past. It is now recognized as endangered, and any further trade in it would have disastrous effects [Baldwin and Baldwin, 1972; Thornback and Jenkins, 1982].

The tamarins (*Saguinus*) have also been taxonomically clarified by Hershkovitz [1977, 1979] and only the red-chested tamarins (*labiatus*, of which there are two subspecies) are traded currently at a high rate. *S. labiatus thomasi* occurs only in a restricted area of Brazilian Amazonia [Hershkovitz, 1977]. It is *S. l.labiatus* that is being exported in large numbers [Mack, 1982; Mack and Eudey, 1984; Mittermeier and Coimbra-Filho, 1982]. Almost all of these animals are exported from Bolivia, where *S. labiatus* appears to be relatively abundant [Heltne et al., 1975], although that is not a guarantee that all of them are actually trapped in Bolivia. Preliminary survey information suggests that *S. labiatus* is also relatively abundant in Peru [Encarnacíon and Castro, 1978; Mittermeier and Coimbra-Filho, 1982].

Several other tamarin species have been exported in large numbers until the recent past, often being identified merely as "tamarins" or "marmosets."

This is, however, no longer the case, although some few hundred saddleback tamarins (*S. fusicicollis*, probably subspecies *weddelli*) may still be exported annually from Bolivia [G. Bejarano, personal communication, 1982; Mittermeier and Coimbra-Filho, 1982]. In addition, small numbers of saddleback tamarins are exported annually from Peru to the United States under the auspices of Proyecto Primates, with Pan-American Health Organization (PAHO) sponsorship [M. Moro, personal communication, 1982; P. Soini, personal communication, 1980]. Peru also exports some few hundred moustached tamarins (*S. mystax*) to the United States under the same arrangement. All the evidence suggests that both of these species are abundant and adaptable and that neither are threatened by this trade [Mittermeier and Coimbra-Filho, 1982].

Other Species That May Be Threatened By the Trade

Lemuroidea and Daubentonoidea. Although these two superfamilies include 28 species, all of them are listed on Appendix I of CITES, and controlling the trade in trapped animals is greatly simplified by the fact that wild individuals are found only in two countries: Madagascar (all species) and The Comoros (two *Lemur* spp.). All domestic and export trading in these animals is strictly limited in both countries. The Comoros permits the export of up to 10 females and 20 males per year, and Madagascar allows no regular exports, making exceptions for a very few special cases [A. Jolly, personal communication, 1979; Tattersall, 1977]. During 1979, Madagascar exported six grey gentle lemurs (*Hapalemur griseus*) and six other *Lemur* spp., according to CITES records [Wildlife Trade Monitoring Unit (WTMU) of IUCN]. According to the same data source, 67 lemurids were traded between other countries: most of them were traded between zoos, many if not most were captive bred, and the numbers traced lead to the conclusion that the live animal trade is not threatening these prosimians, even though many of them are severely threatened for other reasons [Goodwin and Holloway, 1978].

Lorisidae. It seems unlikely that any lorisids are currently threatened by the trade. The lorisine species are hardly traded at all, although the United States did receive 23 slow lorises (*Nycticebus coucang*) from Thailand in 1979, in an apparent exception to the Thai export ban (CITES records in WTMU files). Small numbers of the other sub-family (Galaginae) are occasionally traded. The United States used to import large numbers of lesser bushbabies (*Galago senegalensis*) for the pet trade until this was outlawed in 1975. Eight hundred sixty-two of them (and 30 Demidoff's dwarf galagos, *G. demidovii*) were imported in the period 1978-80, but the numbers have now declined to insignificant levels [Mack, 1982, 1983; Mack and Eudey,

1984]. CITES data for 1979 [WTMU files] indicate that Botswana and Kenya were sources of lesser bushbabies at that time, but this trade appears to have declined along with European and American imports. It should, however, be noted that there are several East African forms that have been lumped together in existing taxonomies. Most authors list six species of *Galago* (including *Otolemur* and *Euoticus*), whereas it may be that nine different species would be a more appropriate total (S. Bearder, personal communication, 1982). If the trade in bushbabies were to grow again, careful attention would need to be given to the conservation status of the varieties involved, making research on their classification all the more important.

Tarsiidae This study yielded no indication of current live-trapping of Tarsiers (*Tarsius*) for export, apart from six pairs that were trapped in Sabah, Malaysia for export to Duke University, United States in 1983. This was a noncommercial exercise carried out under the supervision of the Sabah wildlife authorities (P. Andau, personal communication, 1983; C.W. Marsh, personal communication, 1983).

Callitrichidae Among all the marmosets and tamarins, the cottontop tamarin (*Saguinus oedipus,* or *S. o. oedipus* if considered conspecific with *geoffroyi*) used to be the symbol of the worst side of the primate trade. It is endemic to northwestern Colombia, where it is threatened by expanding human settlement which has had the twin effects of destroying much of its habitat and bringing it into contact with animal collectors. Demand for it was sustained by a local pet trade, by international dealing for pets (until the mid-1970s), and by laboratories (until very recently) [Green, 1976; Neyman, 1978]. It was afforded total legal protection in Colombia in 1970 (A. Donadio, personal communication, 1977), but this failed to stop the trade. Panama became a routine laundering point for smuggled cottontops, with Bolivia and Paraguay acting as supplementary outlets. This continued until the cottontop was placed on the U.S. Endangered Species List in 1976 and Panama and Paraguay tightened up their export controls in 1979–1980. Since then, international trading in wild-caught cottontops has virtually ceased [Kavanagh and Bennett, 1984; Mittermeier and Coimbra-Filho, 1982].

Until very recently, no other callitrichid was ever as seriously threatened by the trade as was the cottontop. However, toward the end of 1983 it emerged that some 50 *Leontopithecus rosalia chrysomelas* had been smuggled out of Brazil during that year, mainly through Bolivia [Jackman, 1983; Mallinson, 1984]. These individuals wound up in Belgium, France, and Hong Kong (via Japan). The largest number, 29, went to Belgium, and became the focus of an international storm of protest led by the Rio de Janeiro Primate Center in Brazil and the Jersey Wildlife Preservation Trust.

However, the recovery of these animals by the Brazilian authorities was complicated by the fact that Belgium did not become a party to CITES until January 1985, and the *L. r. chrysomelas* were being held quite legally under Belgian law [Jackman, 1983]. At the time of this writing, the problem has not been resolved, but it highlights the way in which unscrupulous trading can suddenly pose a threat to a rare species and the disproportionate dangers that can be posed by single countries that remain aloof from CITES.

The other two subspecies of lion tamarin (*L. r. rosalia* and *L. r. chrysopygus*) are also critically endangered in the wild and any commercial trapping of them could be disastrous [Coimbra-Filho and Mitttermeier, 1977], although none came to light during this study. There is, however, one other relatively uncommon callitrichid that may be threatened by live-trapping: this is the emperor tamarin (*Saguinus imperator*). This monkey is of indeterminate status, yet small numbers are still coming out of Bolivia [CITES data in WTMU files; Thornback and Jenkins, 1982]. Bolivia also exports regular but small consignments of silvery marmosets (*Callithrix argentata*), but these appear to be of a subspecies (*melanura*) that is not threatened [B.M.B. Papé, personal communication, 1982; Thornback and Jenkins, 1982].

Callimiconidae. Pook and Pook [1979] report that the high prices offered for Goeldi's monkeys (*Callimico goeldii*) in the United States and Europe have filtered down the supply chain to give an incentive for illegal trapping. The United States and Italy both report receiving a few individuals from Bolivia in 1979–1980 [CITES data in WTMU files] which could be significant in relation to the conservation status of this rare and monotypic primate family from the upper Amazon rain forests [Thornback and Jenkins, 1982].

Cebidae. Several of the 15 cebid taxa that are listed in the IUCN Red Data Book [Thornback and Jenkins, 1982] appear in recent trade statistics, but none of them is presently traded at a high enough level to be significant for their future survival. Woolly monkeys (*Lagothrix lagotricha*) and Guatemalan howlers (*Alouatta villosa*) used to be exported in large numbers, expecially to the United States, but the legislative changes of the early 1970s ended this practice [Mack and Eudey, 1984].

In contrast, the trade in night monkeys (*Aotus*) does appear to threaten some taxa. Night monkeys are widespread from Panama to northern Argentina and it is only recently that the taxonomic relationships within the genus have been clarified. Nine allopatric species in two natural groups have been described [Hershkovitz, 1983]. Unfortunately, most laboratories have failed to record precise identifications of the night monkeys that they use, making it difficult to relate the trade to particular taxa [Brumback, 1976; Hershkovitz, 1983]. When they were available, however, northern Colombian night monkeys were specifically in demand for malarial research [Cooper and Hernandez-Camacho, 1975]. According to Hershkovitz's scheme, these animals would probably be *A. lemurinus*, which appears to have two or three subspecies [Hernandez-Camacho and Cooper, 1976]. However, three other species, *A. brumbacki*, *A. vociferans*, and *A. trivirgatus* [sensu Hershkovitz, 1983], also occur in Colombia. *A. brumbacki* appears to have a very limited distribution, which may make it vulnerable if exploited. There is some indication that Colombian primate exports are restarting [Mack, 1982]. If this is true, it could obviously affect the status of certain *Aotus* taxa, although this would be obscured without precise data on the identity and origin of the animals concerned.

Most laboratory-bound night monkeys currently derive from Bolivia and are thus likely to be *A. nigriceps*, *A. infulatus*, or *A. azarae* [Hershkovitz, 1983; Kavanagh, 1984]. The last named has two subspecies, with *A. a. azarae* occurring in southern Bolivia and northern Argentina. At least in the latter area, this animal occurs at low densities and is of special interest as the only facultatively diurnal night monkey [Rathbun and Gache, 1980]. The ecological and demographic base data for the exploitation of these southern populations do not exist, and it is only in Amazonia, from where few night monkeys are exported, that *Aotus* is still relatively abundant [Mittermeir and Coimbra-Filho, 1982].

Capuchin monkeys (*Cebus*) were heavily traded until the mid-1970s and some few hundred annually appear to be exported currently from Bolivia, Guyana, and Paraguay [CITES data in WTMU files; Mack, 1982]. During the 1970s, the heaviest exploitation shifted from the white-fronted capuchin (*C. albifrons*) to the tufted capuchin (*C. apella*) [Mack and Eudey, 1984]. The latter is widespread between Colombia and Argentina, and although it has at least two threatened subspecies, these are Brazilian forms that appear to be protected from any trapping for export [Mittermeier and Coimbra-Filho, 1982].

Cercopithecidae. Although it involves the largest primate family, the cercopithecid conservation situation vis-à-vis live-trapping for the trade is relatively simple. Stump-tailed macaques (*Macaca arctoides*) formerly were used in relatively large numbers in biomedical research, with Thailand having been the major supplier [Harrisson, 1971; Mack and Eudey, 1984]. The species is poorly known, but its largest populations, with the possible exception of southern China, may have occurred in peninsular Thailand. Excessive trapping, in conjunction with habitat loss and the effects of insurgent activity, has resulted in its becoming very rare in the region [Bertrand, 1969; Fooden, 1976]. However, the Thai export ban of 1976 has virtually eliminated stump-tailed macaques from the international trade [Mack and Eudey, 1984]. Nevertheless, since 1976 there has been some illegal traffic in these animals from Thailand [Eudey, 1984]: the species is known to have been smuggled to Laos for export to Belgium [Anonymous, 1978].

The pig-tailed macaque (*Macaca nemestrina*) may also be threatened by trapping. Recent detailed studies show that it is extremely susceptible to the loss of its primary forest habitat, and (in marked contrast to *M. fascicularis*) it does not persist in agricultural areas [Caldecott, 1983; Robertson, manuscript in preparation]. Populations may therefore be expected to diminish as Southeast Asia runs out of large forest tracts; and trapping around the edges will greatly exacerbate the problem [Caldecott, 1983]. The United States imported 1,540 pig-tailed macaques during 1976–1980 [Mack and Eudey, 1984], but available export figures are either contradictory or unreliable. For example, during 1980–81 Indonesia (the main source of the species) exported either 831 [MacKinnon, 1983] or 3,852 (Indonesian government statistics, W. Kadri, personal communication, 1982). Peninsular Malaysian official statistics (1982) report 3,972 exports of pig-tailed macaques in 1979, a figure that seems highly unlikely in view of the general pattern of Malaysian trade, in which almost all exports are of long-tailed macaques [Nordin and Hasnah Samian, 1981]. The Malaysian export ban can only benefit the species, but the opposite could be true of the high export quotas set by the Indonesian management authority: 7,200 in 1983 [MacKinnon, 1983].

Hylobatidae. Although there is some international trade in gibbons (*Hylobates*), it appears to be at very low levels currently, having declined in recent years [Mack and Eudey, 1984]. The most significant figures are those of Japan, which imported nearly 100 gibbons during 1979–1980 [Nishida and Uehara, 1982], implying that captures could be at levels that are significant for some local populations.

Pongidae. Of the four great ape species, the chimpanzee (*Pan troglodytes*) has been the one to be most extensively traded for entertainment, zoo display, and laboratory research, with the biggest market being that of the United States. Trading in this species did, however, decline dramatically in the late 1970s following its listing on CITES Appendix I in 1977; and in recent years, the majority of CITES-regulated trade has been in animals that have originated in North America and Europe [Mack and Mittermeier, 1984]. Nevertheless, chimpanzees are still exported from habitat countries at the present time, even though the numbers are much reduced. There is a steady traffic in young animals imported into Spain and the Canary Islands for the beach photography trade, which can be fairly regarded as a terminal use [Anonymous, 1982b]. The effects of chimpanzee trading are greatly multiplied by the capture technique, which commonly consists of shooting the mother to obtain the infant [Teleki and Baldwin, 1979]. Nishida and Uehara [1982] calculate that Japanese imports during the period 1973–1977 could alone have been responsible for eliminating the equivalent of two chimpanzee communities (unit-groups) per year. They also report that it is likely that nearly 100 chimpanzees were imported to Japan during 1979–1980. Despite Japan's joining CITES in 1980, its government made a diplomatic arrangement to import 50 chimpanzees from Sierra Leone in 1983 for biomedical research at a pharmaceutical company. Thirty of the animals have already reached Japan and a further 20 were believed to be en route in late 1983 [Anonymous, 1983a]. Sierra Leone is not a party to CITES but the transaction violates the provisions of the Convention for at least two reasons: first, Appendix I animals are being traded for primarily commercial purposes; and second, given the precarious status of Sierra Leone's chimpanzee populations, the wildlife authorities there cannot realistically certify to Japan's management authority that the trade will not be detrimental to the survival of the species. West African chimpanzees have recently come under an additonal threat as a result of an Austrian company's attempt to set up a biomedical research unit in Sierra Leone: the proposed laboratory would use up to 60 chimpanzees per year and could only be sustained if it were permitted to utilize imports to Sierra Leone (i.e., smuggled apes) as well as being allowed to exhaust the local populations. It is currently the object of an international protest campaign [Anonymous, 1983a; G. Teleki, in litt., 1983].

Chimpanzees do not appear to be out of danger from commercial uses that stimulate legal and illegal trading or trapping for local use. In a recent development, the U.S. Public Health Service recommended the use of the species in the United States for research into the human disease, acquired immune deficiency syndrome (AIDS), and already at least nine apes have been brought into the program for terminal research use [J.Held, personal communication, 1983; E.N. Brandt, in litt., 1983].

The pygmy chimpanzee or bonobo (*Pan paniscus*) is altogether a less well-known animal. Since it requires some expertise to distinguish it from the former species, there is always the possibility that some trade in it is hidden in the statistics of *P. troglodytes,* as appears to have occurred among Japanese imports, possibly as recently as 1980 [Nishida and Uehara, 1982]. Exports from Zaire, where the bonobo is endemic, are illegal and thought to be minimal; but as late as 1982, several wild-caught bonobos were being held for sale in a Belgian dealer's compound [Anonymous, 1983b].

The numbers of gorillas (*Gorilla gorilla*) that have been traded in recent years could well be significant in relation to local populations, but most of the relevant data appear to be hidden. Imports from habitat countries rarely appear in CITES statistics, although France reported bringing in five from Cameroon in 1979 [CITES data in WTMU files]. Gorillas may still be brought into some countries that are not CITES Parties even if they have been smuggled out of their countries of origin. One such case came to light in Belgium in 1982. At that time it was quite legal to deal in smuggled gorillas in Belgium and there is some evidence that the ape in question was part of a regular trade in western lowland gorillas from Africa to Belgium [confidential sources, 1982]. Belgium ratified CITES at the end of 1983 and this ought to reduce the trade in all endangered species to that country. As in the case of chimpanzees, the capture of gorillas for the trade routinely involves the deaths of other individuals [Sabater Pí. 1980–1].

Throughout most of the twentieth century, orang-utans (*Pongo pygmaeus*) have been widely traded for zoos, laboratory use, and private collections, with several animals being killed for every one that eventually reaches the market [Harrisson, 1971]. Since the species is only found in Malyasia and Indonesia and both of these counries now forbid its export, any trade in wild-caught animals is both covert and illegal. All CITES-declared orangs traded in 1979 originated from North America or Western Europe and were probably zoo exchanges. However, very few second-generation captive-bred orangutans have been born, so the prospect of the captive population's sustaining itself and even a small trade in the indefinite future is very remote [Perry, 1976]. Although occasional hearsay information suggests that the illicit orang-utan trade has not been completely suppressed, it seems unlikely to be a major threat to wild populations.

CONCLUSION

On the basis of this study, the major exporting countries are Indonesia, the Philippines, and Malaysia in Asia; Kenya, Ethiopia, and Somalia in Africa; Bolivia in Latin America; and the United States and the Netherlands among reexporters. Of these, Kenya and Malaysia are now stopping their exports. The main primate importers are the United States, United Kingdom, Japan, Taiwan, and Canada, to which the U.S.S.R. and France should be added on the basis of indirect evidence.

The clearest theme to emerge from this review is the large reduction that has taken place over the last decade or so in the numbers of primates traded (and therefore trapped) worldwide. The pet trade, expecially in the United States, has diminished from a major to a very minor source of demand and there has been a great reduction in the number of species used in biomedical research. These changes have taken place in response to a network of legal controls that grew rapidly in the early 1970s and that is still expanding. The export bans imposed by Peru in 1973, by Colombia in 1974, and by India in 1978 were immediately effective in reducing the overall number of primates in the market and they were backed up by the imposition of quotas or bans in other source countries.

Import controls have also played a major role, with the prohibition by the United States against the importation of primates as pets being the biggest single influence in this category. The health-related restrictions of the European countries were also important in reducing pet imports.

Most of these legal developments grew out of the same area of concern as gave rise to CITES: unregulated trading in threatened species. The Convention has become the ready-made system that most countries can use to regulate the primate trade. Its effectiveness, however, is reduced by the failure of some Parties to enforce its provisions. For example, Bolivia, a CITES Party, is the major world source of South American monkeys because its wildlife trade is virtually unregulated. Among importers, Japan's membership of CITES has not stopped it from importing chimpanzees for commercial purposes in violation of its obligations under the Convention. And within the United States there are reports of the resumption of a limited trade in New World monkeys for the pet market because of incomplete checks by federal agents (confidential source, 1982). On the other hand, CITES has been very effective in certain instances. For example, Austria ceased to be an open laundering point for smuggled primates within days of becoming a Party [Anonymous, 1982c]. A similar effect should be occurring in Belgium now that that country has acceded to CITES.

The Convention has also led to the provision of trade records that would otherwise be unavailable, although these are by no means complete. In January 1983, when most of the work on this study was completed, 59 out of 74 countries had not submittted their 1981 reports. However, the Convention is relatively new and complex, and all aspects of its implementation may be expected to improve as parties become more familiar with it.

The examples of rhesus macaques in India and stump-tailed macaques in Thailand indicate that excessive trapping can be a significant factor in the depletion and local extinction of primates. Yet there is almost a complete lack of the population survey data that are needed to attempt any sustained-yield harvesting. Basic information is not even available for the long-tailed macaque, the most traded primate of all. The examples of *Saimiri* and *Aotus* demonstrate the importance both of linking population studies to those of taxonomy and of precisely identifying traded animals by taxon and locality of trap-site.

Major areas of concern, for which adequate survey information is required, are the huge harvest of long-tailed macaques, the high Indonesian export quotas for pig-tailed macaques, and the possible over-trapping of hamadryas baboons and patas monkeys, all for biomedical purposes. Exporting countries would do well to consider reducing or cutting off the supplies until such time as the biomedical community (i.e., the beneficiary) arranges for the provision of independent population surveys. Since the mid-1970s, a start has been made on this by some biomedical agencies, most notably PAHO in Latin America [see Mittermeier and Coimbra-Filho, 1982]. the World Health Organisation (WHO) in Indonesia [e.g., Wilson and Wilson, 1975] and in Malaysia [see Kavanagh et al., 1981].

In addition, WHO has adopted the IUCN Primate Specialist Group's "Policy Statement on Use of Primates for Biomedical Purposes" [Anonymous, 1982d] (see Appendix). It remains to be seen how effective this will be. Currently, the development of AIDS research on chimpanzees indicates that the U.S. Public Health Service is not adhering to a provision of the policy statement that threatened primate species should only be used in biomedical research projects if they are taken from self-sustaining captive breeding colonies, a condition that has not been realized in chimpanzees [Teleki and Baldwin, 1975]. The policy statement is, however, being cited by primatologists in opposition to the planned biomedical use of chimpanzees in Sierra Leone (see above). Although not a sponsor of the project, it seems likely that WHO could effectively quash it if it decided to do so in support of its own policy statement [Heneson, 1983].

As with any other trade, that in primates is largely governed by market forces. The development of captive breeding programs in user countries has been in response to the decreased availability of wild-caught primates, but captive breeding is relatively expensive, and as long as wild-caught primates are cheaper, they will continue to be in demand, probably well into the 1990s [Caldecott and Kavanagh, 1984; Eudey and Mack, 1984].

From a conservation point of view, insufficient attention has been paid to the notion of using feral populations that need to be controlled, such as the macaques of Mauritius or the green monkeys of the West Indies, both of which are environmentally damaging populations [Temple, 1974; J. Baulu, personal communication, 1983], Consideration might also be given to the selection of alternative species on the basis on known population data. Marsh and Wilson [1981] suggest reassessing the laboratory potential of langurs (*Presbytis*) on the grounds that they are the most abundant primates in Peninsular Malaysia, they are being killed off in large numbers by habitat destruction, and their husbandry does not seem to be as difficult as previously believed. Live-trapping of langurs in areas where their habitat is already being destroyed would be both feasible and within the guidelines of the IUCN policy statement.

In conclusion, the overall drop in trade volumes is an encouraging sign, but a few species remain under threat either because they are widely trapped for the trade or because they are already rare, vulnerable, or endangered, and a small trade threatens them further. The general lack of survey information and the recent examples of trade-related threats to species as diverse as chimpanzees and golden-headed lion tamarins show that continued vigilance is essential to ensure that the international primate trade does not once again become detrimental to many different populations of primates.

ACKNOWLEDGMENTS

The study was funded by World Wildlife Fund (U.S.A. and U.K.) and conducted with the help of TRAFFIC (U.S.A.), Universiti Pertanian Malaysia, and the IUCN Conservation Monitoring Centre. Acknowledgments have been made in full elsewhere [Mack and Mittermeier, 1984] but special thanks for help with this paper are due to Drs. E.L. Bennett, J. O. Caldecott, C.W. Marsh, R.A. Mittermeier and K. C. Searle, to Mr. C. Huxley, and to Encik Saip Sulong, who drew the figure.

REFERENCES

Anonymous (1978): Laotian-Belgian traffic in mammals. IPPL Newsletter 5:2–5.

Anonymous (1979): Japanese imports of primates. Traffic Bull 1:5–7.

Anonymous (1980): Shrew export ban. Bangkok Post December 29th.

Anonymous (1982a): New Kenyan legislation. Traffic Bull 3:69.

Anonymous (1982b): The beach chimp trade in Spain. Traffic Bull 4:24.

Anonymous (1982c): Stop press. Traffic Bull 4:10.

Anonymous (1982d): World Health Organization (WHO) and Ecosystem Conservation Group (ECG) adopt Primate Specialist Group's "Policy Statement on Use of Primates for Biomedical Purposes." IUCN/SSC Primate Specialist Group Newsletter Number 2:5.

Anonymous (1983a): Sierra Leone's chimps endangered by commercial exploitation. Traffic Bull 5:48.

Anonymous (1983b): Pygmy champanzees found in Belgian basement. IPPL Newsletter 10:3–5.

Asanov SS, Mirvis AB (1972): The breeding of hamadryas baboons at the Sukhumi Primate Centre, In Diczfalusy E, Standley CC (eds): "The use of Non-Human Primates in Research of Human Reproduction." Stockholm: WHO, pp 483–489.

Baldwin JD, Baldwin J (1972): The ecology and behaviour of squirrel monkeys (*Saimiri oerstedii*) in a natural forest in western Panama. Folia Primatol 18:161–184.

Bertrand M (1969): "The Behavioral Repertoire of the Stumptail Macaque." Bibliotheca Primatologica Number 11. Basel: Karger.

Brumback RA (1976): Taxonomy of the owl monkey (*Aotus*. Lab Primate Newsletter 15:1–2.
Burton JA (1978): Primate imports into the United Kingdom 1965–1975. In Chivers DJ, Lane-Petter W (eds): "Recent Advances in Primatology." London: Academic Press, Vol 2, pp 137–145.
Caldecott JO (1983): "An Ecological Study of the Pig-tailed Macaque in Peninsular Malaysia." Unpublished Ph. D. Dissertation, University of Cambridge.
Caldecott JO, Kavanagh M (1984): Use of primates and captive breeding programs outside the United States. In Mack D, Mittermeier RA (eds): "The International Primate Trade." Washington, DC WWF-US/TRAFFIC (U.S.A.), pp 137-152.
Coimbra-Filho AF, Mittermeier RA (1977): Conservation of the Brazilian lion tamarins *Leontopithecus rosalia*. In Prince Rainier HSH, Bourne GH (eds): "Primate Conservation." New York: Academic Press pp 60–94.
Conway WG (1966): The availability and longterm supply of primates for medical research: A report on the conference held in New York. Int Zoo Yearbook 6:284–288.
Cooper RW (1968): Squirrel monkey taxonomy and supply. In Rosenblum LA, Cooper RW (eds): " The Squirrel Monkey." New York: Academic Press, pp 1–29.
Cooper RW, Hernandez-Camacho J (1975): A current appraisal of Colombia's primate resources. In Bermant G, Lindburg D (eds): "Primate Utilization and Conservation." New York : John Wiley & Sons, pp 37–66.
Darsono C (1979): "Utilization (export) of Indonesian Wildlife." Unpublished manuscript.
Dorst J, Dandelot P (1972): "A Field Guide to the Larger Mammals of Africa." London: Collins.
Encarnacíon F, Castro N (1978): "Informe Preliminar Sobre Censo de Primates No Humanos en el Sur-oriente Peruano: Iberia e Iñapari(Departemento de Madre de Dios), May 15–June 14, 1978." Unpublished report to PAHO.
Eudey A (1978): International traffic in primates from Thailand. In Chivers DJ, Lane-Petter W (eds): "Recent Advances in Primatology." London: Academic Press, Vol 2, pp 127–131.
Eudey A, Mack D (1984): Use of primates and captive breeding programs in the U.S. In Mack D, Mittermeier RA (eds): "The International Primate Trade." Washington, DC: WWF-US/TRAFFIC (U.S.A.), pp 153–180.
Fooden J (1976): Primates obtained in Peninsular Thailand June–July 1973, with notes on the distribution of continental Southeast Asian leaf-monkeys (*Presbytis*). Primates 17:95–118.
Gartlan JS, Brian CK (1968): Ecology and social variability in *Cercopithecus aethiops* and *C. mitis*. In Jay PC (ed): Primates." New York: Holt, Rinehart, and Winston, pp 253–292.
Goodwin HA, Holloway CW (1978):" Red Data Book." Morges: IUCN, Vol 1.
Green, KM (1976): The nonhuman primate trade in Colombia. In Thorington RW, Heltne PG (eds). "Neotropical Primates: Field Studies and Conservation." Washington, DC.: National Academy of Sciences, pp 85–98.
Harrisson B (1971): "Conservation of Nonhuman Primates in 1970."Basel: Karger.
Held JR (1982): Nonhuman primates, their use in biomedical research. Talk given at the symposium on the Use of Nonhuman Primates in Exotic Viral and Immuniologic Diseases, Southwest Foundation Forum, San Antonio, Texas, 28 February–3 March.
Heltne P, Freese C, Whitesides G (1975): "A Field Survey of Nonhuman Primate Populations in Bolivia." Unpublished report to PAHO.
Heneson N (1983): Loophole may allow trade in African chimps. New Scientist, October 20th.

Hernandez-Camacho J, Cooper RW (1976): The nonhuman primates of Colombia. In Thorington RW, Heltne PG (eds): "Neotropical Primates." Washington, D.C.: National Academy of Sciences, pp 35-69.

Hershkovitz P (1977): "Living New World Monkeys (*Platyrrhini*)." Chicago: Chicago University Press, Vol 1.

Hershkovitz P (1979): Races of the emperor tamarin, *Saguinus imperator* Goeldi (Callitrichidae, Primates) Primates 20:277–287.

Hershkovitz P (1983): Two new species of night monkeys, genus *Aotus* (Cebidae, Platyrrhini): A preliminary report on *Aotus* taxonomy. Am J Primatol 4:209–243.

Hershkovitz, P (1984): Taxonomy of squirrel monkeys genus *Saimiri* (Cebidae, Platyrrhini): A preliminary report with description of a hitherto unnamed form. Amer. J. Primatol. 6:257–312.

Inskipp T, Wells S (1979): "International Trade in Wildlife." London: Earthscan.

Jackman B (1983): Sale scandal of rarest monkeys. *The Sunday Times,* December 18th.

Kavanagh M (1980): Invasion of the forest by an African savannah monkey: Behavioural adaptations. Behaviour 73:238–260.

Kavanagh M (1983a): Simians for science: The United Kingdom and the worldwide trade in primates. In Harper D (ed): "Proceedings of Symp Conserv Primates and Their Habitats" Vaughan Paper 31. Leicester: University of Leicester, Vol 2, 149-181.

Kavanagh M (1983b): "A Complete Guide to Monkeys, Apes and Other Primates." London: Jonathan Cape.

Kavanagh M (1984): A review of the international primate trade. In Mack D, Mittermeier R A (eds): "The International Primate Trade." Washington, D.C.: WWF-US/TRAFFIC (U.S.A.), pp 49–89.

Kavanagh M, Bennett EL (1984): A synopsis of legislation and the primate trade in habitat and user countries. In Mack D, Mittermeier RA (eds): "The International Primate Trade." Washington D.C.: WWF-US/TRAFFIC (U.S.A.), pp 19–48.

Kavanagh M, Vidyadaran MK, Nordin M (eds) (1981): "Proc Symp Biol Peninsular Malaysian Primates." Malays Appl Biol 10:71–249.

LeCornu A, Rowan AN (1978): Trends in the use of nonhuman primates in biomedical programmes. Lab Anim 12:235–242.

Le Cornu A, Rowan AN (1979): The use of non-human primates in the development and production of poliomyelitis vaccines. ATLA Abstracts 7:10-19.

Long J (1982): Monkey importer sees fair shake from Bangladesh. Oregon J, March 19th.

Mack D (1982): Trends in primate imports into the United States 1981. ILAR News 25:10–13.

Mack D (1983): Trends in primate imports into the United States 1982. ILAR News 26:10–15.

Mack D, Eudey A (1984): A review of the U.S. primate trade. In Mack D, Mittermeier RA (eds): "The International Primate Trade." Washington, D.C.: WWF-US/TRAFFIC (U.S.A.), pp 91–136.

Mack D, Mittermeier RA (eds) (1984): "The International Primate Trade." Washington, D.C.: WWF-US/TRAFFIC (U.S.A.)

MacKinnon K (1983): "Report of a World Health Organization (WHO) Consultancy to Indonesia To Determine Population Estimates of *Macaca fascicularis* (and Other Primates) and the Feasibility of Semi-Wild Breeding Projects of This Species." Unpublished report to WHO.

Mallison, J. J. C. (1984): Golden-headed lion tamarin contraband a major conservation problem. IUCN/SSC Primate Specialist Group Newsletter 4:23–25.

Marsh CW, Wilson WL (1981): "A Survey of Primates in Peninsular Malaysian Forests." Kuala Lumpur: Malaysian Primates Research Programme, Universiti Kebangsaan Malaysia.

Meister G (1977): Laws, regulations and guide-lines relating to the importation, transfer, care and maintenance of monkeys, and experiments on them. Primate Rep 1:10–14.

Mittermeier RA, Coimbra-Filho AF (1982): Distribution and conservation of New World Primate species used in biomedical research. In Hearn J (ed): "Reproduction in New World Primates." The Hague: MTP Press Ltd., pp 3–37.

Morris R, Morris D (1966): "Men and Apes." London: Hutchinson.

Muchenhirn N (1975): "Nonhuman Primates: Usage and Availability for Biomedical Programs." Washington, D.C.: Institute of Lab Animal Resources, National Academy of Sciences.

Napier JR, Napier PH (1967): "A Handbook of Living Primates." New York: Academic Press.

Neyman PF (1978): Aspects of the ecology and social organization of free-ranging cotton-top tamarins (*Saguinus oedipus*), and the conservation status of the species. In Kleiman DG (ed): "The Biology and Conservation of the Callitrichidae." Washington, D.C.: Smithsonian Institution Press, pp39–71.

Nishida T, Uehara S (1982): Japanese ape imports. IPPL Newslett 9:12–15.

Nordin M (1981): Trends in primate exports from Malaysia. Lab Primate Newslett 20:9-11.

Nordin M, Hasnah Samian (1981): The primate trade in Malaysia. Malays Appl Biol 10:183–196.

Perry J (1976): Orang-utans in captivity. Oryx 13:262–264.

Pook AG, Pook G (1979): "A Field Study on the Status and Socioecology of the Goeldi's Monkey (*Callimico goeldii*) and Other Primates in Northern Bolivia." Unpublished report to NY Zoological Society.

Rathbun G, Gache MJ (1980): Ecological survey of the night monkey, *Aotus trivirgatus*, in Formosa Province, Argentina. Primates 21:211–219.

Robertson JMY (1985): "Behaviour and Ecology of the Pig-Tailed Macaque in Sumatra." Ph.D. Dissertation, University of Cambridge (manuscript in preparation).

Sabater Pí J (1980–1): Exploitation of gorillas *Gorilla gorilla gorilla* Savage & Wyman 1847 in Rio Muni, Republic of Equatorial Guinea, West Africa. Biol Conserv 19:131-140.

Schultz AH (1969): "The Life of Primates." London: Weidenfeld and Nicholson.

Tattersall I (1977): The lemurs of the Comoro Islands. Oryx 13:445–448.

Teleki G, Baldwin L (1975): Breeding programs aim to keep this a planet of the apes. Smithsonian 5:76-81.

Teleki G, Baldwin L (1979): "Thc Need for a Systematic Survey of Chimpanzee Populations in West Africa, Focusing Initially on the Republic of Sierra Leone." Unpublished report to IUCN/SSC Primate Specialist Group.

Temple SA (1974): Wildlife in Mauritius today. Oryx 12:584–590.

Thorington RW (1972): Importation, breeding and mortality of New World Primates in the United States. Int Zoo Yearbook 12:18–23.

Thorington RW (1976): The systematics of New World Monkeys. In "First Inter-American Conference on Conservation and Utilization of American Non-Human Primates in Biomedical Research." Washington, D.C.: PAHO, pp 8–18.

Thornback J, Jenkins M (1982): "The IUCN Mammal Red Data Book Part 1." Gland: IUCN.

Warland, MAG (1972): The status of Latin American primates. Int Zoo Yearbook 14:16–18.

Wilson CC, Wilson WL (1975): Methods for Censusing Forest-Dwelling Primates." 5th Int Congr Primate Nagoya 1974. Karger: Basel, pp 345–350.

World Wildlife Fund Indonesia (1980): "The Animal Trade in Indonesia." Unpublished report to the Director of Nature Conservation, Indonesia.

APPENDIX: POLICY STATEMENT ON USE OF PRIMATES FOR BIOMEDICAL PURPOSES

The Ecosystem Conservation Group (ECG; which includes UNESCO, UNEP, FAO, and IUCN) and WHO recognize that nonhuman primates play an important role in biomedical research and testing and that their use as experimental animals has made a significant contribution to advances in human health and disease control. The ECG and WHO are committed to maintaining the current diversity of the Order Primates and to ensuring the survival of representative, self-sustaining populations of all species in their natural habitats.

A total of 76 primate taxa are currently considered *endangered*, *vulnerable*, or *rare* by the IUCN. Since these taxa are either in serious decline or already at very low and precarious population levels, any exploitation of them threatens their continued survival. Therefore, the ECG and WHO strongly recommend that 1) *endangered*, *vulnerable*, and *rare* species be considered for use in biomedical research projects only if they are obtained from existing self-sustaining captive breeding colonies (i.e., in captive breeding, all animals are required to be at least F2 generation); and 2) species categorized as *status unknown* or *indeterminate* also not be considered for use in such research projects until adequate data indicate that they are not *endangered*, *vulnerable* or *rare*.

Members of more than 30 species of nonhuman primates, the majority of them wild-caught, are currently being used worldwide in biomedical research and testing. However, sustained yield trapping strategies for wild primates, based on long-term ecological field studies and adequate demographic data, have not yet been developed for any primate species. Continuing habitat loss in most areas where primates occur makes demographic projections difficult and unreliable in most cases. The ECG and WHO therefore recommend that 1) wild-caught primates be used primarily for the establishment of self-sustaining captive breeding colonies, the eventual goal of which should be to captive-breed most or all (depending on species) of the primates used in research; 2) populations of the apparently common primate species be trapped only in a) special management areas where demographic data are available, where the populations are continually monitored to avoid over-exploitation, and where sustained yield trapping strategies are being developed and tested; b) areas where the animals are living in agricultural or other man-modified environments and have been shown to be agricultural nuisances that would otherwise be destroyed; or c) areas where the habitat is already being destroyed, where the primates would otherwise be killed or would die from starvation or stress, and where translocation is not a viable alternative.

To minimize impact on free-living populations, the ECG and WHO urge that trapping, holding, and shipping techniques be perfected to the point that accidental death, destruction of habitat, disruption of family groups, and other forms of wastage are kept to an absolute minimum.

The ECG and WHO urge researchers and their funding agencies to assist in the control of international commerce in primates by requiring proper export and import documentation on all animals that they purchase or otherwise obtain and to refuse animals obtained in contravention of CITES and/or protective legislation in the source countries.

SECTION II: APPROACHES

Primate Conservation in the Tropical Rain Forest, pages 181–204

7

Objectives, Selection, and Management of Protected Areas in Tropical Forest Habitats

Jeffrey A. McNeely, Kenton R. Miller, and James W. Thorsell

International Union for Conservation of Nature and Natural Resources, 1196 Gland, Switzerland

INTRODUCTION

Earlier papers in this volume have indicated why protected areas are required for the conservation of primates. It is important to manage forests generally in such a way that primates can find suitable habitats, but in today's world, with more "thorough" logging techniques and constantly spreading agriculture, national parks and other sorts of legally established protected areas provide the surest means of habitat conservation in the tropics. Although some species are able to survive, even thrive, in logged forest and other secondary habitats, most arboreal primates—including the bulk of the rarer and more spectacular forms—need mature forests to support their ways of life.

This chapter will present information on the current status of protected areas in tropical forest habitats, suggest an expanded set of objectives that could lead to broad support for a significant enlargement of tropical forest protected areas, outline criteria for selection, and suggest ways and means for those areas to be managed in such a way that their survival is ensured. Case studies from two countries where conservation of tropical forest habitats is being competently addressed conclude the chapter.

It is important to note that protected areas in tropical forests are a relatively new phenomenon, with most units established within the past 30 years. The idea that people have the responsibility to establish specific areas for the conservation of nature is closely related to the expansion of the human ecological niche in the tropics; as the ability of *Homo sapiens* to disrupt rain forest habitats has increased [Myers, this volume], this destructive impulse has been balanced by an opposite (though not equal) desire to conserve.

A major concern is how, in a world where the population of the dominant species is increasing by 2% per year, samples of tropical forests can be preserved in a natural state. We cannot accept the proposition that current trends in tropical forest destruction will continue far into the future; if they are not reversed, then we can hardly hope to protect viable populations of the large trees and animals that many consider the most interesting part of the tropical forest. On the other hand, it is quite likely that today's best protected natural areas will, at some distant point in the future, be seen as "Holocene refugia," those areas saved from the more disruptive effects of industrial man and from which areas of recovered habitats were repopulated by wildlife. This makes the need for protecting these areas even more profound.

AN EXPANDED RATIONALE FOR PROTECTED AREAS

It is apparent that demands on tropical forests are going to continue to grow and that it is unrealistic to expect tropical countries to strictly protect the full portion of their valuable tropical forest resources that would be required to ensure that all species and ecosystems can survive in the long run. The value of the standing biomass in Indonesia's 200,000 ha Kutai Wildlife Reserve in Kalimantan, for example, is some $2 billion at 1982 market prices (based on an estimate of 150 cubic meters of marketable timber per hectare). Therefore, it is necessary to both rationalize existing protected areas ever more convincingly and to establish a range of management regimes for additional areas that would serve both conservation and economic objectives. Through the explicit presentation of appropriate objectives, conservation can be shown to form an integral part of ecologically sustainable development. The "World Conservation Strategy" [IUCN, 1980a] explains the rationale for the linkage between conservation and development and establishes priorities for strategic action.

By enlarging the concept of "conservation," and by explaining the links between human needs and natural resources, ecologically sound development can take on practical meaning that is appealing to governments. To ensure that protected areas can contribute to this linkage, an expanded set of conservation objectives has been suggested (Table I). These objectives carry some hidden dangers. If national parks, for example, start being used for meat or timber production, there will be serious losses to conservation. A system of conservation unit categories linked to management objectives was therefore devised by IUCN [1978] to accomodate a wider range of wildland outputs without giving up any of the important gains made by national parks and other strictly protected categories.

IUCN's system of 10 management categories for protected areas is presented in Table II. Applications from around the world [e.g., Mosley, 1984]

TABLE I. An Expanded Set of Conservation Objectives for Tropical Forest Protected Areas

1. *Sample ecosystems.* To maintain large areas as representative samples of each major biological region of the nation in its natural unaltered state to ensure the continuity of evolutionary processes, including animal migration and gene flow.
2. *Ecological diversity.* To maintain examples of the different characteristics of each type of natural community, landscape, and land form for protecting the representative as well as the unique diversity of the nation, particularly to ensure the role of natural diversity in the regulation of the environment.
3. *Genetic resources.* To maintain all genetic materials as elements of natural communities and avoid the loss of plant and animal species.
4. *Education and research.* To provide facilities and opportunities in natural areas for purposes of formal and informal education and research and the study and monitoring of the environment.
5. *Water and soil conservation.* To maintain and manage watersheds to ensure an adequate quality and flow of fresh water, and to control and avoid erosion and sedimentation, especially where these processes are directly related to downstream investments that depend on water for transportation, irrigation, agriculture, fisheries, and recreation, and for the protection of natural areas.
6. *Wildlife management.* To maintain and manage fishery and wildlife resources for their vital role in environmental regulation, for the production of protein, and as the base for industrial, sport, and recreational activities.
7. *Recreation and tourism.* To provide opportunities for healthy and constructive outdoor recreation for local residents and foreign visitors and to serve as poles for tourism development based on the outstanding natural and cultural characteristics of the nation.
8. *Timber.* To manage and improve timber resources for their role in environmental regulation and to provide a sustainable production of wood products for the construction of housing and other uses of high national priority.
9. *Cultural heritage.* To protect and make available all cultural, historic, and archeological objects, structures, and sites for public visitation and research purposes as elements of the cultural heritage of the nation.
10. *Scenic beauty.* To protect and manage scenic resources to ensure the quality of the environment near towns and cities, highways and rivers, and surrounding recreation and tourism areas.
11. *Options for the future.* To maintain and manage vast areas of land under flexible land-use methods that conserve natural processes and ensure open options for future changes in land use, incorporate new technologies, meet new human requirements, and initiate new conservation practices as research makes them available.
12. *Integrated development.* To focus and organize conservation activities to support the integrated development of rural lands, giving particular attention to the conservation and utilization of marginal areas and to the provision of stable rural employment opportunities.

Modified from Miller [1980], with permission.

TABLE II. Categories and Management Objectives of Protected Areas

1. *Scientific reserve/strict nature reserve.* To protect nature and maintain natural processes in an undisturbed state in order to have ecologically representative examples of the natural environment available for scientific study, environmental monitoring and education, and for the maintenance of genetic resources in a dynamic and evolutionary state.
2. *National park.* To protect relatively large natural and scenic areas of national or international significance for scientific, educational, and recreational use, under management by the highest competent authority of a nation.
3. *Natural monument/natural landmark.* To protect and preserve nationally significant natural features because of their special interest or unique characteristics.
4. *Managed nature reserve/wildlife sanctuary.* To ensure the natural conditions necessary to protect nationally significant species, groups of species, biotic communities, or physical features of the environment when these require specific human manipulation for their perpetuation.
5. *Protected landscapes.* To maintain nationally significant natural landscapes characteristic of the harmonious interaction of man and land while providing opportunities for public enjoyment through recreation and tourism within the normal life-styles and economic activity of these areas.
6. *Resource reserve.* To protect the natural resources of the area for future use and prevent or contain development activities that could affect the resource pending the establishment of objectives based on appropriate knowledge and planning.
7. *Natural biotic area/anthropological reserve.* To allow the way of life of societies living in harmony with the environment to continue undisturbed by modern technology.
8. *Multiple-use management area/managed resource area.* To provide for the sustained production of water, timber, wildlife, pasture, and outdoor recreation, with the conservation of nature primarily oriented to the support of the economic activities (although specific zones can also be designed within these areas to achieve specific conservation objectives).
9. *Biosphere reserve.* To conserve for present and future use the diversity and integrity of representative biotic communities of plants and animals within natural ecosystems and to safeguard the genetic diversity of species on which their continuing evolution depends.
10. *World Heritage site.* To protect the natural features for which the area was considered to be of world heritage quality and to provide information for worldwide public enlightment.

Adapted from IUCN [1978], with permission.

have indicated the value of this approach in expanding the range of management tools available for protecting natural habitats. Nature reserves (Category I) and national parks (II) must be strictly protected against efforts to exploit their natural resources for such purposes as timber production, mining, hydroelectric works, industrial facilities, commercial fishing, sport and commercial hunting, farming, and grazing of domestic animals. But protected landscapes (V), resource reserves (VI), natural biotic areas (VII), and multiple-use management areas (VIII) can be established around the strictly protected areas to prevent them from becoming biologically impoverished islands, or can stand by themselves to make important contributions to systems of tropical forest management.

For example, a project to develop the water resources of a major river should include financial and technical support to conserve the upstream catchment even if the area is not of national park quality or contains a moderate human population; one of the other categories, such as cultural landscape, can address the local situation quite adequately, and a recognized form of conservation can be provided.

By adopting a range of complementary protected area categories, governments can meet their responsibilities for protecting nature while providing for human development on a sustainable basis. The range of protected area categories can address a broad spectrum of conservation issues and objectives, so conservation can be more identifiably linked to water, protein, ecosystem productivity and diversity, pharmaceuticals, agricultural and livestock breeding, wood and wood products, science education, and the human environment in general. Such an approach will enable conservation to obtain a more important place at the development table and earn greater support by governments and the public for significant expansions of protected area systems. A series of case studies that illustrates this concept, including examples from Ivory Coast, Uganda, Indonesia, Malaysia, Nepal, Sri Lanka, Costa Rica, Peru, Ecuador, Brazil, and Venezuela, can be found in McNeely and Miller [1984].

A theoretical framework for assigning the 12 objectives for conservation to the 10 protected area management categories is proposed in Table III. The model suggests that, given an objective, several alternative approaches to management might be followed. From a technical standpoint, the pursuit of each objective usually can be combined compatibly with others, depending on biophysical relationships among variables. The model also suggests that if the entire array of objectives is to be sought then several categories of conservation areas are required.

TABLE III. Decision-Making Guide to Alternative Protected Area Categories to Attain Conservation Objectives

Objectives for conservation	Alternative management categories									
	I Nature reserve	II National park	III Natural monu-ment	IV Game reserve	V Protected landscape	VI Resource reserve	VII Anthro-pological reserve	VIII Multiple-use mgmt area	IX Biosphere reserve	X World heritage site
1. Maintain sample ecosystems	2[a]	(1)	(1)	(1)	2	3	3	3	(2)	(1)
2. Maintain ecological diversity	(3)	(1)	(1)	(1)	(3)	(3)	(3)	(3)	(1)	(1)
3. Conserve genetic resources	(1)	(1)	(1)	(1)	(3)	(1)	(3)	(3)	(1)	(1)
4. Education, research, monitoring	(1)	(2)	(2)	(2)	4	4	4	4	(1)	(1)
5. Watershed and soil conservation	3	3	3	3	3	4	4	4	3	3
6. Fisheries and game management	—	—	—	(1)	4	4	(3)	(2)	(2)	—
7. Recreation and tourism	—	(2)	4	1	3	4	4	4	2	(1)
8. Forest production	—	—	—	—	4	—	4	2	4	—
9. Protect cultural treasures	—	(1)	4	—	4	—	(1)	4	4	(3)
10. Maintain scenic resources	3	(1)	(1)	3	3	4	4	4	4	(1)
11. Maintain options	—	—	—	—	4	(1)	3	(1)	2	—
12. Support rural development	(3)	(3)	(3)	(3)	(1)	(4)	(1)	(1)	(2)	(3)

[a]( ), Major purposes for employing management system; 1, objective dominates management of entire area; 2, objective dominates management of portions of area through zoning; 3, objective is accomplished throughout portions or all of area in association with other management objectives; 4, objective may or may not be applicable depending on treatment of other management objectives and the characteristics of the resources; —, not applicable.

THE CURRENT STATUS OF PROTECTED AREAS IN TROPICAL FORESTS

To meet a wide range of conservation objectives, over 120 countries have created systems of protected areas. A summary of the number and area protected by each country in the tropical forest biome is presented in Table IV, showing that nearly 300 units protect over 40 million hectares.

Ideally, the best way of analyzing how well the diversity of tropical forest ecosystems is being conserved is through biogeography (the science of distribution of species and ecosystems), but far less than half the world's species have even been scientifically described, and the great bulk of the undescribed species live in tropical forests [Myers, 1979], so it is unrealistic to expect mapping of individual species to become a workable tool in sufficient time to be useful. Because time is running out for major land-use decisions in the tropics, it is necessary to make the best use of the available information while a few options are still open.

A system developed by IUCN [Udvardy, 1975] divides the world into eight biogeographical *Realms*, "continent or subcontinent-sized areas with unifying features of geography and fauna/flora/vegetation." The realms are further divided into *Provinces*, defined by significant differences in flora, fauna, or vegetation structure; the provinces are assigned geographic, ecological, or historically established names as appropriate. Each of the provinces is characterized by one of the world's 14 major *Biomes*, major regional ecological communities of plants and animals extending over a large natural area.

Based on this system, IUCN defined 193 biogeographic provinces for the world, including 25 with tropical rainforests. An indication of how well the world's 23 tropical forest provinces that contain primates (two tropical forest provinces, in Australia and on New Guinea, lack primates) are covered by protected areas is presented in Table V.

THE ADEQUACY OF THE CURRENT SYSTEM OF PROTECTED AREAS

Forty million hectares of tropical forest under protection seems like quite a lot, particularly when almost all tropical forest biogeographic provinces are covered. The "big three" tropical forest countries—Brazil, Zaire, and Indonesia—all have substantial areas protected. But is this area—a bit smaller than California—really adequate to conserve the species, ecosystems, and ecosystem functions that the areas are established to conserve? Even setting aside for the moment the question of how effectively the existing areas are

TABLE IV. Protected Areas in IUCN Categories I–V in the Tropical Moist Forest Biome by Countries Containing Primates[a]

Country	No. of areas	Area protected (hectares)
Bangladesh	4	76,675
Bhutan	1	2,100
Bolivia	1	1,884,375
Brazil	17	7,993,166
Cameroon	7	1,125,300
China	8	25,548
Colombia	6	985,000
Congo	11	1,359,100
Ecuador	4	845,000
Gabon	9	1,735,000
Ghana	4	263,623
Guatemala	1	57,600
Guinea	1	13,000
Guyana	1	11,655
India	33	1,863,712
Indonesia	67	7,444,371
Ivory Coast	6	569,000
Madagascar	5	242,718
Malaysia	16	716,203
Mexico	1	1,000
Nepal	1	6,500
Nigeria	2	259,431
Panama	6	660,902
Peru	2	2,920,306
Philippines	18	304,816
Sierra Leone	1	98,000
Singapore	1	2,434
Sri Lanka	1	110,000
Suriname	8	482,400
Thailand	33	2,394,746
Trinidad and Tobago	12	16,567
Venezuela	4	2,220,000
Zaire	4	6,148,000
Totals	296	42,838,258

[a]Source: IUCN Conservation Monitoring Center.

TABLE V. The Coverage of Tropical Forest Biogeographic Provinces by Protected Areas in IUCN Categories I–V (Only Provinces Containing Primates are Listed)[a]

Biogeographic province	No. of areas	Area protected
Afrotropical realm		
Guinean rain forest	14	1,203,054
Congo rain forest	28	7,875,400
Malagasy rain forest	5	242,718
Central African highlands	3	2,492,000
Subtotals	50	11,813,172
Indomalayan realm		
Malabar rain forest	22	1,527,009
Ceylonese rain forest	1	110,000
Bengalian rain forest	13	419,978
Burman rain forest	No areas listed	
Indochinese rain forest	27	2,013,368
Malayan rain forest	13	894,014
South Chinese rain forest	8	25,548
Andaman and Nicobar Islands	4	2,000
Sumatran rain forest	15	3,450,205
Javan rain forest	24	565,102
Celebesian rain forest	12	767,416
Bornean rain forest	26	2,867,659
Philippines	18	304,816
Subtotals	183	12,947,115
Neotropical realm		
Campechean rain forest	2	58,600
Panamanian rain forest	6	660,902
Colombian coastal rain forest	6	860,000
Guyanan rain forest	25	2,155,122
Amazonian rain forest	16	13,894,181
Madeiran rain forest	1	268,150
Serro do Mar	7	181,016
Subtotals	63	18,077,971
Totals	296	42,838,258

[a]Source: IUCN Conservation Monitoring Center.

managed, the answer is still a clear "no." Two main lines of evidence point to this inadequacy.

Biogeography and Protected Area Design

Advances in the theory of island biogeography [MacArthur and Wilson, 1967; Wilcox, 1980] suggest that any time the total area of an ecosystem is reduced species diversity will decline until it reaches a new equilibrium for the size of the ecosystem. It is clear that no reduced area, no matter how well protected, can ultimately retain all its original species when it becomes an "island" of forest surrounded by other forms of land use. The larger the reserve, the less severe and slower the species loss, but the clear conclusion is that *any* loss of tropical forest habitat will lead to a loss in species. A rough estimate is that a single reserve containing 10% of the original forest area will support just 50% of the original species present.

Population Genetics

The science of population genetics has led to a rethinking of what a protected area must do to ensure long-term survival of key species. Questions of short-term fitness, long-term adaptation, relations between genetic heterozygosity and reproductive behavior, and the continued opportunity for evolution of species have been analyzed. It has become clear that large population sizes are required by species that are least dense, such as the large herbivores, carnivores, scavengers, and trees. Avoiding inbreeding depressions requires populations to be maintained at levels greater than 50 or even 500 or 5,000 [Frankel and Soulé, 1981].

Determining the theoretical minimum size of a protected area required to conserve genetically viable populations of all species is a fairly simple mathematical operation, once all the uncontrollable variables are declared controlled. Simply take the least dense species (many trees occur in tropical forest at densities of 1 per ten hectares, and adult male orangutans may have a density of one per 100 hectares in the richest habitats), determine the minimum genetically viable population size, and multiply the density by the population to give the total area required. For wide-ranging species such as tigers or elephants, it might no longer be possible to have reserves of sufficient size to ensure their long-term survival unless active genetic management measures are instituted.

ECOLOGICAL PRINCIPLES OF PROTECTED AREA ESTABLISHMENT

Most ecological research has been carried out in temperate zones, where scientists are plentiful and ecosystems are relatively simple. In the tropics, things are considerably more complex and ecosystems are only very poorly understood. Decisions to protect tropical forests must therefore be based on less than complete information. But what is known, or at least strongly suspected, is already sufficient to suggest a few principles.

First, *multiple reserves are necessary for the maintenance of biological diversity*. A range of natural situations should be incorporated into the protected area network of a country by including areas that are undisturbed, semiwild, and altered. Plant successional stages need to be included, particularly in relation to species that have patchy distributions [Diamond, 1980]. Multiple reserves provide for the transitory nature of patches [Foster, 1980], the replication needed for research, and the capacity of the network to handle large natural or man-induced disturbances. It is also advisable to maintain several populations of particular species to reduce the chances of accidental extinction due to disease or other catastrophe and to permit opportunities for local adaptation that will increase chances for ultimate survival [Franklin, 1980].

Second, *various types and amounts of manipulation of habitats will generally be required to hold or foster succession*. With a series of reserves within each biogeographic province, some areas can be maintained in their undisturbed, natural dynamic state while others are allowed or encouraged to evolve more rapidly (or slowly), thereby covering the various options considered to be desirable.

Third, *the size of each individual protected area will depend on the species to be included, the area required to meet the species' requirements, and the ecological processes that need to be provided special management to ensure their permanent contribution to the area*. Ideally, whole ecosystems should be included in the protected area, with territory sufficient to encompass a large number of the least-dense species. Large, undisturbed reserves in the tropics may require from 200,000 to 1 million or more hectares, whereas the multiple smaller, undisturbed and altered areas may require a minimum of 50,000–100,000 hectares. From data in Tables IV and V, the mean size of current protected areas in tropical forests is about 140,000 ha (Salonga Naitonal Park in Zaire is the largest, with over 3.6 million ha). On the other hand, many areas fall far short of the minimum requirement and are unlikely to meet the long-term conservation objectives for which they were established unless they are significantly enlarged or managed very effectively.

Fourth, *the shape of protected areas is critical in avoiding insularization effects*, the ideal reserve being relatively round. Serrations, peninsulas, and salients tend to lose species representative of the undisturbed core and replace them by species from the zone external to the reserve. Complex predator-prey relationships are disrupted and local extinctions occur.

Therefore, within any biogeographic province, protected areas should be large in size, round in shape, manifold in numbers, and dispersed in space. Such a network of reserves will focus specifically on the first three objectives for conservation areas: the establishment of representative samples of major ecosystems, the maintenance of ecological diversity, and the conservation of genetic resources. It is apparent that few countries with tropical rain forests have protected areas that fully meet these objectives.

GUIDELINES FOR SELECTING A SYSTEM OF TROPICAL FOREST HABITATS FOR PROTECTION

Based on the principles listed above, a series of practical guidelines can be suggested for selecting tropical forest habitats for protection on a countrywide systems basis. These guidelines are generic and can easily be adapted to different regions.

Step 1: Survey

The entire country or region is surveyed, using satellite imagery, aerial photographs, vegetation maps, literature review, and field visits. Biological factors, ecological processes, and social and economic factors all need to be recorded.

Step 2: Establishment of Criteria

Criteria are then established that make the selection of an area for conservation management as objective and defendable as possible. As suggested by Tables I–III, different criteria will apply to different categories of protected areas. The following criteria (adapted from IUCN [1980b]) can be adopted as appropriate to each particular situation.

Ecological criteria

Dependency. To what extent are species dependent on the area? How important are the ecological processes of the area? Areas critical to key species and ecosystems should have high ratings.

Naturalness. The less disturbed an area is, the more important (and rare).

Uniqueness. Areas that contain species, ecosystems, or habitat features that are found nowhere else should have high ratings.

Diversity. Areas that have the greatest variety of species and ecosystems should receive priority; they are likely to be of particular evolutionary importance.

Integrity. Areas that are able to function as an ecologically self-contained unit should receive priority; they are easier to protect.

Representativeness. Areas that are representative of a habitat type, ecological process, biological community, physiographic feature, or other natural characteristic should receive priority so that the system is complete.

Scientific and educational criteria

Convenience. Areas accessible to researchers or students should receive priority for scientific and education criteria, particularly as such areas are likely to have a range of degrees of disturbance.

Monitoring benchmark. Areas that can serve as nonmanipulated areas against which to measure changes occurring elsewhere should receive priority; they will be the basis for assessing any ecological change. They should form part of any national monitoring program.

Research history. Areas where research has been carried out over a long period, or where major field studies have been carried out, should receive priority, because any new research findings will have a strong foundation on which to build.

Demonstration. Areas that can serve to exemplify techniques or scientific methods are particularly important for education purposes.

Process relationship. Areas that represent ecological characteristics of regional value are particularly important; research can yield arguments that can have impacts far beyond the protected area. Water cycles are particularly crucial in many tropical forest areas.

Awareness. Areas where monitoring, research, education, and training within the site can contribute knowledge and appreciation of regional values should receive priority.

Social and economic benefit criteria

Economic benefit. Areas that have obvious long-term economic benefits, such as watershed protection or tourism, should have a high priority (even though their initial establishment might involve short-term economic disruptions).

Social acceptance. Areas already protected by local people should receive high priority; similarly, areas for which official protection by the government (particularly against outside exploitation) would be welcomed or at least accepted should receive high priority.

Recreation. Areas that provide local communities opportunities to use, enjoy, and learn about their natural environment should receive high ratings.

Tourism. Areas that lend themselves to forms of tourism compatible with the aims of conservation should receive high priority; this criterion is often related to those of economic benefit and social acceptance.

Landscape. Areas with features of outstanding natural beauty should receive priority; these are usually also unique, easily destroyed, and attractive to tourists.

Demonstration. Areas that can demonstrate the benefits, values, or methods of protection should receive high priority, particularly when such areas help to resolve conflicts between natural resource values and human activities.

Pragmatic criteria

Urgency. Areas where action is required quickly in order to avert an immediate threat should receive priority (though it should be realized that this is often a "damage control" action; it is usually best to protect far in advance of threat).

Opportunism. Areas where existing conditions or actions lend themselves to further action (such as the extension of an existing protected area or establishment of a buffer zone around an existing park) deserve priority.

Management feasibility. Areas that can be properly managed with reasonable cost and with relatively little conflict should receive priority.

Availability. Areas that can be acquired easily—through inter-departmental transfer, easements, or other legal forms of control—should receive priority.

Step 3: Selection of Areas To Be Included in the System

Once the criteria have been established and applied, the adequacy of the entire system can be assessed and presented in a published form. This document (the "Protected areas system plan") will be of use both for politicians and the public to explain exactly why each area is required and what benefits will accrue to the country by establishing the system.

Step 4: Establishing the System

The areas within the system are established once they have become institutionalized on a secure basis. This normally means legal establishment, ideally through an act of the highest legislative authority [Lausche, 1978], and assignment of management responsibility and budget to one or more government institutions (national park department, game department, watershed protection department, etc.).

Step 5: Managing the Individual Protected Areas

The next step is a more difficult one: implementing the objectives of management. It begins with the process of preparing a management plan for each individual area. A management plan is regarded as an essential tool in ensuring that protected areas function effectively. It is worthwhile discussing the process that results in a plan to demonstrate that planning is a prerequisite for effective management.

PLANNING FOR EFFECTIVE MANAGEMENT OF TROPICAL FOREST PROTECTED AREAS

Although planning methodology can be complex, the basic steps in an ideal procedure are summarized below. Although the management planning process is presented as a list of ordinate steps, it will be necessary to consider many steps on a simultaneous basis. Planners can also find that upon completion of various steps it may be necessary to review earlier decisions as new information becomes available. Similarly, it is recognized that present developments in some areas may prompt alternative approaches to decision-making. In addition, nontechnical factors such as budgets, institutional limitations, and considerations of a political nature can affect the time scheduling of the plan. These factors can be dealt with only at a higher executive level where matters related to the programming of the park system are considered.

The following steps display a thought process, a means of organizing a future based on an assessment of the present. It is not, therefore, the nomenclature or the exact order that is important but the *process* by which one evaluates and addresses the management problems of a protected area. For a detailed outline of what to do on each of the steps the aspiring management planner is referred to Miller [1980].

Step 1: Form the Planning Team

Although a plan can be prepared by one individual, it is most often a team exercise involving a core group of from three to six individuals. It is important that the team members have a mix of capabilities in the methodology of planning, ecology, sociology, economics, forestry, and various other of the resource sciences. As these individuals are seldom available from within an agency in most tropical countries, they may be hired as consultants from local universities, borrowed from regional institutions, or requested through outside assistance programs.

In addition to these specialists, the preparation of a management plan also requires the participation of those who manage the protected area as well as those who will use the unit and others who will be affected by the plan. To be truly effective, the planning process must be responsive to the needs of the managers in the field because the managers are ultimately responsible for the implementation of the plan. If the planning function is centralized at the headquarters office, field managers should still have an integral role in management plan preparation, perhaps through short-term attachments to the planning team. Most important, the individual who is responsible for managing the area should be a member of the planning team because of his knowledge of the area and his role in implementing the plan. The area manager is perhaps best able to assist the planning team in finding practical approaches and solutions to the management problems faced by his unit.

The full range of both technically trained and nontechnical personnel within the management and research arms of a conservation department should be involved in the planning process. In practical terms, this means that the planning team should involve personnel from all levels of management, in addition to consulting with scientists, tourism experts, educators, concessionnaires, and people living in and around the area.

Step 2: Gather Basic Background Information

The planning process now proceeds through a review of all resource material available on the protected area. This includes an overview of the enabling legislation, biophysical features, cultural resources, and socioeconomic data. This information is collected from various sources including a literature review, office files, and interviews with knowledgeable people. A base map and reference file system are usually prepared at this stage.

Step 3: Carry Out a Field Inventory

Although the content and intensity of this step can vary considerably, all planning requires fieldwork to gather new information, to check and update existing data, and to view the area with new perspectives. The purpose is to develop the information base needed to make informed management decisions. Generally, a review is made of environmental resources and visitor use. Attention is also given to archaeological sites and contemporary cultures, regional economics, poaching, transportation networks, and attitudes of local people. Particular attention is devoted to critical environmental areas and potential development sites. The area manager and his staff, being most familiar with the area, play a key role in accumulating the necessary data.

It is important to note that planning does not set out to seek new information as an end in itself. However, it is concerned with compiling sufficient data to address the important management problems. The caution given here is not to dwell on missing information through a long process of data collection, but to address the most important management needs. The gaps can be identified in the next step.

Step 4: Assess Limitations and Assets

Limitations of an environmental, economic, political, administrative, or legal nature should be recognized and analyzed at this point. Senior management at headquarters plays a critical role in defining these limitations. A review of programs outlined in the national plan and in district development plans should be done at this stage, although the absence of such plans should not deter one from proceeding. The purpose of this step is to ensure that options to be developed in later stages will be realistic in the light of current limitations. Assets should also be recognized and analyzed, so that the plan can make effective use of them.

Step 5: Review Regional Interrelationships

A protected area must be integrated as an essential element of the regional land use fabric. The planning team must attempt to review the potential effects of development outside protected area borders as well as the effects of the protected area on the region. The special arrangements that might be required with adjacent inhabitants are outlined here as are the roles of buffer zones and key regional decision-makers.

Step 6: State the Objectives of the Area

With the above steps completed, it will be possible to spell out in detail the values and objectives of the area in relation to its particular set of resources, to the region, and to the country as a whole. A review by senior management should be undertaken at this point to ensure that all factors have been considered and that the objectives identified are appropriate to the protected area.

Step 7: Divide the Area Into Management Zones

Most protected areas will have zones devoted to different objectives and uses. These can range from zones of intensive tourism development to dispersed recreation zones, from controlled resource production zones to full protection zones. Different management practices allowed or prohibited in each zone should be itemized. A standardized zoning scheme for all categories of protected areas in the country should be developed and applied to the area, perhaps based on the system developed by IUCN [1978].

Step 8: Review Boundaries of the Area

Few protected areas have ideal boundaries from an ecological point of view. With the resource inventory, management objectives, regional integration review, and zoning stages of the plan in place, the team might wish to consider boundary modifications.

Step 9: Design the Management Programs

Once the zoning concept has provided the basis for what is to be done where, the task now is to answer the questions of how, when, and who. This is the heart and action-oriented component of the plan that addresses the four major programs of protected area management.

Resource management and protection. This management program focuses on issues that relate to the protection of the biological and physical resources of the area.

Human use. This program deals with all aspects of use by people, including traditional use, recreation, interpretation, and extension.

Research and monitoring. The management of protected area resources frequently requires an understanding of specific ecological dynamics together with supporting data, which may not yet exist for the area. An important aspect of protected areas management thus involves the design and development of research programs to meet these information needs. Similarly, a monitoring program is needed to gauge progress in meeting the management objectives; linkages with other protected areas around the world can be provided through including the area in IUCN's Conservation Monitoring Center protected area data base [Harrison, 1983]. In addition, the role protected areas play as environmental benchmarks makes them attractive field sites for a wide range of research and teaching uses. A separate management program is often devoted to these aspects of research and monitoring.

Administration. Here the operational, manpower, and financial resources needed to run a protected area are outlined. Headquarters facilities, vehicles, equipment, and maintenance requirements are the types of issues of concern in this program.

Step 10: Prepare Integrated Development Options

This step in the management planning process now summarizes all the physical facilities, personnel and manpower, and institutional developments that are needed to accomplish the various management programs. The team might wish to present various development options and present the engineering and construction implications.

Step 11: Outline Financial Implications

No plan can be evaluated if the costs of the planning proposals are not given, at least as general estimates. In some cases, the economic justifications will need intensive treatment in a cost-benefit analysis. In any case, the planning team must now present the cost estimates of their proposals.

Step 12: Prepare and Distribute a Draft Plan

Before proceeding to the final steps, this is a useful point at which to obtain feedback on the team's progress to date. Therefore, a rough first draft of a plan is compiled and distributed to the range of actors who are key to the success of the plan. These will include those within the agency and might also involve many individuals and groups from outside the agency, depending on the public involvement policy followed.

Step 13: Analyze and Evaluate the Plan

After digesting input from the previous step, the team is now in a position to narrow its options. A final review of all development proposals is made and given approval by senior management.

Step 14: Design Schedules and Priorities

As the plan is now set to be finalized and put in motion, the team decides on the timing and priorities of each event. It is important that a formal approval page is then presented to the director for his signature.

Step 15: Prepare and Publish Final Version of Plan

With authorized approval from the director, the final plan is produced, published, and distributed in a form suitable to reach a general audience.

Copies of the document should be given to political leaders, ministry officials, and related government departments, regional councils, international agencies, scientific personnel involved in research and monitoring, and appropriate public interest groups. Such distribution helps to ensure that all interested parties are fully informed about the management steps proposed for the area. This consensus-building exercise can help enlarge budgets, focus research, and encourage managers to carry out the prescribed measures.

Step 16: Monitor and Revise the Plan

Plans soon need revision as new information becomes available and basic conditions change. Thus a 5-year planning horizon is often used as a realistic lifespan for a management plan. A review of the plan at intervals is thus the final step in this idealized process. It is not necessary to wait 5 years for review. Review can take place more frequently, but, in each case, always planning 5 years into the future.

Management plans have been prepared for many tropical forest protected areas, in countries as diverse as Peru, Indonesia, Costa Rica, Brazil, and Thailand, but most areas still lack such guiding documents, and it is our firm belief that management will not be effective until the management planning process has been completed and an approved plan is ready for guiding decisions. This is not to say that a management plan solves all problems. Far from it. Training, public support, appropriate regulations, application of research to management problems, political support, budgets, government policies in surrounding lands, and many other factors are also crucial. But the management plan lays the foundaiton for all these related measures.

CASE STUDIES: HOW TROPICAL FORESTS ARE BEING PROTECTED IN BRAZIL AND INDONESIA

It is perhaps not surprising that two of the countries with the greatest areas of tropical rain forest on their respective continents have taken serious steps to establish protected area systems. The approaches used illustrate the application of protected area planning systems planning methodology and provide guidance for other countries in the tropics.

Brazil

The Brazilian Institute for Forestry Development (IBDF) and the national FAO project, with support from the Brazilian Foundation for Nature Conservation (FBCN), initiated work on a system of national parks in 1975 in response to the mandates of the Second National Development Plan. The conservation problems of the entire nation were reviewed, and a conceptual framework was given for a major effort to examine the Amazon basin as a single unit [Wetterberg et al., 1976]. Six objectives guided the work: 1) to synthesize the published works of various Amazon specialists into a common format from which biologically significant nature conservation priorities can be tentatively identified; 2) to identify and locate the existing and planned nature conservation units in the Amazon; 3) to analyze the potential compatibility or incompatibilities between the Brazilian priority development poles and the preservation of biologically significant areas; 4) to propose a general outline of an Amazon nature conservation program that gives due consideration to the diversity of that region, permits identification of key areas to be preserved, and yet is flexible enough to adapt to future scientific discoveries; 5) to permit those public agencies respsonsible for national parks and equivalent reserves to gain an offensive position from which an Amazon conservation policy can be actively pursued, prior to the elimination of this option by other developments; 6) to contribute, in the particular case of Brazil, to the development of a National Park System Plan.

Existing and planned conservation units were examined in terms of the objectives being pursued and phytogeographic coverage [Prance, 1976]. The analysis showed where gaps existed and demonstrated the need for more detailed classification to ensure adequate coverage of the diversity of the region.

Vegetation maps were refined to locate areas of biological diversity. Particular attention was given to the areas considered "Pleistocene refugia" [Haffer, 1974; Vanzolini, 1970; Prance, 1973]. Information on birds, lizards, butterflies, and various plants was synthetized. Maps were prepared on vegetative formations and theoretical Pleistocene refuges.

A preliminary system of proposed areas resulted from overlaying these maps to determine where putative Pleistocene refugia for two or more plant or animal taxa overlapped. These areas were then examined in terms of their relationship to the proposed development poles. Three criteria were utilized to select priority areas for the proposed system of protected areas. First priority areas were defined as those that two or more authors have identified as likely Pleistocene refuges. Although these sites might not represent the places of greatest plant and animal diversity at the present time, they probably were the centers of evolutionary dispersion. Consideration was given to sites between the refuges where interbreeding must have occurred among organisms from two or more refuges. Second priority was assigned to areas containing samples of several vegetative formations and perhaps also a refuge. Third priority was given to all other specific sites recommended by the first two criteria.

The document that proposed the framework for a system of reserves in the Amazon [Wetterberg et al., 1976] was distributed to authorities in each Amazonian country as well as national planning bodies, universities, research institutes, and individual scientists both in Brazil and abroad requesting critical evaluation. The results of the critical review, and an update on progress towards implementation, was presented to the Second Meeting of the Intergovernmental Technical Committee for the Protection of Amazonian Flora and Fauna. A training workshop was held in 1978 for senior management personnel on planning the management of conservation areas, and the IBDF interdisciplinary planning team proceeded to prepare management plans for areas in the Amazon and elsewhere in the country. A national strategy plan was published in 1979 [Jorge Padua, 1979]. Between June, 1979, and July, 1980, more than 5 million hectares were added to the Brazilian park and reserve system, mostly in the Amazon basin. Through this effort, a protected area was put in place well in advance of alternative land-use pressures.

Indonesia

Indonesia contains the largest remaining areas of tropical rain forest in Asia, along with the greatest diversity of flora and fauna. The exploitation of timber did not begin in earnest until the late 1960s but quickly grew to the stage when many people—both within Indonesia and abroad—were becoming concerned about the rate of forest destruction. A new government policy to establish some 10% (20 million hectares) of the country as protected areas was implemented with the assistance of IUCN, WWF, and FAO, with major support to training provided by the Netherlands government and the World Bank.

The country was divided into biogeographical divisions treating the major islands as distinct units and grouping the 13,000 small islands according to their relationships (with islands sharing more than 80% of their fauna and flora classed as belonging to the same division). For each biogeographical province, a habitat map was prepared showing the distribution of original ecotypes, based initially on physical and climatic categories (detailed vegetation maps are still lacking for most areas).

The maps distinguished 20–30 habitat types for each biogeographic province, providing the basis for including samples of each within the proposed reserve system. However, preference was given to habitats that are species-rich, rare, fast-disappearing, underprotected in the existing system, biologically distinct, or rich in endemic species. A scoring system was devised by Mackinnon [1982] to evaluate the genetic conservation importance of areas and to score each area on objective criteria.

Using satellite imagery and land-use maps, the remaining primary habitats were determined for each province. Existing and proposed reserves and major protection forests were also plotted, revealing what area of each habitat type is already protected and where major gaps exist in the coverage. Extensions or new reserves were then proposed to give fuller representation of all habitat types.

A system of priority rankings was established, with first priority going to areas of major national importance, containing the most valuable or only possible examples of different habitat types. Second priority indicated important or valuable areas that would fill smaller gaps in the reserve coverage, give examples of representative regional variation within habitat types, and make valuable additions to the reserve system. Third priority went to areas of low conservation importance generally not meriting national attention, but that might be of regional interest or merit protection for other reasons (protection forest, recreation area).

These priorities were based on three types of evaluation made for each potential reserve: importance in preserving genetic resources, socioeconomic justification, and management feasibility. For each of these, relatively sophisticated scoring systems were developed that addressed government policies in an objective manner.

As a result of this effort, the area protected by Indonesia has increased from 2.8 million ha in 1978 to 11.6 million ha in 1982. A further growth to 20 million ha is planned for the next 5-year period. It is hoped that management of these areas will soon catch up to the excellent planning work that has been carried out to date.

SUMMARY AND CONCLUSIONS

In a period in the ecological history of the planet when a single species has become sufficiently dominant to threaten many tropical rain forest habitats with destruction, it is important that the same dominant species take steps to conserve the forests and all their species. There is little indication that governments will be able to take the necessary measures on purely ethical, esthetic, moral, or spiritual grounds, but if it can be demonstrated that protection of significant representative samples of tropical forests serves the enlightened self-interest of people (and politicians), then there are real opportunities for conservation.

This requires that all those who are interested in tropical rain forests—scientists, developers, local people, conservation pressure groups, and many others—emphasize the practical value of intact tropical rain forest. "Practical value" should be defined broadly to include the expanded conservation objectives presented in Table I but also to extend beyond those to include specific benefits to local people, university students, local business, international organizations, and many others. In addition, local and international pressure groups can continue to bring the moral and esthetic issues to the attention of the general public, in the expectation that public pressure will help facilitate appropriate decision-making by politicians.

Once the "practical value" of protecting tropical rain forest habitats has been translated into the political will to do so, the steps outlined in this chapter can provide the means necessary for establishing a worldwide network of tropical rain forest protected areas.

REFERENCES

Diamond JM (1980): Patchy distributions of tropical birds. In Soulé ME, Wilcox BA (eds): "Conservation Biology." Sunderland, MA: Sinauer Associates, pp 57–74.

Foster RB (1980): Heterogeneity and disturbance in tropical vegetation. In Soulé ME, Wilcox BA (eds): "Conservation Biology." Sunderland, MA: Sinauer Associates, pp 75–92.

Frankel OH, Soulé ME (1981): "Conservation and Evolution." Cambridge: Cambridge University Press.

Franklin IR (1980): Evolutionary change in small populations. In Soulé ME, Wilcox BA (eds): "Conservation Biology." Sunderland, MA: Sinauer Associates, pp 135–150.

Haffer J (1974): "Avian Speciation in Tropical South America." Cambridge, MA: Museum of Comparative Zoology, Harvard University.

Harrison J (1983): Maintaining a database on the world's protected areas. Parks 7(4):3–5.

IUCN (1978): "Categories, Objectives and Criteria for Protected Areas." Morges, Switzerland: IUCN.

IUCN (1980a): "The World Conservation Strategy." Gland, Switzerland, and Nairobi, Kenya: IUCN/WWF/UNEP.

IUCN (1980b): "Principles, Criteria and Guidelines for the Selection, Establishment and Management of Mediterranean Marine and Coastal Protected Areas." Nairobi, Kenya: UNEP (mimeo).

Jorge Padua MT (1979): "Plano do Sistema de Unidades de Conservacao do Brazil." Brazilia: IBDF.

Lausche B (1978): Guidelines for protected area legislation. IUCN Environmental Policy and Law Papers 16:1–108.

MacArthur R, Wilson EO (1967): "The Theory of Island Biogeography." Princeton, NJ: Princeton University Press.

Mackinnon JR (1982): "National Conservation Plan for Indonesia" (8 vols.). Bogor, Indonesia: FAO/UNDP.

McNeely JA, Miller KR (1984): "National Parks, Conservation, and Development: Defining the Role of Protected Areas in Sustaining Society." Washington, DC: Smithsonian Institution Press.

Miller KR (1980): "Planning National Parks for Ecodevelopment." Madrid: FEPMA.

Mosley JG (1984): Protected areas and environmental planning in Australia. In McNeely JA, Miller KR (eds): "National Parks, Conservation, and Development: Defining the Role of Protected Areas in Sustaining Society." Washington DC: Smithsonian Institution Press, pp 274–282.

Myers N (1979): "The Sinking Ark." Oxford: Pergamon.

Prance GT (1973): Phytogeographic support for the theory of Pleistocene forest refuges in the Amazonian Basin. Acta Amazonica 6(1):5–28.

Prance GT (1976): "The Phytogeographic Subdivisions of Amazonia and Their Consequences on the Selection of Biological Reserves." Bronx, NY: New York Botanical Garden.

Prance GT (ed) (1983): "Biological Diversification in the Tropics." New York: Columbia University Press.

Udvardy M (1975): A classification of the biogeographic provinces of the world. IUCN Occ. Paper 7:1–44.

Vanzolini PE (1970): Zoologia sistematica, geografia e a origem das especies. Inst. Geografia, Univ Sao Paulo, Teses e Monogr 3:1–56.

Wetterberg GB, Jorge Padua MT, Soares de Catro C, Vasconcello JM (1976): Uma analise de prioridades em conservacao de natureza na Amazonia. PNUD/FAO/IBDF/BRA-45 Serie Tecnica 8:1–62.

Wilcox BA (1980): Insular Ecology and Conservation. In Soule ME, Wilcox BA (eds): "Conservation Biology." Sunderland, MA: Sinauer Associates, pp 95–118.

Primate Conservation in the Tropical Rain Forest, pages 205–229

8

Socioecologic Factors in the Conservation of Afromontane Forest Reserves

A. William Weber

Institute for Environmental Studies, University of Wisconsin, Madison, Wisconsin 53706

INTRODUCTION

Throughout the tropics, economic development activities and the myriad subsistence needs of rapidly growing populations have combined to put great pressure on remaining natural ecosystems. Tropical forests comprise what is arguably the world's most threatened biome, if for no other reason than that they contain roughly half of the biota known to our planet. Many authorities on the subject believe that we could lose up to 1 million species by the end of this century, primarily due to tropical deforestation [Myers, 1979, 1981; Ehrlich and Ehrlich, 1981]. Of the numerous taxa thus threatened with extinction, the forest-dwelling primates are among those at the greatest risk.

The principal cause of this impending ecological disaster is widely acknowledged to be our own species. Yet, while scientific studies have greatly increased our understanding of primate behavior and ecology, we have invested comparatively little time or effort to better understand those people capable of rendering such knowledge purely academic.

This chapter examines socioecologic factors of critical importance in conservation planning and management but which are often overlooked in favor of more direct protection activities for threatened parks and reserves. These include demographic, attitudinal, and economic elements as well as the ecological relationships between people and forest environments. Primary emphasis is placed on increasing local support for conservation, without which forest and primate preservation efforts have little hope for long-term success.

The particular biogeographic region under consideration is the afromontane forest zone along the Zaire-Nile Divide (ZND) in Rwanda and Burundi. There, a chain of natural forest "islands," completely surrounded and isolated from one another by areas of intensive human modification of the landscape, provides an extreme context in which to implement effective conservation programs. The human population densities are among the highest in the world, and the forest relics are relatively small, yet they contain more than a dozen primate species including both the chimpanzee and the mountain gorilla. Consideration of the ZND conservation conflicts and ongoing efforts to resolve them is important not only because of the immediacy of the problems; it is also increasingly relevant to conditions elesewhere in the tropics as human populations continue to grow and natural forests are increasingly fragmented.

THE ZND ENVIRONMENT

Rwanda and Burundi cover 53,434 km^2 of equatorial highlands within the Great Lakes region of east-central Africa. From remnant savanna formations to the east, the land rises steadily through the intensively cultivated central zones of each country to culminate in the montane rainforests of the Zaire-Nile Divide. Set against a backdrop of the western rift valley, this range extends along a 300-km front to form the rugged backbone of the region. The ZND environment can be divided into two generalized ecological zones: the montane forest zone of isolated natural reserves and the human-modified landscapes that surround and separate them.

The Montane Forest Zone

High altitude (1,700–4,500 m) and a humid climate (1,400–2,400 mm/year) combine along the ZND to favor the establishment of moist montane forest vegetation. Human exploitation, however, has resulted in the clearing of all but 1,835 km^2 of this forested zone, which once may have covered more than a third of present-day Rwanda and Burundi. The remnant forests now cover less than 6% of Rwanda and only 1.4% of Burundi. The threatened status of the forest domain is further heightened by the fact that the total area is subdivided into six principal reserves: Virunga (150 km^2)[1], Gishwati (280 km^2), Mukura (50 km^2), Nyungwe-Kibira (1140 km^2)[2], Kibira-Teza (200 km^2), and Bururi (16 km^2) (Fig. 1).

The insular reserves of the ZND comprise an important subset of what has been referred to as the afromontane archipelago. These forest islands dotting

[1]The total Virunga conservation area covers over 400 km^2, counting contiguous sectors in Zaire and Uganda.

[2]Includes Rwanda's Nyungwe Forest Reserve (970 km^2) and contiguous sectors of Burundi's Kibira National Park (170 km^2).

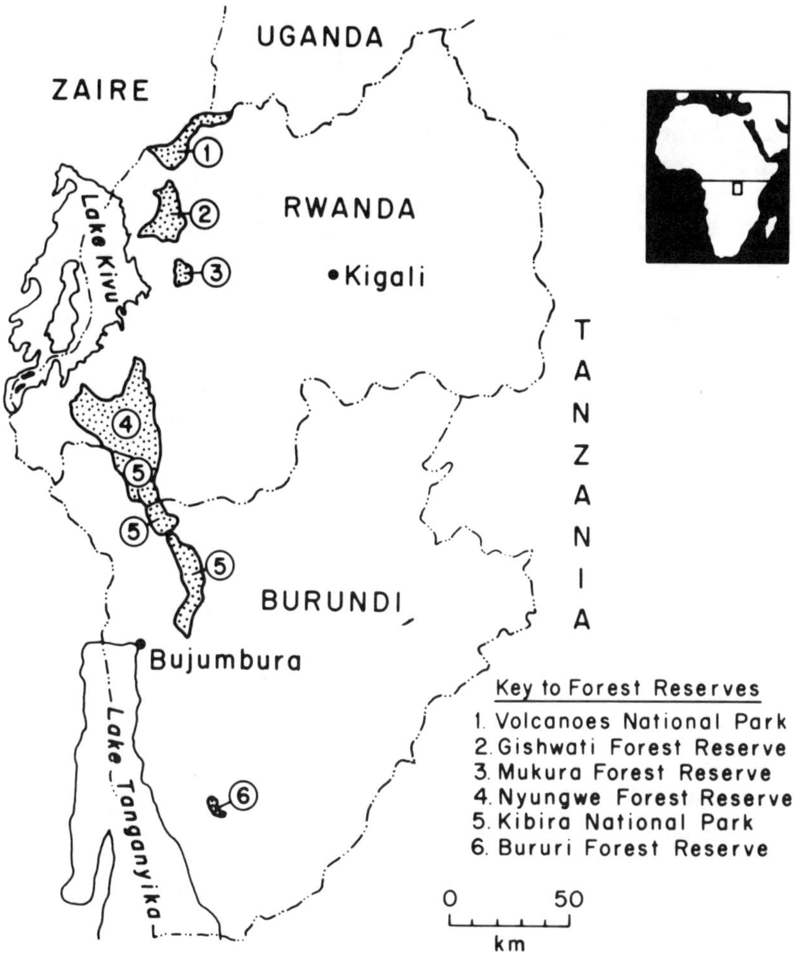

Fig. 1. Location of principal forest reserves along the Zaire-Nile Divide in Rwanda and Burundi.

the African continent can be grouped into several regional associations, with the ZND reserves as part of the larger Kivu-Ruwenzori chain [White, 1978, 1981]. The individual characteristics of three of these forest reserves, selected on the basis of their different sizes, altitudinal ranges, and biological attributes, are briefly described below.

Virunga. The rugged volcanic peaks of the Virunga chain form the international boundary between Rwanda and its northern neighbors Zaire and Uganda. Rising from 2,500 m at its lower limit to more than 4,500 m on

Mount Karisimbi, Rwanda's Parc National des Volcans (PNV) occupies the southern slope of the range along a 45-km front.

The vegetation of the Virungas is generally characteristic of East African montane environments [Hedberg, 1951; Hamilton, 1982; White, 1978]. Vegetation zones are differentiated primarily by altitude in addition to localized factors of exposure, aspect, slope, and microclimatic conditions. Mixed bamboo (*Arundinaria alpina*) forest and *Hagenia-Hypericum* woodland are the principal formations up to 3,200 m. Above that altitude, ericaceous and afroalpine associations predominate.

The lower forest and woodland zones are not only richest in abundance and diversity of plants but also of animal species. More than 100 bird species have been identified from the Virungas, in addition to forest-dwelling leopards, buffalo, elephant, hyena, bushbuck, duiker, and numerous smaller mammals. Four primates are also present (Table I), including the rare mountain gorilla.

Nyungwe-Kibira. Covering roughly 1,140 km^2, the combined areas of the Nyungwe Forest Reserve and the contiguous northern sector of the Kibira National Park represent one of the largest remnant patches of moist lower montane forest in Africa. Largely a function of its altitudinal range (1,700–2,950 m) and greater size, this reserve also contains a far greater number and diversity of animal and plant species.

Vande weghe (in preparation) has divided Nyungwe into four principal ecological zones: moist montane forest, richest in biological diversity; drier, more open, and less diverse *Macaranga* or *Podocarpus* forest; montane marsh forest; and mixed secondary formations. Each can be further subdivided into characteristic plant communities.

TABLE I. Primates of the ZND Reserves

Virunga	
	Gorilla gorilla beringei
	Cercopithecus mitis kandti
	C. mitis doggetti
	Perodicticus potto
Gishwati	
	Pan troglodytes schweinfurthi
	Cercopithecus mitis doggetti
	C. l'hoesti
	Colobus angolensis adolfi-frederici
Nyungwe-Kibira	
	Pan troglodytes schweinfurthi
	Cercopithecus mitis doggetti
	C. l'hoesti
	C. mona
	C. ascanius schmidti
	C. aethiops
	Cercocebus albigena johnstoni
	Colobus angolensis adolfi-frederici
	Papio anubis
	Perodicticus potto
	Galago sp.
Bururi	
	Pan troglodytes schweinfurthi
	Cercopithecus mitis doggetti
	C. ascanius
	C. aethiops

The remarkable diversity of plant associations in Nyungwe provides necessary habitat for an equally great variety of wildlife. Vande weghe (in preparation) has identified 240 avian species, 17 of which are endemic to the Kivu-Ruwenzori highlands. In his inventory of Nyungwe mammals, Storz [1982] lists eight species of carnivores along with his estimates of the remnant elephant, buffalo, wild pig, duiker, hyrax, and other small mammal populations. Among the 11 primates he was able to identify (Table I) are the chimpanzee, five species of *Cercopithecus*, the Angola colobus (*Colobus angolensis*), and the grey-cheeked mangabey (*Cercocebus albigena*).

Bururi. Situated in southwestern Burundi, the Bururi Reserve is the smallest forest relic (16 km^2) in the ZND chain. Its location at the southern terminus of the divide, however, results in certain unique attributes independent of size.

First, the region is subject to a 4–5 month dry season—double the period elsewhere—and natural adaptations to this stress result in a montane semievergreen or mixed deciduous forest. Even more interesting from a biological point of view, Bururi is located in an area recognized as a biogeographical crossroads [Lewalle, 1972]. As a result, it harbors plant and animal species more typically found in the Zambezian region to the south, the drier savanna forests to the east, and even the lowland rainforest of Zaire, as well as a high percentage of species common to the other highland areas of the ZND.

Despite its size and limited altitudinal range (1,900–2,150 m), the Bururi forest is not homogeneous but rather includes a variety of communities differentiated by structure. Among more than 90 species of tree found in the reserve, *Strombosia* and *Myrianthus* are dominant; *Entandophragma, Tabernaemontana* and *Newtonia* are also important. In her forest inventory, Vedder [Weber and Vedder, 1983b] found a comparatively depauperate fauna, which nevertheless contained 87 bird species, five carnivores, one ungulate, and five species of primates including the chimpanzee.

The Human-Modified Zone

The combination of generally adequate rainfall, comparatively good soils and a relatively healthy, high-altitude climate has favored the establishment of one of the world's most densely concentrated human populations in the interlakes highlands. In the more than 2,000 years since Iron Age tools first appeared in the region [Curtin et al., 1978], agriculturalists have cleared the forest from all but the higher reaches of the ZND[3]. In its place, fields of

[3]Evidence from the Ugandan sector of the Virungas indicates that forest clearing had occurred as high as 2,000 m by the year 1000 AD [Morrison, 1968].

sorghum, maize, beans, and bananas form a patchwork across the land. To that mosaic were added extensive pasturelands by the Tutsi, who moved into the region with their distinctive Ankole cattle sometime around the 16th century AD.

Today, agricultural land is at a premium in both Rwanda and Burundi. By 1983, the regional population density had surpassed 200 inhabitants per km^2—the highest level in Africa. Considering only arable land, the occupancy rate doubles to more than 400/km^2: a critical factor in a region where agriculture represents the primary occupation for over 90% of the population as well as the principal source of export revenue [Prefol and Delepierre, 1973; RAPID, 1981, 1982].

Highland zones are generally among the most densely populated areas in both countries. Environmental factors favoring settlement have resulted in densities above 350/km^2 in areas bordering all of the ZND reserves except Bururi, where a predominantly pastoral population is more widely dispersed [Delepierre, 1982; Mertens, 1981]. The regional population of more than 10 million is growing at an annual rate of more than 3% and could therefore nearly double by the year 2000. These demographic factors represent the driving force behind rising pressures on the regional renewable resource base—particularly the remnant ZND forest reserves.

Forest Conversion Trends

One of the first Europeans to travel through the inhabited part of the interlakes region at the end of the last century described a vast area with "neither tree nor bush, only banana trees. . . . it is incredibly populated . . . and very cultivated."[4] With regard to the ZND, the 1907 Mecklenberg expedition found the limit of the Nyungwe forest 9 km farther east than it is today [Lugan, in preparation]. By the late 1920s, however, immigrants from the drought-stricken east were pushing the natural forest back at a rate of nearly one km per year along the entire divide [Scaetta, 1932]. This intensive conversion pressure led Belgian authorities to set aside virtually all remnant forest lands as official reserves in 1933. The Virunga forest had been protected as Africa's first national park several years before.

Some small-scale encroachment continued over the next 25 years, but in the 15-year period between 1958 and 1973 major excisions were made from almost all of the theoretically protected forests. During that time, the Virunga

[4]Hauptmann R: Uha, Urundi und Ruanda, in "Mitteilungen aus den deutschen Schutzgebieten, X" pp 177–181 (cited in Louis [1962]: "Ruanda-Urundi: 1894–1919.", p 112).

and Gishwati reserves saw their areas cut in half, and Nyungwe's forest cover was reduced by 150 km^2 [AIDR, 1981]. Though some farmers did initiate private clearings, most of the montane forest zone was converted for larger-scale, cash crop-production purposes.

Although no major clearing has occurred during the past 10 years, qualitative degradation has continued in all of the ZND reserves. Illegal cattle grazing, uncontrolled prospecting for gold, and hunting for both subsistence and market purposes have all taken their toll on the forests and wildlife.

The principal threat to the remaining forests stems from increasing demand for fuel-wood for cooking purposes. Estimates of per capita consumption center around 1 m^3 per year. Yet even at that relatively low level the sustainable annual yield from existing plantations cannot satisfy more than one-third of current demand. Despite aggressive government reforestation efforts in both countries, the short- to medium-term outlook is therefore quite negative. This is particularly true in Burundi, where some experts project a 60% decline in existing forest cover over the current decade [Brunsfeld et al., 1982].

The people of the ZND are likely to experience painful shortages of wood products—particularly firewood—for the remainder of this century. This situation could lead them to overexploit, degrade, and even destroy much of their natural forest resource base. The ecological implications of this are extremely serious, as the population thus risks suffering not only from a fuel-wood shortage but also from increased erosion, decreased water supplies, and a general degradation of watersheds due to deforestation. In addition, irreplaceable forms of vegetation and wildlife may be lost in the process.

As is frequently the case, it is the larger mammals that have suffered the most from deforestation, habitat degradation, and the direct effects of hunting. In most instances, populations are reduced in number or otherwise stressed by the former, then further diminished or even eliminated by the latter. In Nyungwe, the leopard, buffalo, and possibly elephant are now locally extinct [Storz, 1982]; the leopard and giant forest hog have been killed off in Bururi, and the Virunga elephant population has declined from several hundred to a few dozen individuals.

Most primates have fared comparatively better, probably due to a combination of their arboreal nature, smaller size, and lesser appeal to hunters. It is perhaps significant that the mountain gorilla is larger, more terrestrial, more valuable on international black markets, and considerably more threatened. The most recent data from the Virungas indicate that the gorilla population there experienced a dramatic drop from roughly 450 to 270 between 1960 and 1973, with most of the decline occurring in the Zairian sector of the range and poaching the primary cause. This population has since apparently stabilized at a level above 250, with indications of an increasingly healthy population structure [Weber and Vedder, 1983a]. Though comparable trend data are lacking for the chimpanzee, it is widely recognized to be on the decline across the ZND. Only in Nyungwe does a potentially viable population of several hundred exist [Storz, 1982], whereas the Bururi total is no more than 30 [Weber and Vedder 1983b].

THE ZND CONSERVATION EQUATION

The values of the ZND reserves are many. Because values comparable to these have been widely cited and discussed, however, the principal reasons for their preservation are simply presented in outline form in the paragraphs below. In the following section, more detailed explanations are given as to why many of the claimed benefits of conservation are either unrecognized or greatly devalued by the local populations concerned.

Conservation Values

Scientifically, the ZND reserves are of considerable interest to biogeographers, who see in the afromontane archipelago a network of natural laboratories in which to study issues of extinction, colonization, speciation, and evolution; research that has increasingly applied value as natural habitats become more and more fragmented [MacArthur and Wilson, 1967; Diamond, 1976; Simberloff and Abele, 1976; Diamond, 1981]. Of particular significance is the identification of Nyungwe as a probable Pleistocene refuge during past glaciations [Hamilton, 1982].

There is further considerable interest in individual wildlife species, especially the primates. A great deal remains to be studied about our closest animal relations, the gorilla and the chimpanzee, and some of the ZND monkeys have demonstrated unique variations on the behavior and ecology of their conspecifics elsewhere [T. Struhsaker, personal communication].

With regard to species preservation, the ZND reserves contain many rare and endangered animals as well as plant populations of uncertain status and value. Research on montane forest plants and their uses by the local population might not only document their value to these people but could also reveal new medicines, drugs, and possibly even food crops for wider use [Myers, 1979, 1981, 1983]. More directly applicable research could help to determine native tree species that could be used for plantation and agroforestry purposes by people now inhabiting the former forest zone.

The most immediate and tangible economic benefit associated with forest and wildlife conservation is that represented by tourism revenues. The latter have now reached significant levels in the Virungas, where a tourism-development program centered on the mountain gorilla was initiated in 1979 [Weber, 1979, 1981; Weber and Vedder, 1983a]. Burundi also hopes to profit from increased visitation to its newly created network of reserves and parks [Kabayanda, 1982]. For now, however, tourism outside the Virungas must be recognized as a marginal economic activity.

A final, undeniable value of protecting the remnant ZND forests is their critically important role in water catchment and regulation. High rainfall, protective vegetation cover, absorbant litter, and reduced evapotranspiration all combine to increase water retention in the montane forest. Like a giant sponge, the forest system then gradually releases its load throughout the year, even during the dry season [McCulloch and Dagg, 1965]. The disproportionate importance of montane forested areas in this respect can be seen in the fact that, although the southern flank of the Virungas constitutes only 0.5% of Rwanda's total land area, it represents more than 10% of the country's effective catchment area [Spinage, 1972]. Comparable statistics could be calculated for all of the ZND forests, and therein lies their greatest conservation value to the surrounding human populations.

Devaluation of the Conservation Benefits

The combined weight of the preceeding arguments would seem to represent a powerful force for preservation of the remaining forests and wildlife on the Zaire-Nile Divide. That being the case, why does pressure continue to grow to convert more forest land to fields and pasture while hunting continues to take its deadly toll?

The answer, in brief, is that virtually all the arguments given for preservation above are either unrecognized, devalued, or rejected outright by large majorities of the populations concerned. From presidents to peasant farmers in their fields, residents of Rwanda and Burundi view forest and wildlife conservation issues through a variety of ecological, cultural, educational, economic, and political filters that produce a perspective quite different from that seen by those of us from the outside. The result is not always contrary, just different. Moreover, it would be a mistaken simplification to assume, as some have, that these people have absolutely no interest in forests or wildlife. To derive a successful conservation strategy, however, the initial equation must include factors that accurately reflect the socioecological reality of life along the ZND. Three aspects of that reality are examined below: people-land relations, attitudes toward nature preservation, and the economics of tourism as a regional development alternative.

People-land relations. As was noted earlier, Rwanda and Burundi are, respectively, the first and second most densely populated countries in Africa.

Their collective growth rate is well above 3% (3.7 and 2.9%, respectively), and in both more than half of the population is under 18 years old. Population totals and trends mean little, however, except in relation to land availability and use patterns [RAPID, 1981, 1982]. It is in this context that demographic factors take on added significance in this predominantly rural and agricultural region.

Roughly half of the regional territory is considered to be suitable for cultivation; another 10–20% consists of marginal pastureland. Most of the highland zones bordering the ZND forests have higher percentages of arable land in addition to relatively more fertile soils, but this has only resulted in considerably higher rates of settlement. Virtually every hectare of arable land around the Virungas is now exploited to its fullest by a population that has reached over 900/km^2 in certain sectors. Data from both countries indicate that the highland agricultural zones are now the primary sources of migrants to urban and marginal rural environments. Although discussions of carrying capacities for humans must be tempered by recognition of a highly elastic adaptive capacity, it is nevertheless clear that certain ecological limits have been reached along the ZND.

Opinion surveys of nearly 500 individuals living around the Virunga and Bururi reserves [Weber, 1981; in preparation; Weber and Vedder, 1983b] indicate that most of these people are very much aware of the land crisis in their countries. Whereas majorities of farmers living around the Virungas and the Bururi forest feel that their land holdings are sufficient for current family needs, fewer than half have enough to eventually subdivide for their sons as is traditional practice in the region. Virtually everyone believes that the next generation will suffer from chronic land shortages. Asked what solutions they see to this problem, however, significant minorities of both populations are unable to provide a single suggestion. Emigration is the solution most commonly cited. Historically, this has been a release valve for regional population pressures. The existing political and economic climate in neighboring countries, however, is no longer favorable. With prospects for internal migrations equally limited, this no longer seems a viable option.

A parallel survey of nonfarmers from the Virunga region found that, although 40% still favored emigration, more than 10% also called for increased birth control and more intensified agriculture, significantly higher percentages than among the less well educated farming population. Rwandan university students were also much more inclined to favor these latter solutions, and few saw any potential in emigration. Preliminary analysis of the Bururi survey indicates comparable trends from that region as well.

Although differences in educational and occupational background might cause them to perceive different solutions, virtually all inhabitants of the

ZND highlands are aware of the land availability problem in Rwanda and Burundi. Only against a backdrop of that recognition can their attitudes toward preservation be properly viewed.

Attitudes toward preservation. Although there is nearly universal recognition that it is currently illegal to exploit the resources of either reserve, a majority of those living around Bururi and 42% of the Virunga farming population feel that local residents nevertheless do need to hunt and cut or gather wood in the forest. Smaller percentages admit doing so personally, but the evidence is clear that many local people do exploit each of the reserves in an effort to satisfy their needs.

Asked to consider nonconsumptive values of forests (Table II), most of those in the Virunga farm survey population either state directly that the forest has no value or that they are not aware of any beyond its being a source of products. Highly significant differences again arise, however, with regard to the perceptions by groups with different educational and occupational backgrounds. Nonfarmers and university students are considerably more likely to cite one or more values of forest preservation, with increased rainfall or related climatic benefits at the top of their lists. Students are the only group to cite research and tourism as potential benefits. The Bururi farm population seems considerably more appreciative of microclimatic benefits associated with montane forests, perhaps due to the much drier environment in which they live. Even among the better educated and salaried groups, emphasis in that area is placed on such utilitarian values as opposed to those associated with research and tourism.

With regard to wildlife values (Table III), more than half of the Virunga farmers again fail to mention a single benefit of animal protection. In Bururi, the local population is even less appreciative. Forty-nine percent state frankly

TABLE II. Perceived Forest Preservation Values[a]

	Percentage of:					
	Virunga sample			Bururi sample		
	F	NF	U	F	NF	U
No value	17	9	5	14	5	0
Increases rainfall	19	40	59	45	68	54
Erosion control	7	5	38	8	32	67
Wildlife refuge	19	35	12	10	22	50
Research	0	0	11	0	0	0
Tourism	0	0	11	0	3	0
Don't know	38	9	7	22	3	0

[a]Popular perceptions of nonconsumptive forest values based on surveys in Virunga and Bururi forest areas. F, farmers; NF, nonfarmers; U, university-educated.

TABLE III. Perceived Wildlife Preservation Benefits[a]

	Percentage of:					
	Virunga sample			Bururi sample		
	F	NF	U	F	NF	U
No value	14	16	1	53	22	0
Tourism	39	61	71	5	29	100
Research	1	5	12	0	0	0
Species preservation	0	2	20	0	5	17
Ecologic value	0	0	7	0	0	0
Esthetic	1	5	8	0	0	0
Don't know	44	9	8	34	13	0

[a]Popular perceptions of nonconsumptive wildlife values based on surveys in Virunga and Bururi forest areas. F, farmers; NF, nonfarmers; U, university-educated.

that animals that can't be hunted have absolutely no value; another 32% are unable to provide any reason for their protection. This finding is undoubtedly linked to the fact that 90% of the Bururi sample feels that the forest harbors animal pests—with monkeys and chimps heading the list of nuisance species that eat crops. The gorilla does not have a similarly bad reputation, but half of those living within the shadow of the Virungas are unable to provide even a single descriptive characteristic of this spectacular creature, nor does the gorilla figure in any local myths or legends.

Tourism is the most commonly cited benefit of wildlife, though significant differences arise between areas and subgroups. Bururi has no active tourism program, and this is reflected in a lower recognition of tourism benefits than found in the Virunga region, where a national park has existed for over 50 years. Nevertheless, Virunga farmers are far less likely to perceive values of tourism than nonfarming or student elements in the population. The latter also stand alone in their recognition, albeit by a small minority, of scientific, conservation, and esthetic arguments for species preservation.

Roles are reversed with respect to water catchment benefits, as it is the local farmers who are more aware of the Virunga forest's beneficial contribution than the better educated but more ecologically removed residents of towns. Still, fewer than half of the farming population see the connection between the montane forest above them and their personal supply of water. This awareness appears to be even less widespread in the Bururi region.

Considering the socioeconomic context in which they live, along with their uses of and attitudes toward the forest environment, it is not surprising that more than half of the farmers living around the Virungas would like to see the park opened up for exploitation. Nonfarmers and the better educated support preservation to a much greater degree than do farmers, though level of education also correlates strongly with conservationist attiudes within the farming population. Age, exposure to radio, and recognition of nonconsumptive values of forest and wildlife preservation are also positively correlated with support for park protection. Comparable correlations appear in the preliminary Bururi survey results.

The awareness and attitudinal factors presented above indicate that a limited base of popular support for conservation does exist along the ZND. Still, almost every argument for preservation cited in the preceeding section—with the notable exception of tourism, which is discussed below—is either seriously devalued or simply dismissed at the level of the general population. This is particularly true for the peasant small-holders who live closest to the forest, who perceive significant personal opportunity costs associated with its continued protection and yet upon whom successful conservation is ultimately most dependent.

Tourism. The primary value of wildlife preservation, as perceived by all groups in the survey population is the potential to generate increased revenue through tourism. In this regard, residents of the ZND highlands are in line with many who feel that tourism offers great development potential for less developed countries (LDCs) like Rwanda and Burundi. Peters [1969] cites three principal ways in which LDCs can benefit from the development of tourism: 1) "contribution to the balance-of payments as an earner of hard currency," 2) "the creation of employment opportunities," and 3) "dispersion of development to nonindustrial regions." Evidence from many countries, including Rwanda, however, indicates that much of this classical protourism argument is open to criticism.

First, gross tourism receipts must be balanced by the cost, in foreign currencies, of tourism infrastructure development. For East Africa, Mitchell [1970] has calculated that this reduces the true benefits of tourism to between 5% and 10% of receipts. Data from Rwanda [AGCD, 1980] and the Caribbean [Bryden, 1973] indicate that even a negative balance can occur. With regard to employment, the 1,525 jobs generated by the Rwandan tourism industry as of 1979 represented just 1.2% of all salaried employment. Although this compares favorably with figures of 1.5% for Kenya and 0.8% for Senegal [AGCD, 1980], it is clear that tourism-related employment cannot take significant pressure off the agricultural sector.

Peters' final point, that tourism disperses development to nonindustrial areas, is almost a truism in regard to nature tourism, yet, even if tourists disperse to rural, undeveloped areas, are the economic benefits of their activity equally distributed? The answer is clearly no. Visitors to Africa tend to pay the greater part of their vacation costs to international airline carriers;

they spend considerable money in the capital city for lodging, meals, car rental, and supplies; and then they drive to the parks of their choice, where they pay a minimal entry fee and purchase virtually nothing else. Significant economic development around parks and reserves is unlikely to be stimulated by such low-level inputs.

Rural residents appear to be quite aware of this unequal distribution of tourism benefits. Whereas a large majority of those living around the Volcanoes Park perceive national benefits from tourism, many fewer feel that their region will benefit similarly. Only one out of every four Virunga farmers thinks that he or she will benefit personally in any way. This perception, which accurately reflects the economic reality, could further complicate support for conservation, as those being asked to give up the most are in turn those who perceive the fewest benefits of doing so. And yet few governments have thus far moved toward any form of revenue-sharing with the affected populations.

There is no escaping the fact that there are serious flaws in many of the arguments supporting tourism development. These do not, however, appear to be fatal to the concept. If infrastructure and import costs can be kept down, tourism can be a significant and steadily growing source of foreign revenue in less developed countries [Gearing et al., 1976]. In addition, the regional potential for tourism must be viewed against the lack of development alternatives along the ZND.

Poor, landlocked, devoid of major minerals, essentially agrarian and overpopulated, Rwanda and Burundi have limited possibilites for development. Options in the ZND highlands are even more restricted. Introduced crops such as tea, cinchona, pyrethrum, and potatoes, which spread at the expense of the lower montane forest zone throughout this century, have now encountered serious ecological barriers to further altitudinal expansion. Ultimately, there are also finite physical limits to just how much any form of agricultural exploitation can expand in the montane zone. Even if all the remaining forests could be cleared, cultivated, and settled at average regional density levels, they would provide land for barely more than 1 year's population growth. It is against this reality that continued calls for further forest conversion must be weighed and the value of tourism reconsidered.

Gearing et al. [1976] note that "a tourist industry may be the only available alternative for development in an area that has natural beauty" but little else. Others have argued the more specific case favoring tourism in tropical montane environments for both economic and ecologic reasons [Dasmann et al., 1973; Poore, 1976]. Such is the situation along the Zaire-Nile Divide. Tourism, to the extent that it can be developed, represents the best commercial use of the ZND forest parks and reserves. It is also the most ecologically sound development alternative in that it does not compete with other nonconsumptive values of conservation. At the same time, tourism should be approached on a site- and situation-specific basis, as it is not necessarily appropriate for all natural areas. Although all of the ZND reserves have some tourist potential, careful consideration should be given to the particular attributes and appeal of each before any major investments are made.

Certain conclusions can be drawn from the multiple issues treated in this section. First, it is clear that the ZND regional population is fast approaching, if not already past, a point where the available land base can no longer support it. Although most people are well aware of this situation, they are considerably less well aware or appreciative of forest conservation values. Forest products are valued; forest services are not. A partial exception concerns the large minority's recognition of water catchment benefits. Greater appreciation of nonconsumptive values is generally found among those with higher socioeconomic and educational status.

The one value of forest and wildlife preservation that is widely acknowledged is its potential for generating revenue through tourism. Although the true benefits of the latter can be overstated, it is nevertheless a fact that this potential is real and particularly attractive in a region like the ZND highlands, which has few development alternatives. If some of the parks and reserves concerned can at least partly pay for themselves through tourism, and do so in a manner that distributes benefits to the local as well as the national economy, then it will be much easier for government decision-makers to practice the politics of preservation. If benefits are marginal, slow in coming, not perceived at the local level, or not enhanced by awareness of other values, then policies of expediency may permit continued forest conversion.

INTEGRATED MODELS FOR CONSERVATION

Preserving the montane forest relics of the ZND is a complex task that requires immediate attention. Both governments concerned have given this matter very high priority, as have international donor organizations. Unfortunately, the interventions taken thus far have been of varied form and ecological soundness. The small Mukura reserve is already scheduled to be cut and replanted with exotic timber species, and more than one-hird of the Gishwati forest is to be cleared for a World Bank sylvopastoral scheme. In Burundi, French technical assistance is engaged in reforesting a buffer zone around the Kibira National Park but is paying little attention to its relation to either the natural forest or the surrounding inhabited area. Four of the more comprehensive and integrative ZND conservation efforts, some of which draw on the research and analysis presented above, are described below.

Mountain Gorilla Project

Continued poaching of gorillas and threats of further habitat losses from the Volcanoes National Park (VNP) led a consortium of conservation organizations to launch the Mountain Gorilla Project (MGP) in 1979 [Weber, 1979, 1981]. Its three-pronged strategy, to be carried out in cooperation with Rwandan authorities, was to improve park security, develop tourism centered on the gorillas, and establish a national conservation education program.

Nearly 7 years have passed since the inauguration of the MGP, and a brief review of its accomplishments is now possible. Whereas at least eight gorillas were killed by poachers in the 18 months prior to the inauguration of the project, only two killings have been confirmed in recent years since better equipped Rwandan guards began a program of systematic patrols under MGP supervision. In addition, a 1986 census found over 280 gorillas, the highest total and first increase since 1960 (Vedder, in preparation).

With regard to tourism, park visitation has increased from a few hundred visitors in 1978 to over 7,000 in 1985, due primarily to the habituation of several free-ranging gorilla groups to daily tourists' visits (Fig. 2). This has generated annual revenues of $200,000, which, in turn, has allowed the government to hire more guards, increasing both park security and local employment. Finally, indirect revenues associated with the dramatic rise in nonresident foreign visitors coming to see gorillas have totalled several million dollars over the life of MGP, and tourism now ranks as the country's third major source of foreign revenue [Monfort, personal communication; Siebrecht, 1981; PCBR, 1981].

The education component of the MGP has succeeded in disseminating written and audiovisual materials to every secondary school in the country. These initially concentrated on gorillas but have since been expanded to cover a broader range of conservation issues. A mobile education unit has also allowed presentations to be given to well over 100,000 Rwandans at schools and villages around the country. As a result, several schools have organized wildlife and nature clubs that visit natural areas, carry out local conservation projects, and circulate a newsletter. That the official reaction to all this has been positive is indicated by a recent government decision to increase the number and length of environmental education courses taught in all schools.

Bururi Forest Project

The Bururi Forest Project (BFP) differs from the MGP in several important respects. First, it is not centered on one spectacular animal, like the gorilla,

Fig. 2. Tourist visit to free-ranging gorillas in Rwanda's Parc National des Volcans.

that is capable of attracting considerable outside support. It must instead depend on a more local base of support, and for that reason humans have figured in the conservation equation from the beginning. In addition, the project is a joint venture between the U.S. Agency for International Development and the Nature Conservation Institute of Burundi.

The primary purposes of the BFP are to preserve one of the last remaining montane forests in Burundi and to develop new sources of fuel-wood and construction timber in the surrounding area. Based on the principal findings of a socioecological survey [Weber and Vedder, 1983b], the following recommendations have been made. First, its combination of biological attributes and water catchment functions makes the forest worthy of protection. Due to a lack of alternatives for several traditional forest uses, however, a multiple use management plan is called for. Within this framework, nondestructive activities such as trail use and the gathering of lianas, fruits and medicinal plants would be allowed; fallen fuel-wood could be collected for personal consumption until satisfactory alternative sources existed, but all cutting of live trees would be halted immediately. With regard to reforestation, native species have been proposed for recolonization of forest clearings. Where exotics are to be planted, more varied and flexible agroforestry approaches with an emphasis on fuel-wood species were favored over larger block plantations of predominantly commercial timber. Finally, proposals to initiate tourism and education programs modelled on the MGP were made.

The multifaceted approach of the BFP gives it considerable potential as a model for integrative approaches to conservation and development. In particular, it takes human interests into consideration from the start and seeks to satisfy these while also providing alternatives which complement long-term conservation goals.

Nyungwe Multiple-Use Management Plan

The Rwandan government is currently considering several multiple-use options for the Nyungwe forest based on recommendations by Swiss forestry advisors and IUCN ecologists [IUCN, 1983]. The final plan is intended to serve as a long-term, comprehensive management tool and is likely to include three principle elements.

First, it is recommended that at least 40% and up to half of the 970-km^2 reserve be fully protected as a natural forest wilderness. This would include the most biologically diverse sectors of lower montane forest as well as the entire ZND crest area. In addition, a small part of the existing forest, consisting of no more than 7,000 degraded hectares around the perimeter, would be opened to certain forestry operations. The remaining principal forest block would be closed to all but small-scale, controlled forms of exploitation for a period of roughly 15 years. This would allow research to be conducted in this area to determine the environmental effects of various management alternatives (J. Stebler, personal communication). The final component of the plan would concern the reforestation of the entire forest/ field edge. This would continue the extensive buffer zone already planted by a joint Swiss-Rwandan effort over the past 13 years and help to satisfy local demand for wood.

Cooperative Regional Demonstration Project

The most comprehensive and ambitious approach to conservation in the ZND region is the Cooperative Regional Demonstration Project (CRDP), currently proposed for the entire Virunga watershed in northern Rwanda [SECID, 1983]. Originally conceived as a multimillion dollar, multilateral approach to integrate conservation and development, the future of this project is now uncertain. It is nevertheless of interest to consider its major components.

The CRDP in Rwanda would cover an area of 1,687 km^2 that includes the Virunga forest and all the surrounding highlands that drain into the Mukungwa River system. Following an initial assessment of environmental problems and trends affecting the region's forests, wetlands, lakes, rivers, and various agricultural zones—along with the results of a sociological survey—priority actions would be determined. On-site demonstrations would then be developed that would show both the nature of the problems and possible means of reducing their effects. Wherever possible, cooperation with ongoing projects such as the MGP would be emphasized. The education component would be large, comprising the dissemination of both general conservation principles as well as more specific information such as agroforestry techniques and appropriate farming and energy technologies. And, in a parallel effort, the capacity of Rwandan institutions and personnel would be strengthened through training and experience to confront environmental problems on a continuing basis.

The CRDP approach has considerable potential. It complements the MGP conservation effort through the inclusion of the surrounding human environment in its overall ecosystem approach and is thus able to deal with issues of fuel-wood and water supplies as well as more general land use issues. At the national level, it can cooperate with ongoing efforts in education and even family planning and help to build and strengthen institutions concerned with conservation. Regionally, the model and methodology can be applied, with necessary modifications, to other ZND highland areas.

All the projects and proposals presented above represent creative, comprehensive attempts to deal with the problems of conserving montane forest reserves along the Zaire Nile Divide. All are based on multidisciplinary research or reconnaissance efforts, and all are to be carried out by multiple donors in cooperation with host governmental officials. Both conservation and development organizations are involved with all but the MGP, and the latter would be nicely complemented by the CRDP. Finally, each of the projects represents an attempt to consider the interests of local populations and to satisfy their needs in some way through the generation of tourism revenue and employment, provision of wood and water supplies, and/or permitting some continued traditional uses of the forest concerned.

At the same time, limitations of these approaches must be considered. First, although tourism has been successful in the Virungas, this could be primarily a function of the appeal of the mountain gorilla and the spectacular alpine scenery of the Virunga volcanoes. Extension of the model to other areas is of questionable potential.

Concerning reforestation, agroforestry approaches are more ecologically sound and serve to distribute wood resources better than large commercial plantations. The latter, however, are easier to manage and generate considerably more revenue than do decentralized on-farm plantations; important considerations for understaffed and financially strapped government bureaucracies.

Third, there is a lack of regional coordination with regard to conservation along the ZND. Research on the natural forest systems, appropriate agroforestry and plantation technologies, antierosion techniques, and more general agricultural issues should be coordinated throughout the highland zone of both countries. Tourism could be linked on a regional basis, as well, providing a broader base for development. Cooperation on management issues concerning contiguous reserves that overlap international boundaries, such as Nyungwe-Kibira and the trinational Virunga conservation area, also deserves greater attention.

The one project that could help to stimulate a more regionally oriented approach is the CRDP, and it has not been funded; this leads to a final point. Although most of the projects described above involve cooperation between conservation and development agencies, the former lack the necessary financial resources and the latter lack the necessary commitment to the concept of preserving natural areas. In addition, important elements within the development field still fail to see the essential complementarity of conservation and development interests. As a result, few projects or proposals are as yet sufficiently funded.

CONCLUSIONS

The afromontane forest reserves of the Zaire-Nile Divide represent a rich biological resource of considerable value to scientists and conservationists. Although sharing a number of common attributes, the individual forests are composed of a variety of plant communities, which in turn provide needed habitats for a surprising diversity of wildlife. Among the latter are more than a dozen species of primates, including the mountain gorilla and chimpanzee. In addition, the forests play important ecological roles, attracting and retaining rainfall for slow release to the surrounding human-occupied areas.

Despite these values, however, the regional conservation equation is complex and the future of the forests uncertain. This is partly a function of biological factors. The reserves are generally small and completely isolated from one another. Component species usually exist at low population levels and, without nearby sources of replenishment, are at high risk of extinction. It is nevertheless human factors that have caused this biological predicament and continue to place the ZND forest reserves in jeopardy. Demographic conditions in combination with a limited land base have resulted in ecological imbalances. These, in turn, have led to overexploitation of potentially renewable resources such as forests and wildlife.

Attitudinal factors are of critical importance in the overall conservation equation. Though not technically "forest people," the highland agriculturalists have lived on the forest edge for centuries, primarily moving it back but also exploiting it as a source of basic necessities. As a result, they continue to value forest products but generally ignore or devalue forest services. Their greatest difference with Western conservationists' views concerns values associated with research and species preservation, which are widely unrecognized at the local level. There is more of a convergence of values with regard to water catchment and regulation services, though even the latter are apparently not appreciated by significant proportions of the populations most affected by their benefits. It is perhaps a result of tourism's dual nature—part product, part service—that is the most commonly supported conservation value.

Although it is important to be aware of existing value conflicts, it is equally important not to overlook areas of value convergence. The findings of the ZND attitudinal surveys show that although there is no large popular base of support for conservation nor is there a monolithic mass opposed to it. The average highlands resident fits neither of two generalized behavioral paradigms. He is neither the "noble ecologist," completely aware of his place in nature and living in harmony with the surrounding environment, nor is he the "great destroyer," ecologically ignorant and blindly attempting to satisfy needs with no consideration of environmental impacts. Instead, an image emerges of a more heterogeneous population, with individuals aware of certain conservation values, particularly those of a more utilitarian nature, and indicating a potential for continued attitudinal change in response to socioeconomic, educational, and ecologic factors: in sum, a population not so radically different from our own.

Several projects have been initiated or proposed to deal with forest and wildlife conservation problems on the ZND. Whereas some seem ecologically or socially unsound, others represent creative attempts to resolve very complex conflicts. In the Virungas, tourism helps to provide an economic justification for preserving the last of the mountain gorillas. In Bururi and Nyungwe, reforestation efforts are intended to provide alternatives to continued exploitation of the natural forests, as the latter are given increased protected status. The larger CRDP would add concern for water and soil issues in addition to reforestation and tourism in the Virunga region.

All the projects place a common emphasis on education. Environmental education can not only serve to pass on those traditional conservative values already held by the culture, it can also provide the knowledge necessary to develop and maintain a new conservation ethic more relevant to people's current and changing relationship with their land and natural resources.

The research findings, integrative models, and general methodological approach described in this chapter can be applied, with certain reservations, to a broader range of biogeographic and socioecologic contexts. Conditions on the ZND are relevant to other sectors of the afromontane complex as well as to comparable highland areas of Latin America and Southeast Asia, where forest islands also survive in a sea of humanity. Applications of ZND approaches would appear to be least relevant to lowland tropical forest environments where larger areas, greater biological diversity, and significantly different human pressures combine to alter radically the conservation equation. Yet, even there, increased concern for human factors could help to avoid the forest fragmentation now common in the montane areas of the world.

No matter what the context, there are generalizations that derive from the common situation of people living in or around forests which they exploit to varying degrees. These permit the formulation of multifaceted strategies for conservation centered on the following propositions.

1. As biological research continues to increase our understanding of forests and wildlife, the role of humans as components in the forest ecosystem also demands greater attention. The interests of local populations must then be taken into serious consideration in the planning of any conservation program.

If these people are not involved nor perceive any personal benefits, they are unlikely to support a particular program and its failure is likely.

2. Conservation and development interests must be integrated in a complementary manner and the resources of organizations concerned with both brought to bear on the problems at hand.

3. Regional cooperation on conservation matters affecting particular ecological zones, such as the ZND highlands, would increase efficiency, reduce redundancy, and stretch the value of limited financial and technical resources.

4. Within a regional approach, planning should nevertheless be site- and situation-specific. As each forest has its own unique attributes and exists within a particular socioecologic context, so there are situation-specific strategies of varying degrees of protection and differing management objectives.

5. The capacities of national government institutions to deal with environmental conservation issues need to be strengthened through a combination of training and hands-on experience for personnel.

6. Education programs are needed to increase local populations' awareness and appreciation of forest conservation values, particularly those of a nonconsumptive or service nature.

Finally, education is needed in the other direction, too. Scientists, conservationists, and the general public in more developed countries need to better understand a few basic facts about tropical forest and wildlife conservation. First, although there are excellent reasons to save the ZND forests with their gorillas and chimpanzees, we must admit that for now these are predominantly our reasons. We are the ones who want to carry out research and we are the ones who want to experience tropical wildlife and wilderness. As these reasons—along with more relevant watershed arguments—are widely unrecognized or generally devalued at the local and national level, we are, in effect, asking some of the poorest nations in the world to export benefits of preservation to the wealthiest. Once this situation is recognized, then the problem of preservation becomes one of moving toward a more convergent set of values.

The general program outlined above should help to build a base of popular support for conservation among critically important local populations. At the same time, we must recognize the fundamental linkages between the latter's irreversible drive toward economic development and the current rate of forest conversion. If we put our money where our interests are and support both conservation and sustainable forms of development across the tropics, then there may yet be time for a more hopeful resolution of the present predicament. If we do not, then we should resign ourselves to the disapppearance of perhaps millions of species in our lifetimes, and the more spectacular creatures such as the mountain gorilla among the first to go.

REFERENCES

AGCD (1980): "Evaluation du Projet Belgo-Rwandais: Tourisme et Parcs Nationaux." Brussels: Administration Generale de la Cooperation au Developpement.

AIDR (1;981): "Evolution Entre 1958 et 1979 du Couvert Forestier." Brussels: Geomines-Somirwa.

Brunsfeld SJ, et al. (1982): "Biomass Resources in 66 African, Asian and Latin American Countiries, vol.1: Africa." Boise: University of Idaho.

Bryden JM (1973): "Tourism and Development."London: Cambridge University Press.

Curtin P, Feierman L, Thompson L, Vansina J (1978): "African History." Boston: Little, Brown and Company.

Dasmann RF, Milton JP, Freeman PH (1973): "Ecological Principles for Ecological Development." New York: Wiley & Sons.

Delepierre G (1982): Les regions agro-climatiques en relation avec l'intensite de l'erosion du sol. Bull Ag Rwanda 15:(2):87–96.

Diamond AW (1981): Reserves as oceanic islands: Lessons for conserving some East African montane forests. African J Ecol 19:21–26.

Diamond JM (1976): Island biogeography and conservation: strategy and limitations. Science 193:1027–1029.

Ehrlich P, Ehrlich A (1981): "Extinction." New York: Random House.

Gearing CE, Swart WW, Turgut V (1976): "Planning for Tourism Development." New York: Praeger.

Hamilton AC (1982): "Environmental History of East Africa." London: Academic Press.

Hedberg O (1951): Vegetation belts of the East African mountains. Svensk Bot Tidskr 45:140–202.

IUCN (1980): "World Conservation Strategy." Gland, Switzerland: International Union for the Conservation of Nature and Natural Resources

IUCN (1983): "Conservation et Amenagement des Forets Naturelles de la Crete Zaire-Nil au Rwanda." Gland, Switzerland: International Union for the Conservation of Nature and Natural Resources.

Kabayanda A (1982): "L'Institut National pour la Conservation de la Nautre: Le Passe, le Present, l'Avenir." Bujumbura: INCN.

Lewalle J (1972): Les etages de vegetation du Burundi occidental. Bull Jard Bot Nat Belg 42:1–247.

Louis W (1962): "Ruanda-Urundi: 1894–1919." Oxford: Clarendon Press.

MacArthur RH, Wilson EO (1967): "The Theory of Island Biogeography." Princeton: Princeton University Press.

McCulloch JSG, Dagg M (1965): Hydrological aspects of protection forestry in East Africa. East African Agricultural Forestry J 30:390–394.

Mertens A (1981): "Contribution a la Connaissance des Regions Naturelles du Burundi, No. 5." Bujumbura: Institut des Sciences Agronomiques du Burundi (ISABU).

Mitchell F (1970): The value of tourism in East AFrica. East Africa Economic Rev 1970:1–21.

Morris WHM (1979): "A Report on Agricultural Production, Marketing and Crop Storage in Rwanda" Kigali: U.S. Agency for International Development.

Morrison MES (1968): Vegetation and climate in the uplands of south-western Uganda during the later Pliestocene period. J Ecology 56:363–384.

Myers N (1979): "The Sinking Ark." Oxford: Pergamon Press.

Myers N (1980): "Conversion of Tropical Moist Forests." Washington: National Academy of Sciences.

Myers N (1983): "A Wealth of Wild Species." Boulder: Westview Press.

NRC (1982): "Ecological Aspects of Development in the Humid Tropics." National Research Council, Washington, D.C.: National Academy Press.

PCRB (1981): "Tourisme et Protection de la Nature." Brussels: Project de Cooperation Rwando-Belge.

Poore D (1976): "Ecological Guidelines for Development in Tropical Rain Forests." Morges: International Union for the Conservation of Nature and Natural Resources.

Prefol B, Delepierre G (1973): Disponibilite et Utilisation des Terres au Rwanda. Butare: Institut des Sciences Agronomiques au Rwanda.

RAPID (1981): "Rwanda: Les Effets des Facteurs Demographiques sur le Developpement Social et Economique." Washington, D.C: The Futures Group.

RAPID (1982): "Burundi: Les Effets des Facteurs Demographiques sur le Developpement Social et Economique." Washington, DC: The Futures Group.

Scaetta H (1932): "Les Famines Periodiques dans le Rwanda." Brussels: Institut Royal Colonial Belge.

SECID (1983): "A Proposal for a Cooperative Regional Demonstration Project (CRDP) in the Ruhengeri Prefecture of Rwanda." Chapel Hill: South-East Consortium for International Development.

Siebrecht D (1981): Presentation et interpretation des characteristiques du tourisme au Rwanda. Kigali: Office Rwandais du Tourisme et des Parcs Nationaux.

Simberloff DS, Abele LG (1975): Island biogeography theory and conservation practice. Science 191:285–286.

Spinage CA (1972): The ecology and problems of the Volcanoes National Park, Rwanda. Biological Conservation 4(3):194–204.

Storz M (1982): "La Foret Naturelle de Nyungwe et sa Faune." Kigali: Ministere de l'Agriculture, Cooperation Technique Suisse.

Vande weghe JP (in preparation): L'avifaune de la foret Nyungwe, Rwanda.

Weber AW (1979): Gorilla problems in Rwanda. Swara 2:28–32.

Weber AW (1981): "Conservation of the Virunga Gorillas: A Socioeconomic Perspective on Habitat and Wildlife Preservation in Rwanda." Madison: MS thesis, Univeristy of Wisconsin.

Weber AW, Vedder AL (1983a): Population dynamics of the Virunga gorillas: 1959–1978. Biological Conservation 26:341–366.

Weber AW, Vedder AL (1983b): "Socio-ecological Survey of the Bururi Forest Project Area." Bujumbura: U.S. Agency for International Development.

White F (1978): The Afromontane region. In "Biogeography and Ecology of Southern Africa. Werger MJA (ed): The Hague: Dr. W. Junk bv Publishers, ch 11, pp 465–513.

White F (1981): The history of the Afromontane archipelago and the scientific need for its conservation. African Ecology 19:33–54.

Primate Conservation in the Tropical Rain Forest, pages 231–262

9
Environmental Education in Developing Countries

Rosalind J. Aveling

Frankfurt Zoological Society, D-6000 Frankfurt/Main-1, Federal Republic of Germany, and Zaire Gorilla Conservation Project, B.P. 106, Goma, Kivu, Zaire

> Education, I think is the only answer—if there really is an answer. Governments educated to be responsible to the future, citizens educated to show more respect for other living things. It sounds like a dream, perhaps, but I think we have to try.
> —Lekagul [1979].

INTRODUCTION

In presenting the 1979 Getty Wildlife Conservation Prize to Thai conservationist Dr. Boonsong Lekagul, S. Dillon Ripley described him as "a voice crying in the wilderness." In 1947, Dr. Lekagul founded the region's first wildlife conservation society in Thailand and since then had campaigned against environmental degradation in Thailand, a degradation that was escalating throughout the developing world. Over 30 years later, he reiterated a conviction that conservation cannot succeed without *education*. People cannot be expected to conserve without understanding both the nature of their environment and their dependence on it.

The *World Conservation Strategy* produced by IUCN/WWF/UNEP in 1980 attempted to pull together the basic tenets of environmental conservation and to indicate how individual national governments could incorporate a longer-term approach to the use of natural resources into their development plans. The strategy emphasizes that environmental education is an essential part of this process with the rather all-embracing statement that, "ultimately the behaviour of entire societies towards the biosphere must be transformed if the achievement of conservation objectives is to be assured" [World Conservation Strategy, 1980].

But what is environmental education? Walton and Wood [1982] point out that several authors distinguish between "environmental education" and "conservation education," using the former to describe programs that aim at long-term attitude formation with regard to the natural environment and the latter for those aimed at immediate solutions to specific environmental problems. They propose a definition of environmental education that incorporates both these concepts: ". . . any type of education that is designed to lead to an improvement in the management of natural resources." Walton and Wood further distinguish environmental education from "technology transfer," which they suggest more accurately reflects training of natural resource specialists, who then themselves undertake environmental education. This author, however, includes such technology transfer within the above definition, arguing that the difference is only in the level of education being considered. Increasing the knowledge of a natural resource specialist can in itself lead directly to an improvement in the management of natural resources.

Taking the above as a working definition, is it possible to make any generalizations about how environmental education should be approached in developing countries? Differences between countries of Central and Southern America, Africa, and Asia are involved; their different cultures, political outlooks, degrees of development, economies, sizes, and resources are just a few of the factors that affect not only their environmental problems but also their receptivity to different approaches to environmental education.

One generalization identified by Lieberman and Walton-Wood [1982] is that it is first necessary to identify the conservation problem, be it forest degradation, species rarity, soil loss, or whatever. Then, in looking for solutions to the problem, environmental education can only be considered as one of the available tools. Questions such as the following should be posed.

1. What is the conservation problem?
2. Who is causing it?
3. Are there alternatives to which those causing the problem could turn?
4. If so, could environmental education in any form assist in turning them towards those alternatives?

If environmental education is then pinpointed as one of the methods to use in seeking a solution, the following procedure can determine a suitable approach.

1. Identification of relevant target audiences
2. Assessment of relevant cultural and social aspects
3. Development of appropriate educational content
4. Accurate selection of educational methods

5. Evaluation

Lieberman [1984] has summarized this process in chart form (Fig. 1).

However, in reviewing environmental education projects that have taken place in developing countries, it becomes apparent that this ideal approach is more the exception than the rule. Some programs have evolved in response to a need perceived by governments, organizations, or communities, such as the establishment of regional training colleges for wildlife managers in West and East Africa and Southeast Asia. Others have emerged through, or been catalyzed by, individuals taking advantage of certain situations, such as the promotion of wildlife clubs by teachers or volunteers working with youth groups. Very few of these programs have been evaluated at all, and, where they have, it is often the subjective evaluation of the people involved, which makes comparative assessment impossible.

In undertaking this review, it became clear that sifting out a few examples of successful education programs that could be described in detail and used as models by those concerned with primate conservation was impossible. The author has therefore not attempted a comprehensive review and comparative assessment. Instead, a variety of approaches to environmental education are outlined in the first section from a range of developing countries. This should serve to emphasize the one central factor—that it is important to keep an open mind when approaching environmental education, always considering what one is trying to achieve. What may seem an obvious or attractive approach, which may have been successful in one country, could be a total waste of time given the cultural or political conditions of another.

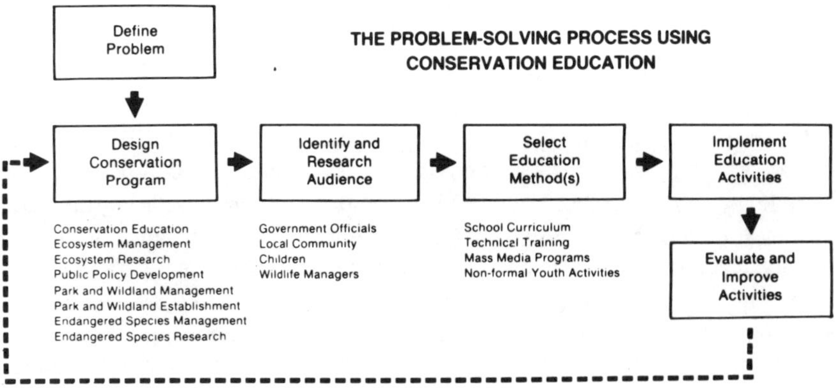

Fig. 1. The problem-solving process using conservation education. From Lieberman [1984], with permission (drawn by Nancy F. Sanford).

The author has not confined the examples given to those directly related to primate conservation, but these have been included wherever possible in the first section. The second section looks at some of the organizations associated with this field, to assist a program planner in orientation with respect to the national and international bodies concerned with environmental education. Finally, in the Discussion, the author returns to the question of program design and discusses in more detail factors that anyone approaching environmental education should consider.

HOW IS ENVIRONMENTAL EDUCATION APPROACHED IN DEVELOPING COUNTRIES?

Approaches to environmental education in developing countries have been through both formal and informal means. Formal education applies to the established educational system—through *primary and secondary schools* as well as *universities and training colleges*. Informal approaches cover a wide range, here subdivided into *public awareness campaigns, clubs,* and *foundations.* A separate category is proposed for education related to protected areas, in particular *education centers* and *training of personnel.*

Formal Education—Primary and Secondary Schools

If environmental education is built into a formal education system it can have a longer-term effect than more informal approaches. Children are exposed to a program over successive years, and they are the people who will be making the decisions and choices about environmental use in years to come. A basic education in biology provides a substrate on which interest in conservation and the environment can grow. Unfortunately, in some developing countries biology is not treated as a priority subject in the formal education system.

In Thailand, Lekagul (personal communication) points to overcrowded curricula as a reason for lack of environmental topics. Lekagul believes that a tendency to make school courses too complicated inhibits the introduction of new topics such as conservation, even though the deficiencies of applied science teaching in schools have been recognized by the administration.

In Paraguay (Darr, personal communication), U.S. Peace Corps Volunteers working with the Ministries of Agriculture and Education produced information on environmental topics as guidelines for teachers. Some difficulties were experienced in getting their use sanctioned as a formal course of study, and, initially at least, it was left to the teachers to sandwich the information and concepts in between conventional subjects.

In francophone East and Central Africa, a relative lack of tradition in biology teaching permeates the educational system, resulting in few trained biology teachers to feed back into primary and secondary schools. In Rwanda, however, a complete revision of school curricula in the early 1980s has provided the opportunity for considerable input of environmental conservation concepts.

By contrast, in anglophone East Africa there is an established tradition of biology teaching, although in some areas political and economic problems of recent years have reduced the availability of good teachers and materials. Curricula need to be revised to reflect the current ecological situation; forests are often mentioned in old textbooks only as a widespread and plentiful resource. A study by Connor [1982] pinpointed the need for more materials, and one response has been the collaboration of the African Wildlife Foundation and CARE in producing environmental issues ("Rainbow Environmental Special") of a children's magazine. These have now been distributed to 17,500 primary schools in Kenya and Uganda, reaching an estimated 2,700,000 upper primary school children (Hetz, personal communication). Although not part of a formal course of study, this novel method of getting the message across takes advantage of the formal education system to reach that age-group. This project is undergoing evaluation and hopes to consolidate the effects of the Rainbow Environmental Specials by strengthening contacts with local educational institutions, such as the Institutes of Education in Kenya and Uganda. The aim is to facilitate the use of the supplement within teaching schedules and multiply its effect at little extra cost.

Formal Education—Universities and Training Colleges

It can be argued that the quickest return on investment in environmental education can be achieved by concentrating on input at the university level. It is usually the graduates from national universities who feed into the upper echelons of administration and can have considerable influence on critical decisions affecting the environment. In many developing countries, field biologists are a rarity; field studies are not widely encouraged nor widely sought by the small elite of graduate scientists. This is unfortunate, as it is this first-hand experience of the problems of conservation that can turn graduate scientists into active conservationists.

This is partly the rationale behind the Primate Ecology course integrated into the Masters Degree program of Mahidol University in Thailand [Prempree and Baimai, 1978]. The program encourages Thai graduates to undertake further studies in primate biology, with an emphasis on field work and conservation of remaining natural habitats. The approach involves collaboration both within Thailand (for example, with the Royal Thai Forest Department, to initiate primate management and conservation programs) and internationally (both to give Thai students the opportunity to study abroad and to attract funds and personnel to improve possibilities for primate study and conservation in Thailand).

Also at Mahidol University, an MSc program in Technology of Environmental Management has, since 1974, taken a multidisciplinary approach to the training of environmental managers. Graduates retain their original specialization, as engineers, economists, educators, etc., but a broad-based course of 2 years enables them to integrate environmental and resource ideas into their professional work [Tuntawiroon, 1982]. According to Tuntawiroon, this program was established from the bottom up, growing out of an environmental movement that had its roots in student nature clubs and other such groups, and was initially established by popular demand, against some reluctance from the academic estabishment. Apparently, this initial reluctance has been reversed, and both subject and approach have become fashionable. Tuntawiroon considers that the impact of the course is much greater than the numbers involved indicate. Although only 15–20 students graduate annually, they feed into the top of the hierarchical system and have considerable influence on the planning and management of natural resources in Thailand.

In Kenya, an MSc program in the Biology of Conservation was started in 1965 at the University of Nairobi, and through this African ecologists from Kenya and other countries have received training at the graduate level in subjects directly related to wildlife conservation and management. Since 1981, an experimental postgraduate program in environmental studies has been under development at the University of Nairobi. The program takes an interdisciplinary approach to environmental studies, combining the interests of departments such as Economics, Sociology, Geography, and Urban and Regional Planning [Marwick, 1983].

A novel development in Indonesia has been the establishment of "Environmental Study Centers" (Pusat Studi Lingkungan or "PSLs") attached to 17 universities around the country, an initiative of Indonesia's Minister for the Environment and Development Supervision. The task of PSLs is to generate qualified manpower in the field of environmental management and at the same time to monitor local development [Pattinasarany et al., 1980]. Each PSL is assigned a particular field of study related to the dominant environmental features in its area. Although it is difficult to assess their effectiveness, it is encouraging that the PSLs are sited throughout the country and have a brief to become directly involved with local conservation issues. This is particularly important in a country like Indonesia, where the outlying islands tend to suffer as a result of central government and administration being at some remove on the island of Java.

Informal Education—Public Awareness Campaigns

> First we must educate the adults. Our children will not have an impact on our economic development for 5, 10, or even 20 years. The attitudes of the adults, on the other hand, have an impact now
>
> —President Nyerere of Tanzania [Mangubuli, 1983].

President Nyerere's statement reflects a view of educational priorities that can apply equally to environmental education. It presupposes, however, that adult education can permanently change both attitudes and behavior and also does not recognize that adults are often reached quite effectively through education of their children. Most public awareness campaigns are aimed at adults—both at the general population and more specifically at community leaders—but there is a widespread view that their effect on attitudes can be temporary. Their success can lie more in publicizing and gaining support for specific local issues and in influencing decisions directly.

Some sort of attempt to increase public awareness is often the first, and even more often the only, education component of many conservation projects. This may range from dissemination of leaflets or posters to audiovisual presentations and discussions in local communities and use of the mass media.

National government departments concerned with protected areas often produce posters and leaflets, usually connected with specific reserves or endangered species. Some have undertaken more extensive campaigns, sometimes with financial and technical input from outside groups. For example, the Botswana Department of Wildlife, National Parks, and Tourism has for some years had an Education Section, established in the early 1970s with the aid of U.S. Peace Corps volunteers and various conservation groups. The education unit tackled the problem of getting conservation topics to the general community through radio programs, newspaper and magazine articles, and purpose-made films [Halloran, 1975]. To reach sparsely distributed rural populations, they capitalized on seasonal congregations around certain large villages during the winter months and transported a display about game law enforcement, wildlife education, and wildlife research to regional agricultural shows. Another approach was to contribute reading cards on wildlife-related topics for a program of the Botswana Extension College in 1974 to help increase English Comprehension, a program that was directed at all segments of the population.

The Mountain Gorilla Project (MGP) in Rwanda, run by several conservation organizations working with the government's department of National Parks and Tourism, considered environmental education an integral part of its program from its inception in 1978 along with concurrent emphasis on park protection and park development [Weber, 1979]. Initially, the education component took the form of a public awareness campaign, particularly in the region around the park that is the mountain gorilla's only habitat in Rwanda. Using posters and audiovisual material and equipment, MGP personnel and park officials saturated the area around the park with publicity about the gorillas, their conservation, and the ecological/economic importance of the park to people (Fig. 2). In a small country like Rwanda, it was thought feasible to have a deeper educational impact, and not only was the public awareness campaign extended to the whole country but, with the involvement of the Ministry of Education, also into the formal education system. Studies by Weber (personal communication), who started the Rwanda program, provide some of the few data available on the effectiveness of public awareness campaigns. In 1984, he conducted a follow-up survey in the area around the park. This indicated that the local farming population around the park had become more aware of the values of park preservation, with the eco-

Fig. 2. Children in Rwanda, in central Africa, enjoy an outdoor audiovisual presentation about conservation of mountain gorillas and their habitat. Photo: Rosalind J. Aveling.

nomic benefits of tourism and the ecologic benefits of forest protection most commonly cited. A minority who still saw no value in preservation was composed mostly of the elderly, the less educated, and those unexposed to any MGP education effort. Whereas in 1980 52% believed that the park should be opened for exploitation, by 1984 only approximately 25–35% still expressed this belief. However, the region currently supports 700 people per km^2 and that population pressure is increasing. As Weber rightly points out, there is no guarantee that the observed attitudinal changes will translate into behavioral changes when people are faced with increased demand for firewood and farmland.

Turning to Southeast Asia, there are also examples of public awareness campaigns initiated and carried out by nongovernmental organizations (NGOs). Sometimes these are emergency reactions to critical conservation situations; for example, Malaysian NGOs alerted the general public to illegal logging in the Endau-Rompin reserve, which was causing the rapid destruction of one of the few remaining habitats of the Sumatran rhinoceros. After a massive press campaign, sufficient public pressure was brought to bear to get high-level action to stop the logging, at least temporarily.

A Sri Lankan Society called "March for Conservation," which started by organizing public processions to attract interest in conservation, has expanded its activities to include lecture series, seminars and workshops, wildlife film festivals, radio and press publicity, and specific research and conservation projects [Jansen and Raheem, 1981].

In South America, conservation education campaigns using endangered primates as their foci have been launched in Brazil (with the muriqui [*Brachyteles arachnoides*] Fig. 3) and Peru (with the yellow-tailed woolly monkey [*Lagothrix flavicauda*]), largely promoted by national and foreign scientists studying the endangered primates concerned. These campaigns have concentrated on increasing public awareness using audiovisual materials and media publicity to stimulate interest in conservation of the primates and their forest habitats [Mittermeier et al., 1984].

In some instances, NGOs are asked to provide expertise on a consultancy basis for education programs promulgated by government or international institutions. In 1980, the Green Indonesia Foundation (YIH) was asked to plan and undertake a 2-year education program as part of the implementation of proposals for the Gunung Leuser National Park in northern Sumatra. YIH decided to concentrate on a critical area of conflict between local communities and the integrity of the National Park—the Alas valley, which almost separates the park into two sectors. They planned a multifaceted approach, particularly aimed at illegal cutters, nonformal leaders in society, civil ser-

Fig. 3. Carlos Alberto Machado Pinto, a member of the joint WWF-US Brazilian team working to save the muriqui, distributes posters in the state of Minas Gerais, Brazil. Photo: Russell A. Mittermeier.

vants of all types who were concerned with local communities, and youth. Two important aspects were an emphasis on working closely with other NGOs in carrying out the program (those concerned with social welfare and family planning, for example), and the inclusion of a socioeconomic survey to assess possible alternative sources of income in the valley (Frey, personal communication).

Informal Education—Clubs

An encouraging development in recent years has been the phenomenon of "wildlife clubs," groups of schoolchildren and students who have formed themselves into clubs to extend and share an interest in their environment.

Wildlife Clubs of Kenya (WCK) developed from an initial gathering of 12 clubs and interested patrons in December, 1968, into a national network of 1,060 registered clubs with over 60,000 members and associate members in 1984. A national headquarters in Nairobi, run by a small Kenyan staff, coordinates activities and provides back-up services to the countrywide clubs. In view of their apparent success, it is worth taking a more detailed look at how WCK evolved and at the activities and approaches they take and to try and assess how effective they are in influencing attitudes to conservation in Kenya.

The WCK Association was the first conservation education program of its kind on the continent of Africa and takes pride in the fact that it came about through the interest and enthusiasm shown by Kenyan youth themselves [Chumo, 1982]. However, organizing and developing that enthusiasm in the initial years depended on considerable input of funds and personnel from other organizations. Initial coordination by the National Museums of Kenya's Education Section was taken on in 1969 by the first permanent staff member, Sandra Price. Price [1978] stresses the importance of a "personalized" approach, each member participating directly in his club. Each club pays a small membership fee to the national organization. A 50-page wildlife magazine is produced by the national organization for the clubs three times each year, and half of each issue is devoted to material submitted by members. The WCK education officer also makes personal visits to as many clubs as possible each year.

Most of the club activities are centered around the immediate environment of the school or college. They include monitoring of local habitats, collections for club museums, and a variety of activities to raise funds to provide transport for club members to visit national parks. All club members are admitted free to national parks, and WCK will loan them camping equipment and binoculars. WCK also has hostels at Nakuru National Park and near Nairobi National Park, which students can visit for short courses. The headquarters staff organizes workshops for club leaders and advisors.

"Learn to Conserve for a Better Tomorrow" is the motto of WCK, but their role in conservation is not a passive one. In 1974, a thousand WCK members marched down Nairobi streets with placards to support the government in stepped-up antipoaching measures. They have petitioned the president, marched publicly to protest other wildlife issues, and lobbied for changes in the laws. Moreover, ex-WCK members often maintain their interest after leaving school or college, keeping in touch through associate membership, and continue to act as part of an informed pressure group in society. There is a marked trend towards wildlife-related careers being filled by former WCK members, and most Kenyan MSc students enrolled in the Biology of Conservation course at the University of Nairobi attribute their initial interest in studying wildlife to earlier participation in school wildlife clubs.

WCK have also had a ripple effect, contributing to the formation of wildlife club organizations in other African countries—Uganda, Sudan, Zambia, Cameroon, Rwanda, Malawi, and Tanzania—many of which are set up in a similar way and undertake the same sort of activities (Fig. 4). Are these likely to enjoy equal success? In assessing this, it is helpful to examine some of the factors influential in the early years of WCK. 1) From the beginning, the interest and enthusiasm shown by Kenya's youth in forming themselves into clubs was channeled through a national organization, backed up by the National Museums of Kenya. Although initially only one person, the national headquarters played a vital role in maintaining momentum, providing support to the clubs, and lobbying for more funds and facilities. 2) Although not part of the formal education system, WCK took advantage of a well established network of secondary schools and colleges within which there was already considerable emphasis on teaching of the natural sciences. 3) Kenya was also fortunate in having a history of conservation efforts, a network of National Parks, and several active conservation organizations based in the country. The spread of national parks meant that it was possible for most clubs to envisage field trips, and conservation groups helped in developing facilities for the embryo clubs.

Obviously Kenya had a substantial base from which to build wildlife clubs, and some of these factors are not apparent elsewhere. New club organizations need dedicated personnel, transport and communication facilities, sufficient

Fig. 4. Members of Wildlife Clubs of Uganda demonstrating against poaching at a World Environment Day Rally, 1984. Photo: Wildlife Clubs of Uganda.

funds and support—from both NGOs and government—and a certain measure of political stability. WCK was not an overnight success, and support must be maintained through a period of years to get a club's organization well established.

Of course, Africa is not the only continent with wildlife clubs. Similar groups exist elsewhere, such as the Nature Clubs of India. This movement was started in 1976 by WWF–India and now comprises over 1,000 clubs in 22 of 25 states [Wakankar, 1984]. WWF–India produces quarterly newsletters for the clubs with national and international conservation news, promotes practical club projects specifically adapted to Indian conditions, and organizes nature camps for students in areas with different ecological problems. An annual camp held for club organizers at the National Defence Academy also exposes cadet officers for all the armed forces to conservation education. In Suriname, STINASU, the Foundation for Nature Preservation, supervises the "Suriname Wildlife Rangers Club," whose members can attend study camps and undertake other activities during school vacations. Various scattered "nature lover" groups already exist in Indonesia, and the Green Indonesia Foundation is trying to get them into contact with each other to provide the basis for a national organization.

Informal Education—Foundations

Recent years have seen the emergence of "nature foundations" in countries of the developing world. Mostly, these are nongovernmental groups whose broad aim is to promote a greater awareness of the environment and its protection in countries where such awareness is minimal. In general, these foundations seem to be undertaking an essential catalyctic role in stimulating and supplementing the effectiveness of often low-status government conservation departments.

STINASU—Stichting Natuurbehoud Suriname (Foundation for Nature Preservation in Suriname). One of the older foundations, STINASU was established in 1969. According to Education Officer Gisla Sommer (personal communication), the foundation works in close cooperation with the Suriname State Forest Service to realize three main aims: 1) to make the public aware of environmental problems, 2) to make hunting rules accepted, and 3) to create facilities to use the existing nature preserves for research, study, leisure, and education.

A tourism program contributes to the foundation's income, but the research and education programs also depend on grants from outside organizations. The education department, set up in 1976, undertakes both public awareness campaigns (media publicity and film shows in the interior) and work with schoolchildren (lectures and audiovisual presentations in schools at every level combined with production of literature and posters).

Fundación Natura (Nature Foundation of Ecuador). Fundación Natura is described by executive director Yolanda Kakabadse (personal communication) as a private, nonprofit organization dedicated to the conservation of nature. Created in July, 1978, the foundation undertakes contracts for different organizations to supplement government conservation efforts. It is funded both through these contracts and through selling sponsorship of nature films to private enterprise. Fundación Natura started an educational program—EDUNAT I, funded by U.S. A.I.D.—in September, 1980, which concentrated initially on assessing the country's most important ecological problems and then getting these across to the public (through media publicity) and leaders (through roundtable disussions and conferences). Fundación Natura tackled the problem of how to get its message over objectively, undertaking at an early stage a study of communication channels within the major government departments and associated organizations and then using this information to adjust its own communication methods accordingly. The second phase of Fundación Natura's educational program—Edunat II—places a greater emphasis on formal education (curriculum modification and development).

SLENCA (Sierra Leone Nature Conservation Association). SLENCA is the only nongovernmental conservation organization in Sierra Leone. The honorary secretary, Daphne Tuboku-Metzger [1979, and personal communication], reports that dissatisfaction with the lack of progress of conservation in the country was the main reason behind its establishment in December, 1976. Efforts to generate public awareness of conservation through media campaigns (radio, film shows, and newspaper articles) brought a favorable public response, but SLENCA realized that little progress would be made while the authorities remained apathetic to conservation. The association made direct presentations to the relevant government departments on conservation issues, backing them up with hard-hitting press coverage. One issue, the extensive and illegal traffic in juvenile chimpanzees, they took to the President of Sierra Leone, using the occasion to draw his attention to other crucial conservation matters. Apart from continuing such pressure, the association is undertaking an environmental awareness program in schools and villages.

Yayasan Indonesia Hijau (YIH)—Green Indonesia Foundation. YIH, a nongovernmental organization established in January, 1978, is dedicated to nature conservation education and generation of public awareness. Its immediate aims [Pattinasarany et al., 1980] are: 1) production and effective dissemination of educational and public awareness material and 2) close cooperation with youth and nature-lover groups in order to establish a nonformal network of conservation-conscious groups that could act as an environmental monitoring system as well as a conservation lobby.

YIH seeks some funding directly from organizations and companies for specific projects, such as its bimonthly magazine *Suara Alam* (*Voice of Nature*). 2,500 copies of *Suara Alam* are distributed free to high schools, teachers colleges, university and public libraries, and youth/nature-lover groups throughout Indonesia. Sales, subscriptions, and advertisements allay production costs. YIH also undertakes environmental education programs around some of Indonesia's protected areas, tackling the problem of illegal forest exploitation through a combination of education and promotion of alternative sources of income.

YIH differs from the other foundations described in two main respects; firstly, it is not the only NGO concerned with conservation in the country, although it is the only one concerned specifically with environmental education at the grass-roots level throughout Indonesia. Secondly, there are additional problems associated with a country comprising 1,000 islands and 700 ethnic groups. Problems of distance, communication, and stretching of meager resources add to the difficulties of national campaigns and of undertaking local projects in regions remote from the national headquarters.

However, the experience of the YIH does illustrate well problems also encountered by other foundations in trying to achieve their goals, in particular *personnel* and *funding*.

Personnel. Most foundations are attempting to counteract a lack of public awareness of environmental issues at all levels. In general, a lack of public awareness can be partially attributed to a formal education system that places low emphasis on ecology and the environment. This situation results in few qualified teachers and even fewer graduate scientists to feed back into the system. The few who do result tend to form the core of an organization like YIH—a small group of dedicated people. However, the general lack of qualified personnel (and where there are so few there is often competition for their services) also acts as a brake on what the organization can achieve. To undertake some projects, YIH has had to train personnel itself—a difficult and time-consuming task detracting from the other work of the organization. The only solution to this dilemma is in the long term and depends on the success of attempts to integrate environmental education fully into all levels of the formal education system.

Funding. Whether a foundation concentrates on obtaining funds from the public at large or through large businesses or conservation organizations, fund-raising takes time. With the shortage of personnel described above, a foundation may have to sacrifice valuable manpower to fund-raising, taking people away from practical projects. This can be a double disadvantage, as the skills required for the two activities are different, and are not always found in the same people. This highlights a conflict that can arise particularly in the vital formative stages of a new foundation: A certain minimum of funds is necessary to achieve good results, which in turn are essential to encourage and attract donors. The organization always has to find a compromise between spending too little on fund-raising, resulting in too few funds to be effective, and too much effort, which can absorb the working capacity and cause the main practical tasks to be neglected.

Such administrative conflicts can be a brake on the effectiveness of non-governmental foundations. Counterbalancing this, however, is their independence and conviction, which give them a freedom from complacency and an enthusiasm often lacking in the more secure government institutions. The foundations also compare favorably with other conservation groups in their approach to environmental education. Springing from the culture of the country itself, they often come up with appropriate culture-based methods of getting the message across. For example, use of traditional drama by YIH to attract attention to conservation topics stimulated more interest and discussion than many of their more modern approaches.

Education Related to Protected Areas—Education Centers

Many protected areas in the developed world feature visitor centers, with interpretive displays to increase the area's educational value. These can be supplemented by guided walks to demonstrate distinctive natural features and are often used by parties of schoolchildren and students. Similar facilities are appearing in protected areas of developing countries but in some cases are aimed more at the foreign tourists than at national visitors and function more as status exhibitions than educational centers. However, where protected areas are accessible (in terms of distance and travel costs especially) to national visitors, they can be invaluable in attracting local interest, which can then be developed by appropriate education facilities.

Chitwan National Park in Nepal boasts an informative visitors' center at the park headquarters. The exhibits describe the history, geology, and ecology of the park and its flora and fauna; several are "activity exhibits" manipulated by the observer to match up questions and answers. However, this is a prime example of a park used mostly by foreign visitors, and very few Nepalese benefit from either the park or its visitors' center. With captions in both English and Nepali it has education potential, but this is likely to be limited to Nepali administrators on official visits and not be available to a wider audience.

By contrast, the Nature Study Center in Malaysia's Taman Negara National Park was purpose-built for short study courses by Malaysian teachers and students. The initiative of an Australian volunteer in the Forestry Department, its budget for construction was provided by the Wildlife and National Parks Department, and funds to run the courses were raised by the children themselves, often with help from local companies. The facility was small (for less than 100 people) and was constructed of natural materials. Weekly courses during school holidays were run with the cooperation of the park staff, with emphasis not directly on conversation but rather on increasing knowledge of the forest and thus overcoming fears. Course organizers (Rubelli, personal communication) found that it was far easier to catch the interest and imagination of the children than their teachers, although the cooperation of the latter was important. Although this Nature Study Center was well received, sustaining its operation, and proposals to extend the idea throughout the country, have met with problems. These include financial support—inflexible government budgets with no slot for such new developments—and a lack of qualified and enthusiastic personnel to nurse the plans through to fruition. It is often the case that difficulties occur in trying to maintain the momentum of successful developments like the Taman Negara Study Center through unwieldy and complex government administration systems.

Attempts at rehabilitation of captive orang-utans have attracted a great deal of interest in Indonesia and East Malaysia, an interest which has been turned to good purpose for environmental education at the two rehabilitation centers nearest to areas of high human population [Aveling and Mitchell, 1982]. At Bohorok in northern Sumatra, about 4,000 Indonesian visitors make the trip out to the Gunung Leuser Reserve annually. They are attracted by the orang-utans and during supervised visits have the opportunity to see the semi-wild orang-utans in their natural habitat. Visitors are usually encouraged to discuss with center personnel the conservation issues underlying the project and have the chance to study a photographic exhibition on conservation of tropical rainforest (Fig. 5) accompanied by illustrated leaflets on the same theme. The orang-utans are the reason why visitors go to the forest, but, once there, their interest can be developed and chanelled to convey a broader conservation message [Aveling, 1982].

A similar education policy has been followed at Sepilok Rehabilitation Center in Sabah, East Malaysia, during the last few years. A nature study building has been constructed at Sepilok, and from that base, environmental

Fig. 5. Students visiting Bohorok orang-utan rehabilitation center in Sumatra, Indonesia, learn about conservation of tropical forest. Photo: Rosalind J. Aveling.

education is being extended into the schools of the province by Canadian volunteers working in collaboration with the Game Branch of the Forest Department (Mitchell, personal communication). Other rehabilitation centers, at Tanjung Puting in Kalimantan and Ketambe in north Sumatra, are not suitable for popular environmental education because they are too remote and have long-term research projects under way. These two centers serve instead as field study sites for both Indonesian and foreign students—environmental education at the upper university level.

Elsewhere in Indonesia, on the more densely populated island of Java, attempts to use popular recreation areas for environmental education have had some setbacks. The Cibodas Reserve would seem ideal for education purposes due to its accessibility from urban centers, and it already attracts many national visitors. However, the reserve is not sufficiently well patrolled and guarded to cope with such an influx, and interpretive structures as well as guardposts and campsites have been rendered useless by vandals. Effective administration and control would seem to be essential prerequisites for any education program involving large numbers of people in protected areas. A further problem is that areas with different status can come under different authorities. The Mentawai Islands in Indonesia, for example, are a complex of reserves and logging concessions that are under the control of different Ministries.

An encouraging note is that provision for education centers is sometimes made at the planning stage for new protected areas. Development of new parks and reserves in Malawi during the late 1970s included plans for a Nature Center at the Lilongwe Nature Sanctuary, with the education program organized by the Environmental Education Unit, a separate entity within the Ministry of Agriculture and Natural Resources [Morris, 1977].

Mexico boasts an example of a park established purely for educational reasons [Crespo, 1979]—Belgica Lagoon, 42 hectares of varied topography but predominantly lush semitropical vegetation. An "island" of natural vegetation surrounded by cultivated and pasture lands, the park was established by the Chiapas Nature Protection Council. To fulfill its goals (to improve understanding of natural sciences for present and future generations and to stimulate understanding of the value of natural resources), the park has three principal trails featuring different ecological aspects of the habitat. These are used for field excursions by parties of schoolchildren as part of their state education in natural sciences. The idea is to take natural science teaching out of the classroom into the field and to try to instill a respect for ecological balance.

Protected areas provide good opportunities for environmental education, which are not always fully realized. If an area contains a rare species, particularly one endemic to that region, it can be used to focus both national and international attention on the wider issue of habitat conservation. This is particularly true of primates, which are obvious and generally appealing elements of an ecosystem and, along with other large mammals, are among the first species to suffer from habitat destruction.

Education Related to Protected Areas—Training of Personnel

The national personnel in charge of a country's protected areas have a crucial role. The management decisions they make have both immediate and long-term effects on the conservation value of these areas.

Realization of the importance of immediate-level wildlife managers was the rationale behind the founding of the African Wildlife Leadership Foundation (now AWF), which was instrumental in establishing the College of African Wildlife Management at Mweka, Tanzania, in 1963. This college serves a regional function, which to date has been utilized by 17 different African countries in training wildlife and park managers [Mosha and Thorsell, 1983]. The courses offered by Mweka College are at three different levels and last for 1 or 2 years. Between 1963 and 1984, almost 1,000 students were awarded a diploma or certificate from the college.

In 1970, a similar establishment began operating to serve the countries of francophone Africa. The School for the Training of Wildlife Specialists at Garoua in Cameroon has since graduated just over 400 wildlife and national parks managers from 20 different countries and in addition has provided training for over 250 game and park guards for Cameroon itself [Allo and Ngog Nje, 1983]. The Garoua school experimented with different course lengths and contents and now offers a 2-year training program.

The School of Environmental Conservation Management at Ciawi in Indonesia was established in 1978 as a joint endeavor by the governments of Indonesia and the Netherlands and was modeled on the two regional African colleges. The principal objective of the Ciawi school was to train intermediate-level personnel for the Directorate of Nature Conservation in Indonesia, but steps are being taken to develop it into a regional training center for Southeast Asia [Duryat and van Lavieren, 1982]. Between 1978 and 1982, there were 127 participants in four courses of 9–10 months' duration; of these, 124 were Indonesian and three Malaysian.

The curricula of all three schools are a combination of background sciences, management techniques, and supporting subjects both academic and technical. Duryat and van Lavieren [1982] comment that "Intermediate level personnel involved in the management of nature reserves should be able to understand the complexity and functioning of ecosystems, and to monitor and evaluate conditions and trends in components of the system. They should be able to evaluate the impact of human activities on the ecosystem and to take measures to curb undesirable changes. This requires a basic knowledge of biology, ecology, soil and water conservation and habitat and wildlife management techniques." It can be argued that this is the ideal situation for a wildlife manager but that the practical daily problems of running a protected area require a greater emphasis on administrative and technical skills. The courses of all schools have received some criticism for too heavy an academic loading at the expense of more practical training [Cobb, 1979]. All the schools, however, stress the importance of fieldwork, and endeavor to maintain this as an important part of their curricula. The Mweka College has always organized visits to various Tanzanian National Parks and Reserves for fieldwork and management exercises and in 1983 was granted control over part of Arusha National Park, which it plans to use as a practical management exercise. The Garoua School is situated in a National Park and field courses are an integral part of the curriculum. The Ciawi course of 10 months includes 15 weeks of fieldwork in protected areas of Java, Bali, and Sumatra, emphasis being placed on the practical application of class or field instruction.

A major problem that all the regional colleges face is a tremendous variation in previous education and experience of their students. This is exacerbated by the fact that to date they have all concentrated on training protected area personnel already in service, who range from relatively new recruits to experienced officers. A partial solution tried at Garoua was to limit participation to students between 17 and 30 years of age. The Ciawi School is considering recruiting candidates from school-leavers and making the School's Diploma a prerequisite for entering the service of the Directorate of Nature Conservation in Indonesia.

The training offered by the schools described above is increasing the ability of wildlife managers to manage the environment. But what of training managers to undertake environmental education of others? Wetterberg and Milne [1983] describe recent advances in this sphere in Central and South America, such as the "Central American Workshop on Environmental Interpretation and Education" held in Costa Rica in 1981 and the "International Workshop on Environmental Interpretation" held in 1982 in West Virginia. The latter used Spanish as the workshop language and included participants from a variety of countries who were, or would be, undertaking environmental education programs. The idea was to provide a forum for dissemination of practical information on techniques and discussion of appropriate application of concepts and techniques in the different countries concerned. Such workshops are useful in bringing participants together but could be more useful if the technical aids presented were geared more to the conditions prevailing in the developing countries concerned. Also, their effect could be multiplied by providing a follow-up service to assist participants in developing programs once they have returned to their own countries.

The training courses offered by the WWF/IUCN International Education Project (now the International Center for Conservation Education, ICCE) to developing-country nationals involved in environmental education also aim to increase the knowledge and practical skills of participants in the techniques available to the environmental educator. By bringing together trainees from different countries, the courses intend to promote interchange of ideas on suitability of different methods in different cultural situations as well. The course organizers maintain close contact with, and continue to provide assistance to, trainees after the course period.

WHO IS INVOLVED IN ENVIRONMENTAL EDUCATION IN DEVELOPING COUNTRIES?

Environmental education in all its many forms has been attempted by an equally diverse variety of organizations. These have here been subdivided into national and international organizations, the former comprising both *governments* and *local nongovernmental organizations* (NGOs). In the brief descriptions below, the author has separated international organizations into those of the United Nations, aid agencies, and international NGOs.

National Organizations

Governments. A variety of government departments are concerned with environmental protection. In many developing countries, wildlife authorities are subsections of departments of forestry or agriculture, which itself can cause conflicts of interest in deciding how best to manage the natural environment. Environmental education activities of government departments concerned with wildlife/forestry/agriculture vary from production of leaflets or posters about specific parks or endangered species to more wide-ranging activities of education extension units within those departments. However, where government funds are limited, education seems to take low priority. Many of the education activities that emanate from government departments are in fact funded from outside aid or conservation organizations.

Local nongovernmental organizations (NGOs). NGOs, such as wildlife clubs or foundations and other public pressure groups, usually spring from the community and regard community education as a high priority. Many are therefore very active in environmental education. By bringing public attention to specific environmental issues, it is often the NGOs that generate pressure on government departments.

The Cipko movement in Nepal has had some influence in reducing central exploitation of dwindling tree resources, and promoting community participation and investment in developing stocks at the village level.

In Ecuador, Fundación Natura carefully studied how it could be most effective in lobbying on conservation issues, by investigating the communication channels of different government departments and then adjusting its material and approach to best effect for each one.

In Indonesia, the Ministry of the Environment actively sought the assistance of environmental NGOs, using their knowledge and contacts to communicate with the public at large and to undertake specific projects.

NGOs can be thorns in the sides of government departments, but in many cases the two have a symbiotic relationship. The grass-roots support of the NGOs is the very support on which some governments depend politically for their own survival, so they cannot be easily ignored. In fact, they can serve to increase attention on wildlife departments and improve their status in government hierarchical structures.

International Organizations

United Nations. Two United Nations organizations are specifically involved in environmental education—the United Nations Educational, Scientific, and Cultural Organization (UNESCO) and the United Nations Environment Program (UNEP). UNEP is the coordinating agency for all environmental activities undertaken by the United Nations group [NRDC, 1978] and is responsible for acquainting the world public with the need to conserve the human environment [WCOTP/IUCN, 1977].

An intergovernmental conference on environmental education was organized by UNESCO in cooperation with UNEP in October, 1977, at Tbilisi in the U.S.S.R. This conference formulated 41 recommendations providing a framework, principles, and guidelines to promote and develop environmental education at the national, regional, and international levels, for all age groups, both inside and outside the school system [CONNECT, 1982] and resulted in the "Tbilisi Declaration."

In that the United Nations is made up of member states, one should be able to assume a closer commitment to its programs from those member states than to those of an external organization. However, attempts to translate recommendations at the conference level into action in the field are often obstructed by complex international and national bureaucracies. As the UNEP representative to the Fourth Extraordinary Session of UNESCO's General Conference put it (in reviewing the progress of the joint UNESCO–UNEP International Environmental Education Program—IEEP), "The challenge now is to translate the results of the past decade [since the IEEP was established in 1975] into concrete action in the second decade" [CONNECT, 1982]. This depends not only on even closer cooperation between UNESCO and UNEP in the years ahead but on the implementation of IEEP's fourth phase—"with a stress on the practical application of the results of the earlier phase among target groups in member states."

UNESCO–UNEP performs a valuable service in stimulating reviews of environmental education, through funding of workshops and conferences on this subject. UNESCO publishes an environmental education newsletter, *CONNECT*, in conjunction with UNEP, which is printed in English, French, Spanish, Russian, and Arabic. This serves as a means of information exchange between environmental education initiatives in countries throughout the world. However, it can be argued that the United Nations programs are having proportionately less effect in the developing countries despite the fact that it is these same countries where the need is greatest. Other United Nations organizations involved to some extent with environmental education are the Food and Agriculture Organisation (FAO), the United Nations Development Program (UNDP), and the United Nations Children's Fund (UNICEF).

Aid Agencies. Bilateral agreements between a number of developed countries and those of the developing world bring a flow of "aid in development" from the former to the latter. Under this umbrella gather both the direct government-supported and the volunteer organizations—the U.S. Agency for International Development (U.S. A.I.D.) and the U.S. Peace Corps, for example. The extent to which the field programs of these organizations contribute to environmental conservation or destruction, and the consequent extent of the environmental education component, depends largely on the vagaries of the politics and economics (and hence perceived priorities) of the source countries. Most, however, have been involved to some extent in environmental education in the developing world in recent years, often responding to a specific request from the host country.

The volunteer organizations are particularly active in this respect as their personnel are most closely connected with people and problems at community level. There are instances in which such assistance is of direct relevance to primate conservation. For example, Canadian volunteers have recently developed environmental education programs in Sabah, East Malaysia, based at the Sepilok Orang-utan Rehabilitation Center.

On a larger scale, U.S. A.I.D. recently sponsored a program, Environmental Training and Management in Africa (ETMA), that is assessing regional environmental training and management needs and establishing integrated projects to address these needs in selected areas of eastern and western Africa.

International nongovernmental organizations. Several international conservation organizations are active in the field of environmental education. There has been an increased awareness of the importance of such education to direct conservation efforts in recent years, and an environmental education/public awareness component is now built into more conservation projects from their inception. Among some project planners in conservation groups there remains a lingering reluctance, where budgets are limited, to use funds for education rather than what is seen as direct conservation action. This may be connected with the difficulty of showing "results" from environmental education, but the difficulty of maintaining the success of "direct action," such as anti-poaching work, without the back-up net of an effective education program is becoming more widely recognized.

The International Union for Conservation of Nature and Natural Resources (IUCN) is a union of 520 governments, government agencies, and NGOs formed outside the United Nations system. IUCN has a Commission on

Education that aims to provide the stimulus and technical expertise to members of IUCN and other international agencies to formulate, promote, and undertake environmental education programs in their own countries. The commission is currently in the process of reorganizing and redirecting its efforts into a more practical approach. It has over 300 volunteer experts as members around the world and a number of national committees, particularly in Europe, and is becoming more active in the developing countries. Sakya [1982] reports that in Nepal a National Committtee of the IUCN Education Commission has been established. This is in the context of a national conservation strategy with government participation but has yet to succeed in its aim to reform the education curriculum to include conservation. More recently, the IUCN Commission on Education set up a "tropical forest working group," composed of recognized experts who have worked in the field, with the brief to assess how practical environmental education programs can contribute to conservation of tropical forests.

Some conservation organizations have a special emphasis on environmental education. RARE Inc. (previously the Rare Animal Relief Effort and now the international conservation education affiliate of WWF-US) is particularly involved in environmental education in Central and South America. This ranges from input to the formal education system (for example, RARE's Resource Management Education Program, which has proved so successful in Costa Rican schools that it is currently being extended to Colombia, Honduras, and other countries of the region) to support for conservation clubs and public awareness programs from Costa Rica to Ecuador, professional training, and NGO development [RARE Inc., 1982].

The African Wildlife Foundation (AWF) has concentrated on assistance to training of wildlife managers and general environmental education in Africa over the past two decades and currently maintains a full-time education officer at its African headquarters in Nairobi.

From 1975 to 1984, World Wildlife Fund supported an International Education Project, based in the United Kingdom. This project was established as a "special project for the promotion of conservation education in developing countries," and although budget constraints dictated that it took on contract work for countries of the developed world also, this was used to subsidize assistance to developing countries, which remains the primary objective. Support mainly took the form of providing the specialized equipment and educational materials that are some of the working tools of environmental education programs. Project staff are sensitive to the needs and constraints of field programs, as all have field experience. The custom-designed mobile audiovisual units that the project has put together for use in Gambia, Senegal,

Cameroon, Rwanda, Sri-Lanka, Madagascar, Uganda, and elsewhere have more than proved their worth in bringing environmental education/public awareness to audiences of thousands [Boulton, 1982]. Audiovisual series (slide packs with tapes) and resource filmstrips on approximately 300 topics have been produced and distributed for worldwide use. To date, Africa has been a major target for support, with programs in Gambia, Senegal, Sudan, Kenya, Uganda, Rwanda, and Zambia and support services in Tanzania, Ethiopia, Sierra Leone, Malawi, Swaziland, and elsewhere. Support in Asia has included India, Malaysia, Indonesia, Nepal, and Sri Lanka, but little work has as yet begun in or for the Americas [Yorath and Boulton, 1982; Boulton, personal communication].

The International Education Project has also developed 12-week training courses at its base in the United Kingdom for national staff running education programs in developing countries. It hopes to put more emphasis on this "training of trainers," both in the United Kingdom and in the field overseas, with the eventual aim of establishing resource/production centers in the developing countries themselves, first on a regional and then on a national basis.

The work of the WWF/IUCN International Education Project has now been subsumed into the wider brief of the International Center for Conservation Education (ICCE), established in 1984 to undertake all aspects of conservation education—training, advisory and consultancy services, production of educational materials, and provision of equipment—operating as an independent registered charity.

In developed countries, zoological gardens often have extensive education programs, most of which draw attention to wildlife conservation. In general, zoos in developing countries are not so advanced in this respect, although there are some developments along these lines, as seen in the education program of Kuala Lumpur Zoo in West Malaysia. However, some zoos in Europe and America make a considerable input to conservation education programs in developing countries through their zoological societies, such as the Frankfurt Zoological Society (FZS), the Smithsonian Institution, and the New York Zoological Society (NYZS). For example, a conservation-education program for the golden lion tamarin in Brazil has received support from the Smithsonian Educational Outreach Program and the Frankfurt Zoological Society account "Help for Threatened Animals," as well as its core funding from WWF-US [Dietz, 1984]. Working from a reserve in the last remaining habitat of the golden lion tamarin, this program is saturating the people around the reserve (landowners, schoolchildren, the general public, and officials) with conservation information using the tamarin as a focus. By targeting landowners around the reserve, it hopes to extend the protected habitat available to the tamarins and through the local people to enlist long-term support for these conservation measures.

DISCUSSION

As the examples given here indicate, environmental education is not always initiated as a considered response to a problem, whether that problem be a specific conservation issue or a more general lack of environmental awareness. In practice, educational efforts usually spring from another starting point—funds becoming available within a project or the appointment of an individual to a new position in which he or she can promote environmental education activities.

Undoubtedly, the many disparate efforts and approaches described here have had an impact on conservation in the countries concerned. However, asessing them critically is difficult (could another approach have been more effective, for example) and comparing them impossible (variations of culture, environment, and politics present different matrices of problems), so perhaps it is more constructive to presume that an educator has the opportunity to start at stage one in designing an environmental education program and consider in more detail some of the factors involved.

In general, it would seem important to try to assess the problems of conservation in any one place at the outset and then develop a program that suits the problem. The ultimate goal must be carefully considered, and the target audience carefully selected, as a means of achieving that goal. Is it important to influence general attitudes or those of a certain group of policy-makers?

Identification of Relevant Target Audiences

Unless the environmental education program is directed at the people who can, by taking different decisions or altering their approach, provide a solution to conservation problems, it will not have a demonstrable impact on the environment. Potential target groups include national governments, the general public, the business community, landowners or land-dwellers, schoolchildren, and older students. It might be necessary to direct efforts at more than one group simultaneously. A public awareness campaign around a forest containing an endangered primate might not even reach the government officials responsible for assigning logging concessions in that area. Conversely, government decrees can be ineffective without the support of the land-dwellers affected by them.

Walton and Wood [1982] include international assistance agencies as a target group for environmental education. Many agencies are becoming more aware of the need to consider the environmental impacts of development projects. Few, however, include an educational component in either preparation or implementation of their projects. If, in project design, host country officials have negligible input, or are not adequately educated in the field, they do not develop the skills and commitment necessary for successful implementation. There are many examples of such project plans gathering dust, unrealized, on shelves in government offices.

Assessment of Relevant Cultural and Social Aspects

It should be obvious that an understanding of the cultural and social background of the target audience is an essential prerequisite of any environmental education attempt. Education requires effective communication, and communication is per se a two-way process. Environmental educators, imprinted by the influence and beliefs of Western culture, will not get very far if they assume the same background for most target audiences in developing countries. The same applies to a national from one sector of a community trying to influence those from a different sector. Nor are the solutions to environmental problems designed by highly educated minds the only, or necessarily the most appropriate, solutions available. Local attitudes to environmental conservation are not always negative, as seen from the developed world; they may just be different. Taken into account and harnessed, local beliefs, taboos, and priorities can contribute to more effective environmental management schemes in given communities.

Development of Appropriate Educational Content

Having determined the target audiences and looked at the problem from their point of view, the next step is to develop a message that presents suitable alternatives to the problem at hand. The alternatives must appear relevant and viable for the target population, and the educator must be able to convince the appropriate people that those alternatives are in their best interests. If such alternatives are not available, it may be that environmental education is not the tool to use in that instance.

Accurate Selection of Educational Methods

It is natural for educators to turn first to the methods with which they are most familiar, but, in a new situation, these might not be the most appropriate, and educators should keep an open mind about how best to communicate their message. Different target groups will probably respond best to different approaches. Whereas available funds and time will also narrow the selection of methods, it must be borne in mind that more time-consuming approaches (such as personal contact with community leaders) can in the end prove more cost-effective than general public awareness campaigns, if they achieve the desired goal.

Evaluation

A program can be evaluated from many different angles. From a purely educational viewpoint, has it been effective in terms of altering attitudes, skills, and level of knowledge? Is it a financial success—were the approaches taken cost-effective? But perhaps most crucial is evaluation in terms of ecological effects—has it had a positive influence on conservation? Results should ultimately be measured in terms of effect on the environment rather than in the quantity of attractive publicity material produced, but of course this is an ideal of environmental education and is not always feasible. Even if there is not a demonstrable effect, environmental education might have provided a suitable climate for that to occur. It may be possible to demonstrate a change in attitudes, but is that only relevant if it can affect behavior? Villagers around the Parc National des Volcans in Rwanda may now understand the reasons behind protecting gorillas and their habitat, but will they therefore stop poaching and destroying the habitat when they personally need wood and land? And, if not, has the education effort been a waste of time? Programs can also be evaluated in terms of achievement of their original objectives; but were those the "right" objectives? If that approach has failed, is it possible to say that another would have succeeded?

It may be that a more carefully considered approach, as outlined above, would lead to more effective programs, and certainly most initiatives would benefit from attempts at objective assessment of their efforts by those involved. Some form of evaluation should be built into every program from the outset.

The range of environmental education programs illustrated in this paper is at least evidence that efforts are being made. However, environmental education, particularly in developing countries, is very much an experimental field. Developing environmental education programs is a learning process, and learning from the experience of other initiatives can be relevant even though conditions may not be the same. Several programs have demonstrated that primates, obvious and attractive features of their habitats, can be used to highlight wider environmental concerns. In Indonesia the conservation problems of the orang-utan attract attention to forest conservation. In central Africa gorilla conservation efforts are having the same effect, and in South

America the golden lion tamarin conservation program is focusing interest on conserving remaining forest habitat. As most remaining tropical forests occur in the developing countries, it is encouraging that tropical forest protection is a major focus of environmental education efforts in those countries. But perhaps those efforts are not adequately supported by campaigns aimed at changing the behavior of societies consuming tropical forest products in the developed nations.

As with other aspects of conservation, environmental education could benefit from more coordination of initiatives and funding from different sources. However, in general, the current, piecemeal approach can be expected to continue. It can still have its successes, particularly if each program takes a critical look at the problem it is attempting to alleviate and gears its approach to that goal. The ideal educational process has been concisely described by Keating [1981] and applies particularly well to environmental education: "Go to the people, live among them, learn from them, serve them, plan with them, start with what they know, build on what they have."

ACKNOWLEDGMENTS

The author would like to thank the many people who provided the information on which this paper is based, and also the following for commenting on the draft manuscript: D. Allen, M.N. Boulton, J.V. Connor, V.C. Gilbert, I. Hattingh, P.E. Hetz, G.A. Lieberman, J.S. Marwick, A.J. Potterton, S.A. Price, A.L. Vedder, and A.W. Weber.

REFERENCES

Allo A, Ngog Nje J (1983): School for the Training of Wildlife Specialists at Garoua, Cameroon (mimeo). In: "Proceedings of the 22nd Working Session, Commission on National Parks and Protected Areas. IUCN/January, 1984." pp 42–46.

Aveling RJ (1982): Orang utan conservation in Sumatra, by habitat protection and conservation education. In de Boer LEM (ed): "The Orang-utan—Its Biology and Conservation." The Hague: Dr. W. Junk Publishers, pp 219–315.

Aveling RJ, Mitchell AH (1982): Is rehabilitating orang utans worthwhile? Oryx 16(3): 263–271.

Boulton MN (1982): Conservation on wheels. In Cowan ME, Stapp WB (eds): "Environmental Education in Action—V: International Case Studies in Environmental Education." Columbus: Ohio State University, pp 90–94.

Chumo N arap (1982): Wildlife clubs of Kenya. In Cowan ME, Stapp WB (eds): "Environmental Education in Action—V: International Case Studies in Environmental Education." Columbus: Ohio State University, pp 95–100.

Cobb S (1979): Are the right people being trained the wrong way? AWLF News (Winter, 1979) 14(3):17–19.

CONNECT (1982): UNESCO–UNEP Environmental Education Newsletter VII(4):3.

Connor JV (1982): "Educating Upper Primary Students on Wise Use of Natural Resources and Renewable Energy." Proposal to UNICEF/Tanzania/Uganda/Kenya.

Crespo WLH (1979): Belgica Lagoon, First Mexican Educational Park. Parks 1979:11–12.

Dietz LA (1984): "Conservation Education Program for the Golden Lion Tamarin. WWF Project 3148." Progress Report for funding and collaborating agencies, January, 1984.

Duryat HM, van Lavieren LP (1982): Training personnel for Indonesia's protected areas. Parks 7(2).

Halloran D (1975): Conservation Education in Botswana. AWLF News (Winter, 1975) 10(3):15–18.

Jansen MAB, Raheem R (1981): Promotion of Conservation in Sri Lanka. Progress Report 2 of March for Conservation, September, 1980–August, 1981 (mimeo). Department of Zoology, University of Colombo.

Keating R (1981): The harmonious valley. In: "MAB After Ten Years." Paris: UNESCO, p 54.

Lekagul B (1979): Interview in Asiaweek (September 21, 1979) p 31.

Lieberman GA (1984): Step by step: How conservation education works. Focus/WWF-US (July–August, 1984) 6(4):2.

Lieberman GA, Walton-Wood D (1982): Evaluation of operational program grant to Fundacion Natura in Ecuador. Report to United States Agency for International Development, September, 1982.

Mangubuli MJJ (1983): Adult education and environmental conservation education in Tanzania. In: "Proceedings of the first national conservation education seminar, Tanzania, August, 1983."

Marwick JS (1983): North-south co-operation in the development of a university school of environmental studies: Canada and Kenya. Working Papers in Environmental Studies. Faculty of Environmental Studies, York University. November, 1983."

Mittermeier RA, Macedo-Ruiz H de, Luna ML, Young A, Constable ID, Prado CP del, Luscombe BA (1984): Conservation education campaign for the Peruvian yellow-tailed woolly monkey to be launched in Peru. IUCN/SSC Primate Specialist Group Newsletter (March, 1984) 4:19–22.

Morris DK (1977): Malawi: A progress report. AWLF News (Spring, 1977) 12(1):12–14.

Mosha GT, Thorsell JW (1983): The College of African Wildlife Management, Mweka—After 20 years. In: "Proceedings of the 22nd Working Session Commission on National Parks and Protected Areas. IUCN/January, 1984. (Mimeo)." pp 38–41.

Natural Resources Defence Council (1978): Tropical Moist Forests Conservation Bulletin 1. Washington, DC: NRDC International Project.

Pattinasarany C, Hehuwat F, Frey AR (1980): Nature conservation education in Indonesia. In: "International Symposium on Conservation Inputs From Life Sciences, Universiti Kebangsaan Malaysia, October, 1980."

Prempree P, Baimai V (1978): "Study and research in the biology of primates." Department of Biology, Faculty of Science, Mahidol University, Bangkok, Thailand (mimeo).

Price SA (1978): Wildlife clubs of Kenya: Are they worth it? AWLF News (Winter, 1978) 13(3):12–15.

RARE, Inc. (1982): "Annual Report." Washington D.C. 20009.

Sakya K (1982): Environmental education in Nepal: An urgent challenge. In Cowan ME, Stapp WB (eds): "Environmental Education in Action—V: International Case Studies in Environmental Education." Columbus: Ohio State University, pp 67–74.

Tuboku-Metzger (1979): Nature conservation in Sierra Leone. AWLF News (Winter, 1979) 14(3):12–16.

Tuntawiroon N (1982): Philosophy, methodology and history of environmental integrationist training–An experience of a developing country. In Cowan ME, Stapp WB (eds): "Environmental Education in Action—V: International Case Studies in Environmental Education." Columbus: Ohio State University.

Wakanker C (1984): "Conservation Education in India." WWF Monthly Report, (February, 1984).

Walton D, Wood D (1982): Working draft on environmental education as applied to tropical forest management. Washington, DC: U.S. Office of Technologies to Sustain Tropical Forest Resources, 1982.

WCOTP/IUCN (1977): "Pilot Seminar of Environmental Education Methodology in East Africa, September, 1974." (Final Report published by World Confederation of Organisations of the Teaching Profession, 5 Avenue du Moulin, 1110 Morges, Switzerland).

Weber AW (1979): Gorilla problems in Rwanda. Swara 2:28–32.

Wetterberg GB, Milne RC (1983): International workshop of environmental interpretation: Reflection of trends in personnel training. Parks (October-December, 1983) 8(3):8–10.

World Conservation Strategy (1980): Produced by IUCN/UNEP/WWF in collaboration with FAO/UNESCO.

Yorath S, Boulton MN (1982): A case study account of the development of audiovisual materials in conservation. In Cowan ME, Stapp WB (eds): "Environmental Education in Action—V: International Case Studies in Environmental Education." Columbus: Ohio State University.

Primate Conservation in the Tropical Rain Forest, pages 263–299

10

Captive Propagation as a Component of Conservation Strategies for Endangered Primates

Thomas J. Foose, U.S. Seal, and Nathan R. Flesness

AAZPA Conservation (T.J.F.) and International Species Inventory System (N.R.F.), Minnesota Zoological Garden, Apple Valley, Minnesota 55124 and IUCN Captive Breeding Specialist Group, VA Hospital, Minneapolis, Minnesota 55417 (U.S.S.)

INTRODUCTION

Captive propagation can and must be a part of the strategy to preserve endangered primates from extinction. The more traditional philosophy and strategy of conservation by protecting the habitat, and presuming its inhabitants will also be preserved, are becoming increasingly difficult or unfeasible. Other chapters in this book document the accelerating rate of destruction of the tropical habitats of primates. However, even where some natural habitats and populations can be protected, they may be so reduced and fragmented that they are not viable genetically and demographically over the long term. Their gene pools are being converted into gene puddles that are vulnerable to evaporation in an ecological and evolutionary sense.

Genetically, small populations lose variation necessary to enable adaptation to changing or changed environments [Soulé and Wilcox, 1980; Frankel and Soulé, 1981] and to sustain the vigor of individuals against inbreeding depression [Ralls and Ballou, 1982]. Hence it is important to preserve as much as possible of the heritable diversity that has evolved and exists in wild gene pools.

Demographically, small populations are susceptible to stochastic perturbations that can cause extinction [Goodman, 1986]. Obvious examples are disease epidemics and natural disasters that can obliterate remnant populations. More subtle perturbations include the sex ratio of the few births that would occur in a very small population being too distorted or the density, and therefore the encounter rate of potential mates, being too low for adequate reproduction to occur. If populations of primates are going to survive and continue to evolve, conservation strategies must not only protect animals but preserve genetic diversity and maintain demographic stability.

These objectives can be achieved if, but only if, principles of population biology and management are applied to conservation problems and programs.

MINIMUM VIABLE POPULATIONS

To a great extent, preservation of genetic diversity is a function of the size of the population being maintained (protected). The smaller the population, the faster genetic diversity is lost, principally because of random genetic drift (Fig. 1). This phenomenon is true in situ or ex situ. However, the population size of relevance is not merely the total census of all the animals alive (N). Instead, the genetically effective size (N_e) of a population is a measure of the way in which animals are actually contributing to the next generation. An animal that does not reproduce during its life does not contribute at all. A male that dominates breeding for a long time can contribute excessively.

Effective population size is thus a function of the number, sex ratio, and family sizes (variance in life-long offspring) of the animals that reproduce [Crow and Kimura, 1970], i.e., the structure and dynamics of a population. Depending on what these characteristics are, the genetically effective proportion of a population (its N_e/N ratio) can vary from a small fraction, e.g., one-tenth, to twice the actual number of animals (Fig. 2).

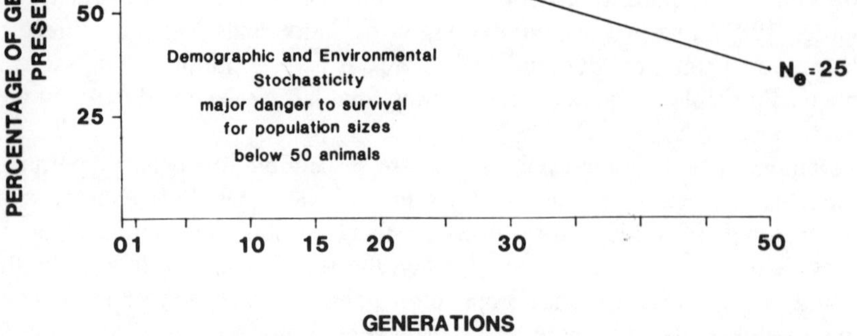

Fig. 1. Decline of genetic diversity resulting from random drift for various effective population sizes (N_e) possible for a total population (N) of 250.

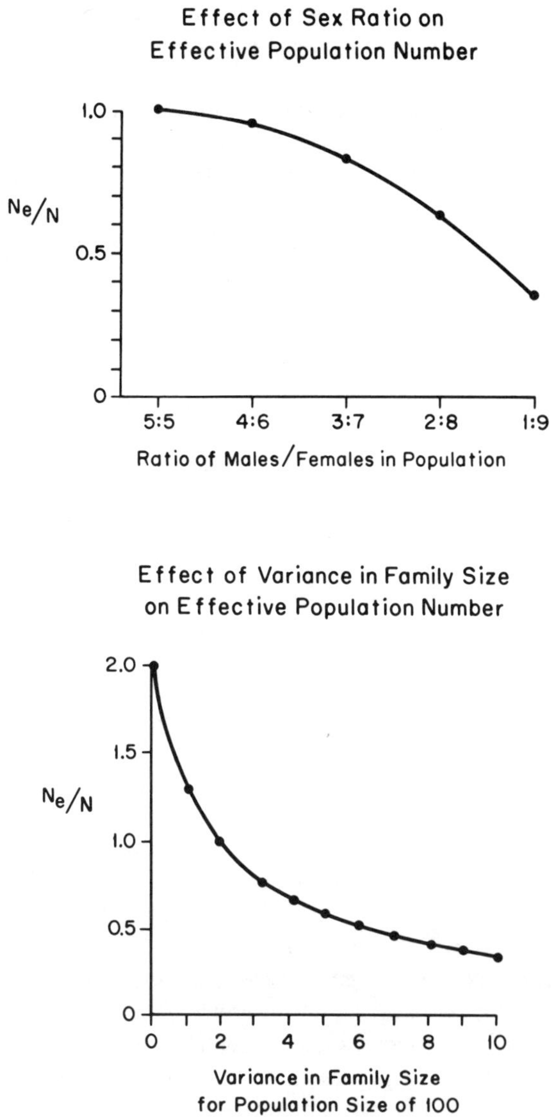

Fig. 2. Effects of sex ratio and family size on N_e/N ratios.

In natural populations, the social dynamics are usually such that the genetically effective size is less than the total number. In ex situ programs, genetic management will potentially be able to produce much higher effective sizes for the population maintained. One of the reasons captive propagation is so appropriate and important for conservation strategies is that zoos can manage intensively to maximize N_e for a given number of animals. However, it is only in the last 5 years that zoos have really become cognizant of genetic management. Hence the recent attempt by Flesness [1986] to estimate N_es for captive primates reveals that the effective population sizes of most populations are unviably low; only nine populations in ISIS faciltities have developed an $N_e > 100$.

Fluctuation in population size from one generation to the next also depresses N_e/N ratios over the long term. Generations with lower N_es have a disproportionate effect on the average N_e for several generations.

There are other reasons for demographic regulation of managed populations. Oscillations in population are not only detrimental to genetic variation, they can cause population explosions and extinctions for the stochastic reasons noted above or because of degradation of the habitat. Thus it is important to stabilize populations around a carrying capacity greater than or equal to the minimum viable population (MVP) size recommended by genetic considerations.

MVPs depend on a number of variables. One is the N_e/N (the effective size to the total number) ratio that can be expected from the population or produced through some management. Another is the amount of the gene pool it is acceptable to preserve. The ideal amount is probably all of it. A more practical amount is some high percentage, although there might be a point of diminishing returns in terms of resources required for the conservation effort. Preserving 95% vs. 90% of the genetic variation will require an N_e twice as large. Recently, some population biologists have suggested 90% of the original gene pool as an acceptable level [Soule et al., 1985]. A third variable is the time scale over which conservation is to occur. How long must the diversity be preserved? The ideal response to this question probably is "forever"; i.e., the gene pool should be protected to continue its evolution immune from extinction by anthropogenic factors.

It is difficult to develop programs that would last forever. However, intensive programs to preserve animal genetic diversity might have a less formidable task. It is only the shrinkage and fragmentation of natural systems that necessitates these intensive attempts to preserve diversity. Were human population expansion and development not encroaching on these ecosystems so severely, diversity could probably preserve itself. At some time in the future, human population and development might equilibrate at a level that permits patches of ecosystems to be large enough that conservation measures can be relaxed. However, it does not appear that such ecological stabilization will occur for rather a long period, what Soule et al. [1985] have described as a "demographic winter" for the earth.

Therefore, a more realistic time scale than "forever" is until the demographic winter is over. The best estimates on its duration are 500–1,000 years. For most vertebrate taxa, preservation of 90% of the genetic diversity in the gene pool for 500–1,000 years would require population sizes from 500 to several thousands depending on the generation time and social organization of the taxon (Fig. 1). Maintaining populations of several hundreds of most primate taxa for the next several centuries will be a formidable endeavor.

However, there might be even more help on the horizon in the form of cryogenic germplasm preservation. Repositories of stored germplasm could contain significant amounts of genetic diversity and hence significantly expand the N_es represented by living populations. From another perspective, large germplasm reserves could reduce the size of the N_es that must be maintained as living populations. However, the viability and applicability of these techniques depends on the capability of regenerating living animals from the germplasm. Currently, this is possible for only a very few, mostly domestic, species, and specialists in animal reproductive technology advise that much more research and hence money will be necessary before such techniques become generally applicable. One estimate is that this research could necessarily extend over the next two centuries. Even with this long development, the critical time period for conservation could be reduced from 500–1,000 years to 200 years, and so minimum viable N_es could be substantially reduced. Nevertheless, two centuries is still a long time for humans to maintain programs.

Yet another variable affecting MVPs is generation time. Loss of genetic diversity resulting from drift occurs generation by generation not necessarily year by year. Assuming that their N_es are equal, gorillas and tamarins will be expected to lose exactly the same amount of genetic diversity over ten generations. However, ten generations for gorillas is probably 150 years, for tamarins probably 50 years (Fig. 3).

As a consequence of these phenomena of loss of genetic variation and demographic instability, conservation strategies for primates must determine MVP sizes. Actually, more is almost always better for preservation of genetic diversity, so populations larger than the MVPs are preferable. However, MVPs demarcate what should be the lowest or last line of defense.

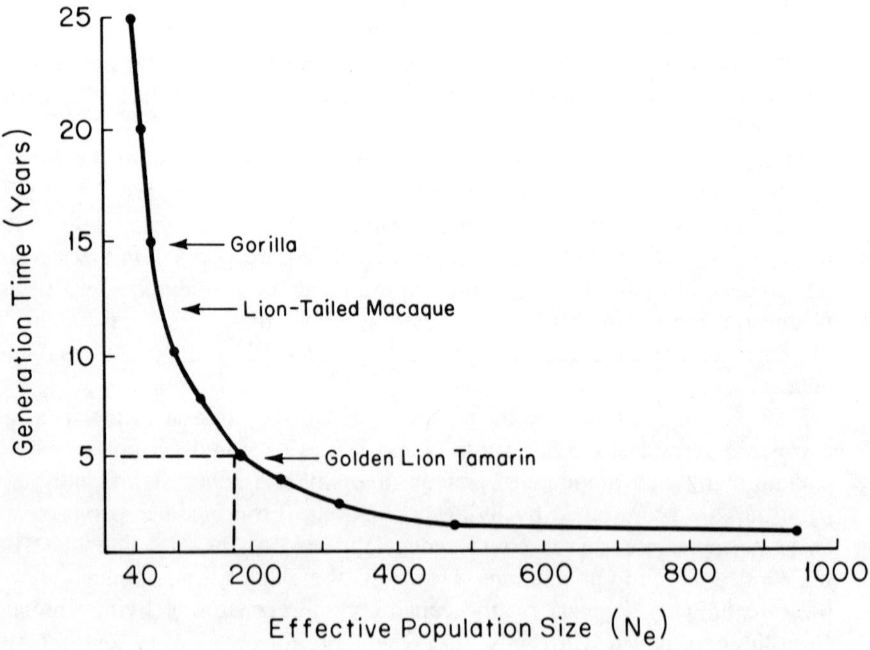

Fig. 3. Effective population sizes necessary to preserve 90% of original genetic diversity for 200 years for taxa with various generation times.

Captive propagation can assist significantly in maintaining minimum viable populations and hence genetic diversity for endangered primates. However, it must be emphasized that the primary objective of captive propagation is the reinforcement, not replacement, of wild populations. Captive populations are not an end in themselves. Even if the worst occurs in the immediate future, there must be optimism that over several centuries it will be possible to restore or even reconstruct ecosystems, but only if we have kept most of the parts. Captive propagation is a way to help preserve options for the future. Moreover, the opportunities will not always be so distant in the future. In fact, two scenarios are relevant.

The more dismal scenario is the situation in which the habitat will be totally untenable for some period of time. In other words, the taxon will become extinct and recolonization would be necessary. The period of untenability could extend from a few years to several centuries, with the latter becoming more probable for many of the so-called charismatic megavertebrates (i.e., the big and beautiful species such as the gorilla), which often disappear before their habitat does.

The more hopeful, though no less challenging, scenario is that some remnants, the gene puddles, will persist in a natural habitat, but not be viable for the long term unless reinforced by genetic material from captive propagation. This scenario thus envisions a system of interactively managed wild and captive populations both of which are critical for the taxon's survival (Fig. 4). In such a system, the captive and wild components are equally important. The wild population permits the species to continue to be subjected to some semblance of natural selection but cannot sustain the genetic diversity necessary for this process or ultimately for survival of the species. The captive programs can provide a reservoir of genetic variation that can be periodically infused into the natural population, or, to use the imagery of G. Evelyn Hutchinson [1965], into the ecological theater so that the evolutionary play can continue.

There are critics who contend that reintroduction from a captivity to the wild will not be feasible even if tenable wildlands are available. One of their

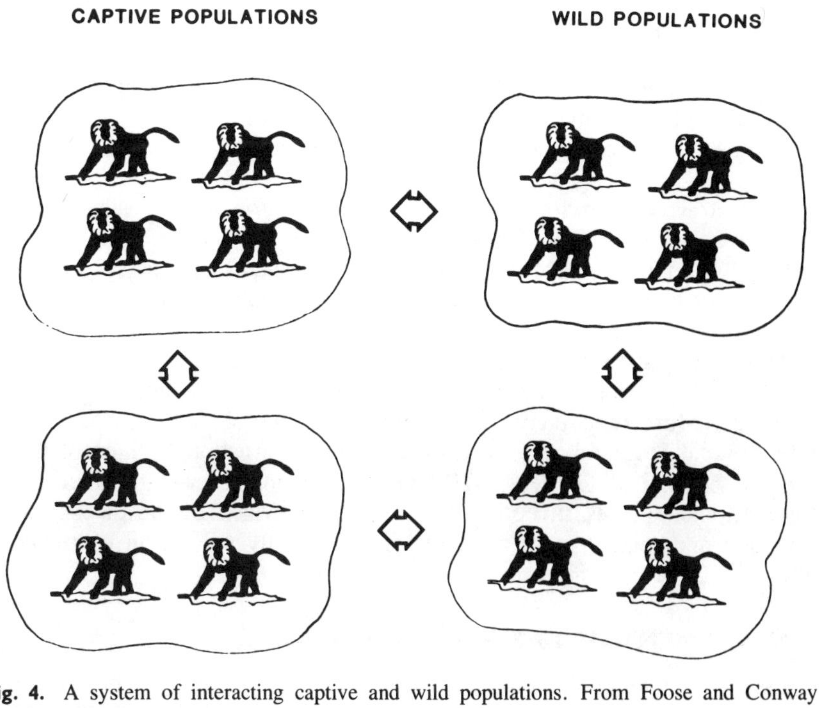

Fig. 4. A system of interacting captive and wild populations. From Foose and Conway [1985], with permission.

major concerns is the possible disruption these "invaders" could cause to wild populations still resident in the natural habitat. This concern would be most applicable to the second scenario described above in which some remnants of the species still survive in the natural habitat.

The other major concern often "articulated" by skeptics of the conservation value of captive propagation is the doubt that animals born and bred in captivity, perhaps for many generations, will be able to readapt to a natural environment. In response to these negative speculations, some limited but encouraging data can be cited from the several projects past and present that have attempted reintroductions with some success. Among primates, the notable example is the golden lion tamarin *Leontopithecus rosalia* whose case will be discussed below [Kleiman and Beck, 1986].

CAPTIVE PROGRAMS FOR PRIMATES

There are two basic problems for programs to propagate species in captivity as part of conservation strategies: first, strategic assignment of priorities and allocation of the limited space and resources (the captive habitat) to taxa in need of assistance, and, second, management of the collections of the selected taxa as biological populations. Captive facilities and programs can contribute significantly to conservation only if they manage their collections as biological populations. This entails both genetic and demographic management as well as all the husbandry that keeps animals alive and permits or promotes their breeding. It also requires extensive compilation and exchange of data.

Such strategic activities by captive facilities are only possible through collective and coordinated endeavor. Some of the zoos of the world are indeed organizing for such programs. In the United States and Canada, the American Association of Zoological Parks and Aquariums has established its Species Survival Plan (SSP) for coordinated and scientific propagation of endangered species. Similarly, in the United Kingdom, zoos are developing the Anthropoid Ape Advisory Panel and the Joint Management of Species Schemes. More tentatively, other zoos of the world do participate in the International Studbooks under auspices of the International Union of Directors of Zoological Gardens (IUDZG) and to an increasing extent in the International Species Inventory System (ISIS). Much of the discussion of selection and management of primates in captivity will be in reference to efforts of these programs.

MANAGEMENT OF CAPTIVE POPULATIONS

Based on the preceding considerations, a general protocol can be delineated for genetic and demographic management of captive populations.

1. Acquire an adequate number of founders. Since no more diversity can be preserved in captive populations than has been obtained from the wild, more founders are probably better in most cases, although there is a point of diminishing returns. Indeed, the number of founders is another factor that can affect the MVP (Fig. 5).

In reality, usually a relatively few founders will be available, especially for rare taxa. Moreover, with such forms, there must be care not to decimate the natural populations. Fortunately, if prudently selected, relatively few founders (20–30) can provide an astonishingly significant sample of the average diversity of the pertinent gene pool [Seal and Flesness, 1979; Senner, 1980; Soule et al., 1986]. A simple guideline is that founders should be unrelated, noninbred, and of course interfertile. More important than the actual number of founders is their proportional representation through time (see below).

2. Establish a carrying capacity equal to or preferably greater than the MVP prescribed by genetic considerations. For managed populations, the carrying capacity will be an analytically determined number that in a sense represents a compromise between maximizing the genetically effective size and providing assistance for as many taxa as feasible with the space and resources available.

3. Expand the population to carrying capacity as rapidly as other components of genetic management permit. Rapid expansion will avoid losses of genetic diversity because of lower N_es during the growth phase.

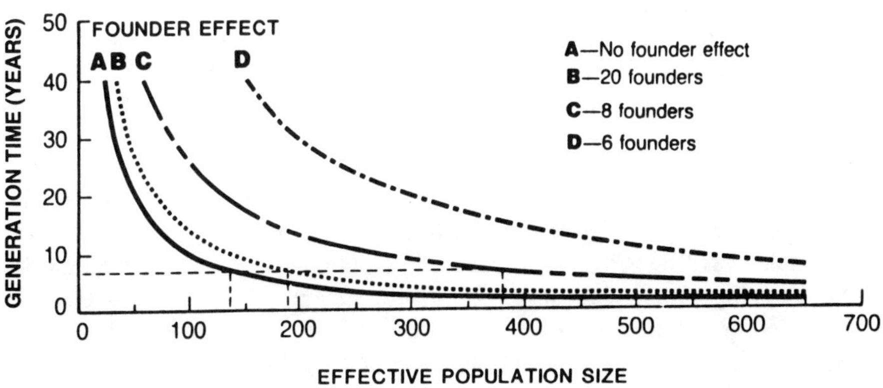

Fig. 5. N_es required to preserve 90% of original genetic diversity for 200 years if population is established by various numbers of founders assuming immediate expansion of founder groups with no loss of diversity. Adapted from graphics prepared by M. Soule and J. Ballou. From Conway [1986], with permission.

4. Stabilize the population at the carrying capacity to reduce oscillations that can cause extinction [Roughgarden, 1979] and avoid depression of N_e resulting from fluctuating size. Demographically, a population can be characterized by and hence managed through control of three basic parameters: a) the age and sex structure of the population at particular times, that is, how many animals of each sex are in each age class; b) the age- and sex-specific survivorships, that is, how long animals live on the average, or, equivalently, what the chances are that an animal will survive to a certain age; c) the age- and sex-specific fertilities, that is, how well animals of various ages usually can be expected to reproduce.

Basically, there are two ways to stabilize a population demographically: a) Prevention—the fertilities can be modified by regulating reproduction through various kinds of birth control, thereby preventing more animals from entering the population. b) Removal—survivorship can be modified by removal of individuals from certain age classes. It will be ideal to place the removed individuals in natural habitats, but in many cases, euthanasia might be an inevitable method of regulation. Normally, demographic management will require some combination of regulation of reproduction and removal of animals.

Whatever the management program, the animals specifically selected for removal or reproduction will best be identified by genetic criteria so as to maximize preservation of genetic diversity. Naturally, both genetic and demographic management must be applied within constraints imposed by the sociobiology of the species.

Stabilization requires that age structure as well as total numbers be regulated. Constancy of numbers is impossible without stability of the age distribution. The nature of a stable age distribution has been discussed in detail elsewhere [Foose, 1980, 1981; Goodman, 1980]. It will suffice here to state that a stable distribution will have a broad base of young animals and a gradual decline in numbers through the successive age classes. Populations without stable age distributions will behave erratically, often oscillating drastically and detrimentally. Hence, to equilibrate a captive population at the carrying capacity, management should attempt to produce a stable age distribution at approximately this desired size. In demographic terminology, a population stable in both total numbers and age distribution is designated "stationary."

For those species whose reproductive husbandry has been mastered, the potential for growth can be explosive and must be regulated (Fig. 6). However, there are other types of demographic problems in captivity. The Western lowland gorilla (*Gorilla gorilla gorilla*) is an eminent example of a

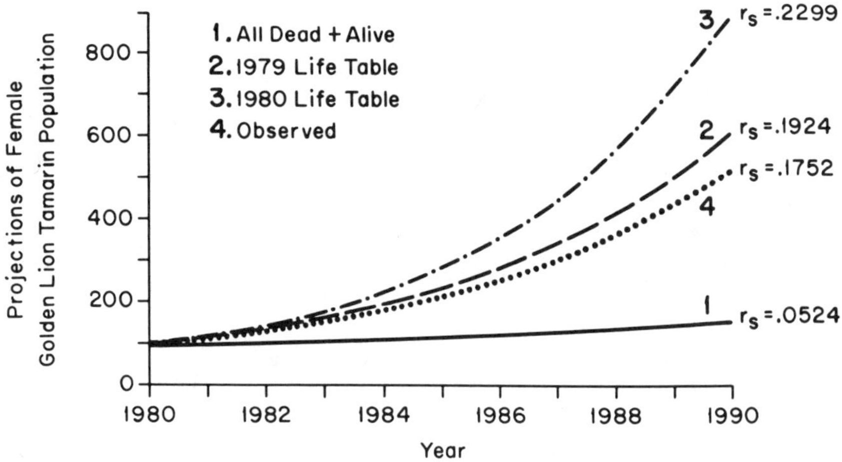

Fig. 6. Projected growth of captive population of golden lion tamarin if certain measured rates of survivorship and fertility pertain. From Kleiman et al. [1982], with permission.

species for which captive reproductive husbandry has not yet been mastered to the point where the population is self-sustaining. As a consequence, the worldwide captive population of gorilla is not self-sustaining and would decline to extinction in about 30–40 years were the problems not solved (Fig. 7)

A particular difficulty for the gorilla seems to be the failure of many males to breed either at all or after early adulthood. However, a few institutions seem to be especially successful in propagating gorillas. Indeed, data for just the North American population seem to suggest a recent turnaround. Consequently, an intensive research and management program is now underway to both master and transfer the art or science of gorilla breeding.

A comparison of the age structures for golden lion tamarins, lion-tailed macaques (*Macaca silenus*), and gorillas reveals the healthy distributions of the first, the moderately stable configuration of the second, and the very unhealthy, top-heavy structure of the third (Fig. 8).

5. *Subdivide the population.* The justification for this measure derives from another phenomenon of population genetics. All small populations will lose genetic diversity because of drift, but different populations will lose and complementarily retain different alleles. Hence a system (or metapopulations) of several small subpopulations might ultimately retain more of the original diversity than one large population managed as a unit [Foose et al., 1986; Chesser et al., 1980].

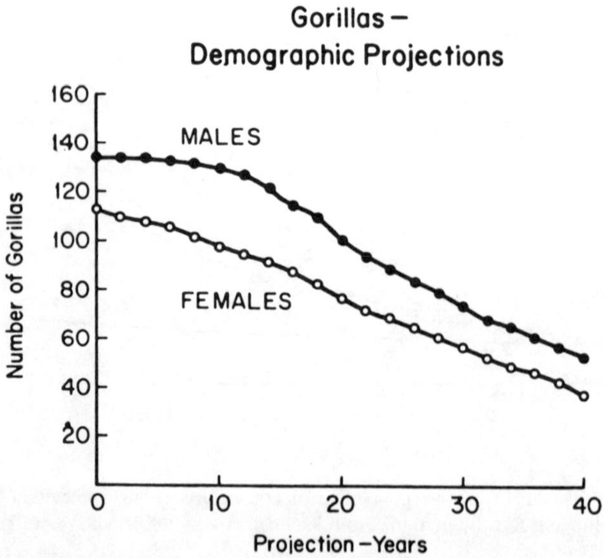

Fig. 7. Projected changes in captive population of lowland gorilla in North America if historical rates of survivorship and fertility continue. From Seal et al. [1986], with permission.

Population subdivision will also protect against epidemics or other disasters that might annihilate any one unit or subpopulation. Logistics of management programs may also be appreciably facilitated, especially if the subpopulations are geographically localized, as most likely will be the case.

The number and size of the subdivisions will vary with the biology of the species, but a general guideline might be that, when the total panmictic effective population size (N_e) > 100, subdivision into units of N_e = 50–100 is recommended. When the overall N_e < 100, smaller subdivisions would be necessary.

The major constraints on subdivision will be genetic and demographic. Subdivisions smaller than a few tens of individuals will be particularly vulnerable to inbreeding depression caused by recessive deleterious mutant alleles [Falconer, 1981] and perhaps even more seriously to stochastic decimation and extinction caused by demographic and environmental factors [Goodman, 1986]. Considering all these factors, the size of subdivision appropriate for a particular taxon could range from a few (e.g., ten) collections to a continental program like the AAZPA SSP.

6. Exchange animals or genetic material periodically between the subdivisions, to achieve some desired compromise between subdivision and panmixia [Foose et al., 1986].

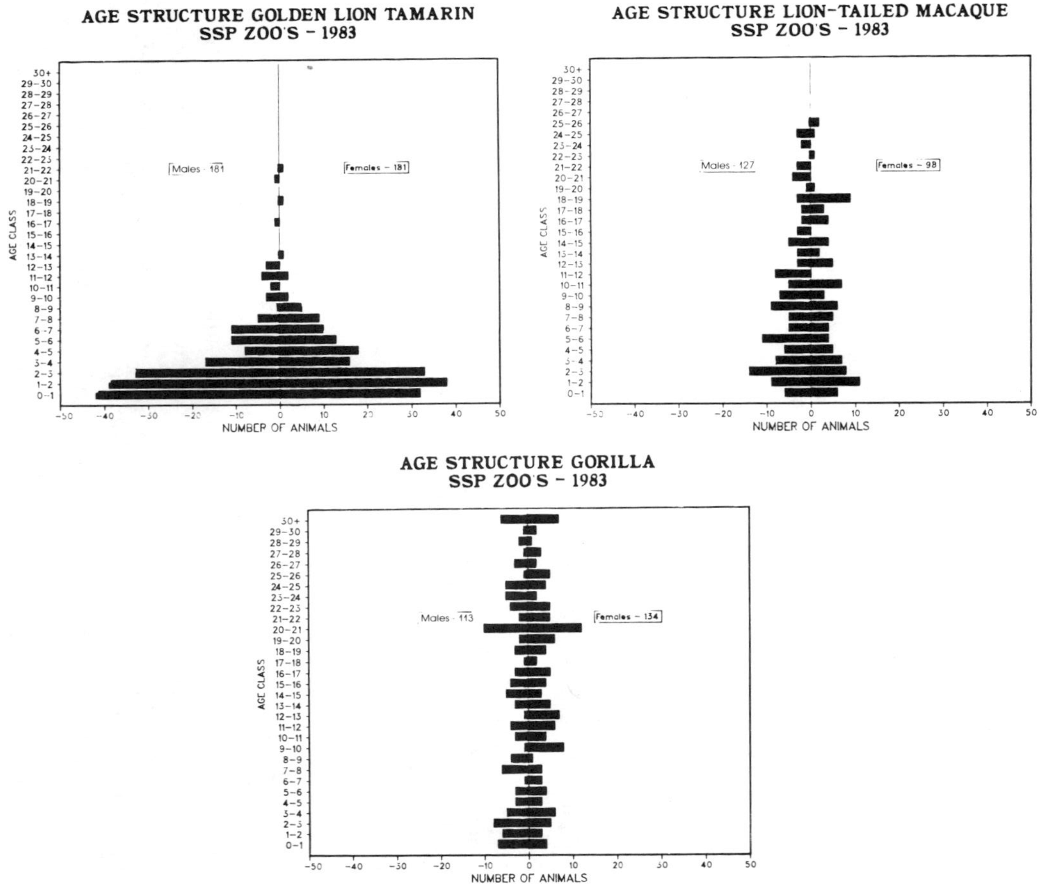

Fig. 8. Age structures of SSP populations of golden lion tamarin, lion-tailed macaque, and lowland gorilla.

7. Within each subdivision, maximize N_e by equalizing family sizes (i.e., life-long progeny production) contingent on 8.

Once the population is at the carrying capacity and parity of founder representation has been achieved, each individual should produce one son and one daughter that are recruited to the next generation. Before the advent of genetic managment in the last 5 years, family sizes in captive populations were very disparate. (Figs. 9 and 10). Equalization of progeny number represents one of the most serious problems for genetic management of captive populations.

8. Redress imbalances in founder or bloodline representation. For most primate taxa in captivity, previous lack of attention to genetic problems and management has produced unequal representation of founder lineages (and presumably alleles; Fig. 11). Available information on pedigrees should be used to rectify these disparities as far as possible. In other words, the representation of founder lineages should be adjusted to be proportional to the probable distribution of alleles surviving from the founders at the time the program of population management is initiated. A number of methods being referred to as "gene drop" programs have been developed to treat this problem [MacCluer et al., 1986; J. Ballou, personal communication; M. Gillpin and O. Ryder, personal communication]. This prescription is a more refined version of what has been a prevailing objective of "equalizing founder representations" [Foose, 1983; Seal and Foose, 1983] and recognizes the realities of allelic losses because of complicated pedigrees, common in captive populations [cf. Ryder et al., 1982; Ballou, 1984]. Reproductive technology can significantly facilitate adjustment of founder representation [Ballou, 1984]. Rectifying disparities in founder representation will entail a period of unequal family sizes.

9. Manage inbreeding by avoiding most matings between closely related individuals. It is important to distinguish loss of genetic variation resulting from genetic drift and inbreeding. *Inbreeding* is actually a term with multiple meanings [Jacquard, 1975]. In its strictest sense, inbreeding is the probability that both alleles at a locus in a diploid individual are identical by descent, i.e., are autozygous. Autozygosity is a special case of homozygosity. Inbreeding is a measure of the lack or loss of diversity at the individual level.

In small populations, inbreeding increases, or alternatively the proportion of heterozygosity decreases, below Hardy-Weinberg levels. However, inbreeding *sensu lato* by itself does not change genic frequencies or cause loss of alleles [Futuyma, 1979]. Inbreeding does reduce heterozygosity and subjects many populations to what is known as *inbreeding depression*, i.e., reduced fertility, survival, or vigor [Falconer, 1981; Ralls and Ballou, 1982].

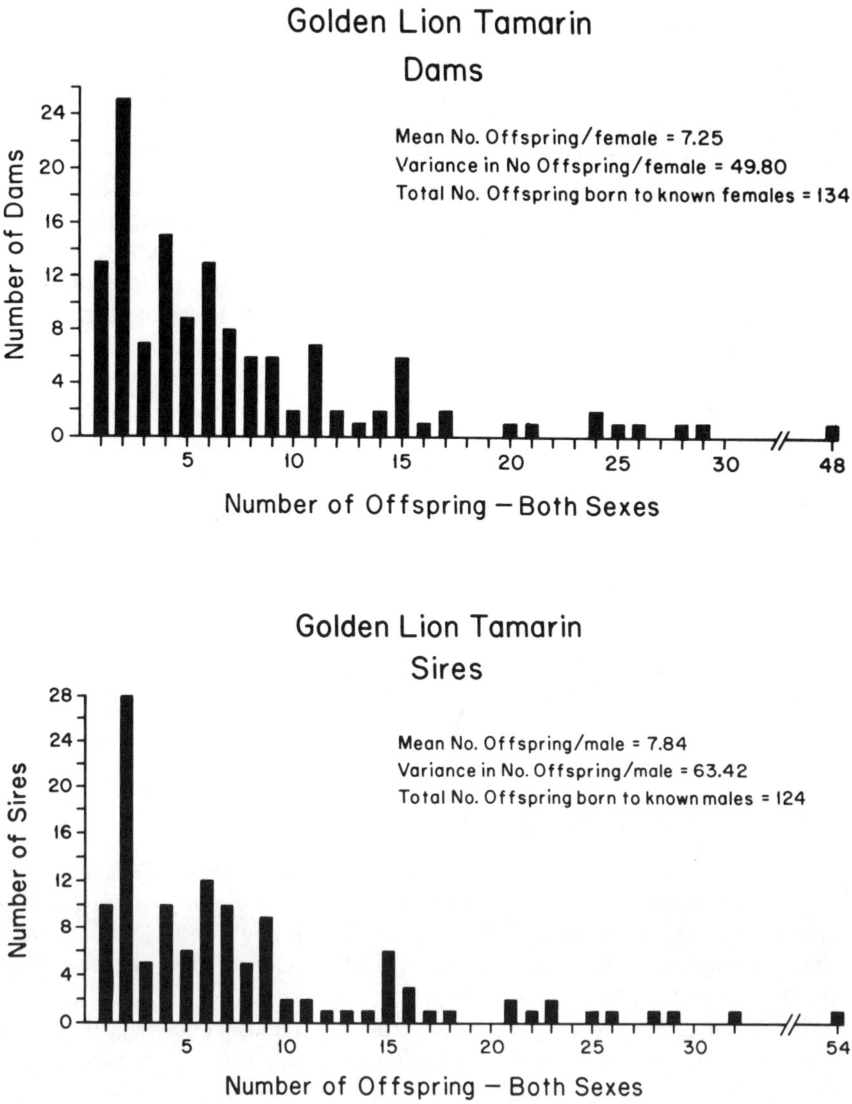

Fig. 9. Family sizes of captive golden lion tamarins. Courtesy of J. Ballou.

Genetic drift does change genic frequencies and causes loss of alleles as some are not transmitted in the sample from one generation to the next. As the frequency of some alleles decreases (ultimately to zero, i.e., lost), the proportion of heterozygotes also declines.

Thus both inbreeding and drift are manifested by a decrease in heterozygosity, but inbreeding entails only rearrangement of the allelic diversity among individuals, whereas drift causes actual loss of variation from the population. As long as more than one allele still remains, inbreeding can potentially be reversed by matings between individuals with different alleles. Loss because of drift can be reversed only by creation of new material by mutation. In this discussion, *inbreeding* will be used in its narrow sense to refer to loss of diversity and its consequences at the individual level. Decline of variation because of genetic drift will refer to actual loss of alleles from the population.

Although not the most important aspect of genetic management, inbreeding sensu lato can cause serious problems in terms of individual survival and fertility of primates, as the Ralls and Ballou [1982] studies indicate. Genetic management has been attempting to reduce levels of inbreeding in recent years, but it still persists even for well managed species.

10. Once the population is at the carrying capacity, maximize generation time to retard loss of genetic diversity over absolute time.

11. Perhaps impose some selection against specific deleterious traits. However, any artificial selection should be applied to managed populations with extreme caution.

12. Periodically exchange genetic material between wild and captive populations.

SELECTION OF TAXA AND THE CAPACITY OF ZOOS

The number of primate taxa whose survival might depend on support from captive propagation is already large (67 or 34% of the approximately 200 species) and rapidly growing [Mittermeier, 1982]. Soule et al. [1986] project that virtually all primate taxa will be critically endangered by the middle of the twenty-first century.

In contrast, the capacity of zoological institutions concerned with and capable of significant captive propagation for conservation purposes is very limited. With a few exceptions, these institutions are the established "zoos" of the world. If zoos are to utilize their space and resources effectively and efficiently, there must be: 1) assessment of what space and resources, i.e., the captive habitat, are available for various kinds of animals, e.g., large

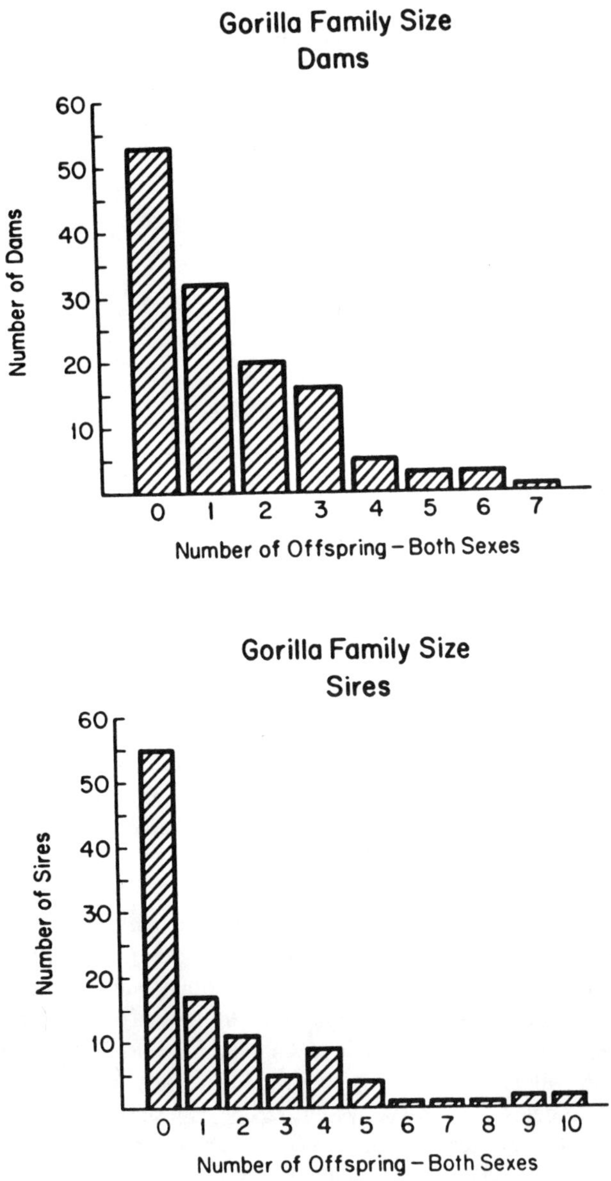

Fig. 10. Family sizes of captive lowland gorillas. From Seal et al., [1986], with permission.

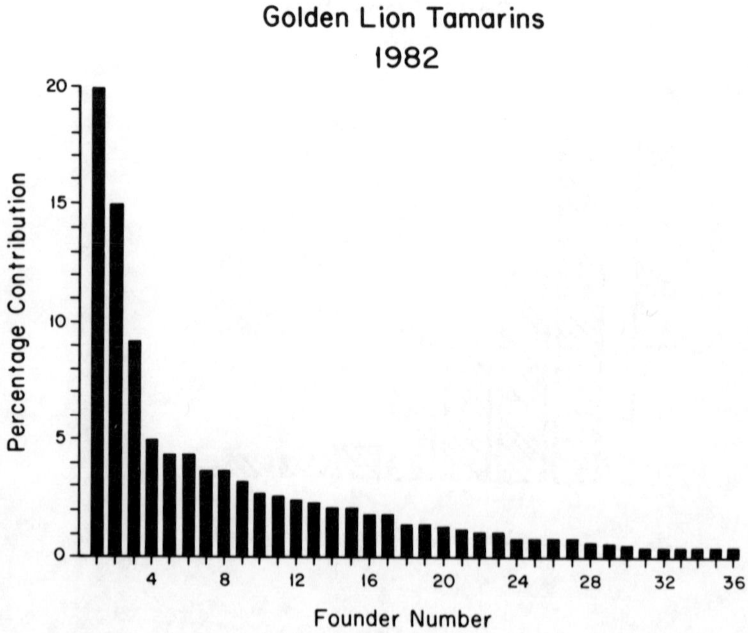

Fig. 11. Relative representation of founder genes in captive population of golden lion tamarins. Courtesy of J. Ballou.

cats, ungulates, tropical forest primates; 2) ascertainment of what taxa are most in need of assistance from captive propagation; and 3) allocation of the captive habitat according to some rational scheme that ensures that viable populations are being accommodated.

A crude measure of the capacity of captive facilities for primates can be provided by the numbers and kinds currently being maintained. Three major sources are available for data on primates in captivity: 1) International Zoo Yearbook (IZY) [Olney, 1984]—Although this annual survey attempts to include all the zoos of the world, its data are compiled only for selected rare and endangered species. Moreover, it is only an inventory; no individual vital statistics are collected. Also, it is always at least three years out of date. 2) International Species Inventory System (ISIS) [Flesness et al., 1984]—This compilation is more current (about 3–6 months out of date) and attempts to collect both inventory data and individual vital statistics for every taxon in captivity. Its major limitations are the restricted number of institutions that participate (virtually all in North America, about 20 elsewhere in the world, especially the United Kingdom, for primates) and the absence of much pre-

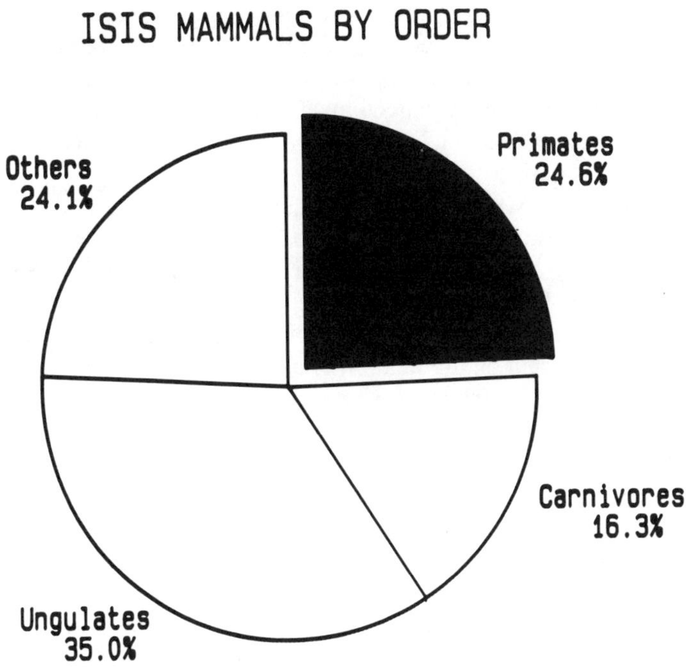

Fig. 12. Living mammals in ISIS.

1974 historical data (ISIS has been operating only for the last 11 years). 3) International Studbooks—These are the most complete in terms of the data and institutions included but exist for only 10 primate taxa. Because many studbook keepers are not yet computerized, currency of the reports has been a problem (e.g., for gorillas).

Strategic and biological management of species requires that there be basic data on the populations and their constituents (individual identification, sex, parentage, birth date, death date, and tracking of animals between institutions). Toward this end, both ISIS and the International Studbooks potentially can complement each other. ISIS serves as a centralized repository of data and can automate much of the logistic drudgery of producing needed reports. Studbook keepers could then serve as data managers in the sense both of ensuring that complete and current data are provided to ISIS and of analyzing information for population management. Moreover, studbook keepers can also then concentrate on being population managers.

Analyses in this paper will apply mostly to information from ISIS, because it represents the most comprehensive, cohesive, and current set of data for a definable collection of zoos committed to captive propagation and population management. The variable coverage in all these data bases applies exclusively or largely to "zoos." The Regional Primate Centers in the U.S., other research centers, and private collections are for the most part not included. The only significant exceptions are the great apes, especially with respect to ISIS, for North America.

Primates constitute 8,974 or 24% of the 35,467 living mammals registered in ISIS (chimpanzees in the biomedical research facilities have been excluded for this survey; (Fig. 12). Table I summarizes data on the status of primates in ISIS facilities by family and genera. These data provide perspective on the "captive habitat" available to primates of similar resource requirements, geographic origins, and public images. (Public perceptions and expectations are still important determinants of what is maintained in a zoo.)

In terms of biotic diversity, Figure 13 compares the relative representation of families and genera in existence and in ISIS. Several important groups are clearly receiving very poor coverage by ISIS (and all) captive facilities. Particularly notable are the Indriidae, the Daubentoniadae, the Tarsiidae, and the Cebidae. Of *Red Data Book* endangered taxa, eight genera (*Allocebus, Phaner, Lepilemur, Daubentonia, Indri, Lichanotus* [*Avahi*], *Brachyteles, Chiropotes, Pygathrix*) and 36 species are not represented at all in ISIS facilities (Table II). (However, a new facility and program for *Brachyteles* are being developed at the Rio de Janeiro Primate Center [WPTI, 1985].) At least nine other species are tenuously represented by singletons or all specimens of the same sex, hardly a secure foothold. Conversely, the best represented genus is *Cercopithecus*, none of whose 19 species is currently classified as endangered by the IUCN *Red Data Book*.

To develop a better assessment of the capacity of captive facilities for primates, the last three columns of Table I compute the number of taxa of each family and genera that might be accommodated if populations of three sizes (100, 250, 500) are maintained. These population sizes extend over the range of several hundreds recommended as MVPs by population geneticists. The values in each column are calculated by dividing the total number of specimens by the population size. Extrapolation based on comparison of several forms considered by both ISIS and IZY suggests that the worldwide capacity of reasonably managed captive facilities for primates might be about two times the values in Table I. It is evident that the captive habitat will be inadequate for total number of taxa that are likely to be endangered in the near future.

Taxonomic categories for Table I are based on the recent classification of mammals published by the Association of Systematic Collectors [Hornacki et al., 1982]. The ASC classification recognizes more families, but frequently fewer species, than other classifications of Primates [cf. Hershkovitz, 1977]. This pattern applies especially to the Callitrichidae. Hence the capacity of captive facilities might be even more limited in relation to the extent radiation of primate species than Table I suggests. In reality, many subspecies might be legitimate entities for conservation. Benirschke [1983] has depicted the significant variation in chromosomes among geographically different populations of what appear to be "phenotypic" species of several South American primates: *Saimiri sciureus*, *Aotus trivirgatus*, and *Ateles* spp. Relative to need at the subspecies level, the amount of captive habitat currently available is exiguous.

There are two major ways in which the limited capacity of captive facilities can be and is being expanded. One way is technological, as discussed above. The other is territorial. Zoos are developing more and larger facilities. Moreover, beyond the established zoos, many wildlife departments in the countries of origin of endangered primates are realizing the potential and importance of captive propagation and are developing captive facilities and programs of their own to complement their field efforts. In the future, these programs could become an important, perhaps the most important, part of captive conservation activities.

In the meantime, beause of the limited capacity of existing facilities, selection of wild taxa for captive propagation or conservation/intervention must be based on rational criteria. The zoo community in North America has formulated a set of three criteria.

1. *Endangerment in the wild*. Certainly some cases are more urgent than others; some taxa still are not classified as endangered. However, at least for the megavertebrates, most will be endangered by the middle of the next century if not before. The best source of data on status and trends of wild taxa is the Species Conservation Monitoring Unit (SCMU) of the International Union for the Conservation of Nature and Natural Resources. SCMU is based near Cambridge University in England but has been limited in resources to perform its function in a really extensive and timely manner.

2. *Feasibility in captivity*. This criterion encompasses a number of considerations: Are there enough animals available for a viable foundation? Are the facilities and staff interested in committing to an intensive management program for the taxon? Can the taxon be kept alive and reproduced in captivity? If not, are the prospects reasonable that a concerted effort will master these skills? (A spectacular example of an unfeasible taxon is the blue whale).

3. *Uniqueness in relation to other taxa*. There are six endangered subspecies of tiger. However, once one or two subspecies have been selected for

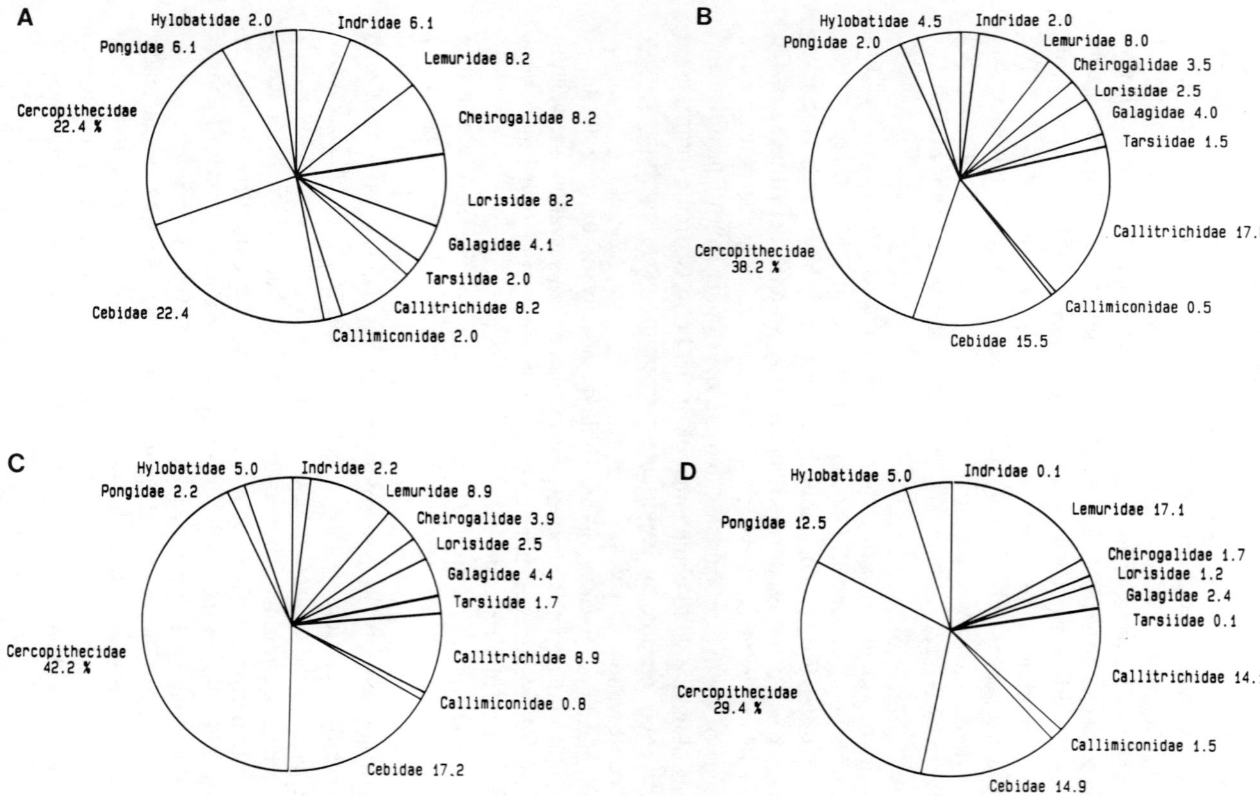

Fig. 13. Primate taxa extant and in ISIS. A. Extant primate genera by family. B. Extant primate species by family ("split" Callitrichidae species). C. Extant primate species by family ("lumped" Callitrichidae species). D. ISIS primates by family.

TABLE I. Captive Status of and Capacity for Primates in ISIS Facilities by Family and Genus

	Extant				ISIS status								Capacity for taxa if MVP		
									Captive born		Species with pop.				
Family	No. Genera	Genus	No. Species	RDB species	Genera	Species	RDB Species	Specimens	No.	%	> 100	> 50	100	250	500
Prosimians															
Cheirogeladae	4		7	4	2	4	2	151	141	93	0	2	2	1	0
		Allocebus	1	1	No	0	0	0	0	0	0	0			
		Cheirogaleus	2	1	Yes	2	1	65	60	92	0	1			
		Microcebus	3	1	Yes	2	1	86	81	95	0	1			
		Phaner	1	1	No	0	0	0			0	0			
Lemuride	4		16	8	3	8	4	1,531	1,371	90	4	5	14	6	3
		Hapalemur	2	2	Yes	1	1	9	5	56	0	0			
		Lemur	6	3	Yes	6	3	1,134	1,003	88	3	4			
		Lepilemur	7	3	No	0	0	0	0		0	0			
		Varecia	1	0	Yes	1	0	388	363	94	1	1			
Indriidae	3		4	4	1	1	1	11	1	9	0	0	0	0	0
		Indri	1	1	No	0	0	0	0	0	0	0			
	(Avahi)	*Lichanotus*	1	1	No	0	0	0	0	0	0	0			
		Propithecus	2	2	Yes	1	1	11	1	9	0	0			
Daubentoniidae	1	*Daubentonia*	1	1	No	0	0	0	0		0	0			
Lorisidae	4		5	0	3	3	0	108	40	37	0	1	1	0	0
		Arctocebus	1	0	No	0	0	0			0	0			
		Loris	1	0	Yes	1	0	17	11	65	0	0			
		Nycticebus	2	0	Yes	1	0	75	26	35	0	1			
		Perodicticus	1	0	Yes	1	0	16	3	19	0	0			
Galagidae	2		8	0	2	5	0	215	143	66	1	2	2	1	0
		Galago	7		Yes	4	0	168	107	64	1	1			
		Otolemur	1		Yes	1	0	47	36	77	0	1			

(continued)

TABLE I. Captive Status of and Capacity for Primates in ISIS Facilities by Family and Genus (Continued)

	Extant				ISIS status								Capacity for taxa if MVP		
									Captive born		Species with pop.				
Family	No. Genera	Genus	No. Species	RDB species	Genera	Species	RDB Species	Specimens	No.	%	> 100	> 50	100	250	500
Tarsiidae	1		3	3	1	3	2	12	0	0	0	0			
		Tarsius	3		Yes	3	2	12	0	0	0	0	0	0	0
Totals	19		44	20	12	24	9	2,028	1,696	84	5	10	19	8	4
New world monkeys															
Callitrichidae	4		16 (35)[a]	8 (11)	4	12 (16)	6	1,265	1,083	86	3	6	14	6	3
		Callithrix	3 (8)	3 (4)	Yes	2 (3)	2	244	214	88	1	1			
		Cebuella	1 (1)	0	Yes	1	0	71	58	82	0	1			
		Leontopithecus	1 (3)	1 (3)	Yes	1 (2)	1	394	384	97	1	1			
		Saguinus	11 (23)	4	Yes	8 (10)	3	556	427	77	1	3			
Callimiconidae	1	*Callimico*	1	1	1	1	1	136	121	89	1	1	1	0	0
Cebidae	11		31	15	9	21	7	1,334	679	51	4	7	13	5	3
		Alouatta	6	2	Yes	3	0	70	29	41	0	1			
		Aotus	1	0	Yes	1	0	142	84	59	1	1			
		Ateles	4	4	Yes	4	4	342	178	52	1	2			
		Brachyteles	1	1	No	0	0	0	0	0	0	0			
		Cacajao	2	2	Yes	1	1	8	2	25	0	0			
		Callicebus	3	1	Yes	2	0	53	34	64	0	0			
		Cebus	4	0	Yes	4	0	201	107	53	1	1			
		Chiropotes	2	2	No	0	0	0	0	0	0	0			
		Lagothrix	2	2	Yes	2	1	45	11	24	0	0			
		Pithecia	4	0	Yes	2	0	66	25	40	0	1			
		Saimiri	2	1	Yes	2	1	411	209	51	1	1			
Totals	16		48 (67)	4 (27)	14	34	14	2,735	1,883	69	8	14	29	11	6

Old world monkeys															
Cercopithecidae	11		76	19	11	50	7	2,638	1,873	73	9	19	25	10	5
		Allenopithecus	1	0	Yes	1	0	10	0	0	0	0	0	0	0
		Cercocebus	4	1	Yes	3	0	97	59	61	0	1	1	0	0
		Cercopithecus	19	0	Yes	15	0	568	361	64	2	7	5	2	1
		Colobus	9	6	Yes	3	0	301	223	73	1	1	3	1	0
		Erythrocebus	1	0	Yes	1	0	127	93	73	1	1	1	0	0
		Macaca	16	3	Yes	13	2	902	705	78	3	6	8	3	2
		Nasalis	2	2	Yes	2	2	11	5	55	0	0	0	0	0
		Papio	3	2	Yes	3	2	418	324	78	2	2	4	2	1
		Presbytis	16	3	Yes	7	0	132	91	69	0	1	1	0	0
		Pygathrix	4	2	Yes	1	1	10	8	0	0	0	0	0	0
		Theropithecus	1	0	Yes	1	0	65	50	77	0	1	0	0	0
Totals	11		76	19	11	50	7	2,641	1,919	73	9	19	25	10	5
Apes															
Hylobatidae	1	*Hylobates*	9	4	1	7	3	450	2	52	2	3	4	2	1
Pongidae	3		4	4	3	4	4	1,120	NA	<50	3	4	11	4	2
		Gorilla	1	1	Yes	1	1	276	133	48	1	1	3	1	0
		Pan	2	2	Yes	2	2	503	NA	<50	1	2	5	2	0
		Pongo	1	1	Yes	1	1	341	228	67	1	1	3	1	0
Totals	4		13	8	4	11	7	1,570		<50	5	7	16	6	3

[a]Figures in parentheses are for "split" Callitrichidae.

captive programs in zoos, it becomes difficult to justify adding more kinds of tigers when there are so many other forms desperately in need of help, e.g., a kind of leopard or a type of jaguar. The captive programs with their limited capacity must try to incorporate a gross representation of as much of the taxonomic diversity of the planet as possible. Thus species that are the only representative of their genus, family, or order will receive particular priority.

In North America, six primate species are among the 37 taxa that have been designated to date for the SSP program (Table III). In the United Kingdom, a number of primate taxa have been selected for the Joint Management of Species Schemes (Table IV). Studbooks, which are a necessary and therefore often a preliminary development for population management, have been established on an international or regional basis for 12 species, including four not yet officially designated for SSP-type programs (Table V).

Clearly, many more primate taxa need to be managed by SSP-type programs. Indeed, there are more urgent cases than the species already designated. However, designations have proceeded rather conservatively to ensure that programs are organized and operating well before designation is extended to many taxa. Initial designations have concentrated on taxa that represent a diversity of management problems and taxonomic groups and hence that can serve as models for the entire program.

What is needed next is to consider entire groups or radiations possessing phylogenic affinities and similar captive requirements. For example, two groups desperately in need of comprehensive and systematic evaluation for allocating captive resources are the Callitrichidae and the Prosimii, especially the Malagasy community.

There also needs to be a more rational system for deciding when to intervene for intensified management in situ or establishment of ex situ programs. Recent attempts to apply decision analysis seem promising [Maguire, 1986; Maguire et al., 1986]. However, the alarm bell has been ringing very late or not at all for many taxa. The decision to establish ex situ propagation programs should occur long before the numbers of a taxon are reduced to levels below viable N_es and indeed often so low that the situation is deemed virtually hopeless by those who would prefer to preserve in situ. Moreover, many taxa currently considered to be secure in situ reserves clearly are in jeopardy by population biology criteria and must be managed genetically and demographically if they are to survive for the long term. Toward this goal, there needs to be more consideration and collaboration

TABLE II. Red Data Book Taxa not Currently in Captivity in ISIS Facilities

Family	Scientific name	Common name
Cheirogalaidae	*Hapalemur simus*	Broad-nosed gentle lemur
	Allocebus trichotis	Hairy-eared dwarf lemur
	Phaner furcifer	Fork-marked mouse lemur
Lemiridae	*Lepilemur mustelinus ruficaudatus*	Red-tailed sportive lemur
	Lepilemur mustelinus dorsalis	Nossi-be sportive lemur
	Lepilemur mustelinus leucopus	White-footed sportive lemur
Indriidae	*Indri indri*	Indris
	Propithecus diadema perrieri	Perrier's sifaka
	Avahi laniger occidentalis	Western woolly avahi
Daubentaniidae	*Daubentonia madagascariensis*	Aye-aye
Tarsiidae		
Callithricidae	*Callithrix aurita*	Buffy tufted-ear marmoset
	Cellithrix humeralifer	Tassel-eared marmoset
	Saguinus oedipus oedipus	Cotton-top tamarin
	Saguinus bicolor	Bare-face tamarin
	Leontopithecus chrysomelas	Golden-headed lion tamarin
Cebidae	*Saimiri oerstedii*	Central American squirrel monkey
	Callicebhus personatus	Masked titi
	Chiropotes albinasus	White-nosed saki
	Chiropotes satanas satanas	Southern bearded saki
	Cacajao calvus	Uakari (red and white)
	Cacajao melanocephalus	Black-headed uakari
	Alouatta fusca	Brown howler monkey
	Alouatta villosa	Guatemalan howler monkey
	Lagothrix favicauda	Yellow-tailed woolly monkey
	Lagothrix lagotricha	Woolly monkey
	Brachyteles arachnoides	Woolly spider monkey
Cercopithecidae	*Cercocebus galeritis galeritis*	Tana river mangabey
	Macaca pagensis	Mentawai Islands macaque
	Colobus badius rufomitratus	Tana river red colobus
	Colobus badius gordonorum	Uhehe red colobus
	Colobus badius kirkii	Zanzibar red colobus
	Colobus verus	Oliver colobus
	Colobus satanas	Black colobus
	Presbytis johni	Nilgiri langur
	Presbytis geei	Golden langur
	Presbytis potenziani	Mentawai Islands langur
	Simias concolor	Pig-tailed langur
	Pygathrix roxellanae	Snubnosed langur
	Pygathrix nemaeus	Douc langur
Hylobatidae	*Hylobates klossi*	Kloss's gibbon
Pongidae	*Gorilla gorilla beringei*	Mountain gorilla

TABLE III. Taxa Designated for the AAZPA Species Survival Plan (SSP)

Puerto Rican crested toad	Ruffed lemur	Asian elephant
	Black lemur	Indian rhino
Chinese alligator	Golden-lion tamarin	Sumatran rhino
Radiated tortoise	Lion-tailed macaque	Black rhino
Aruba Island rattlesnake	Gorilla	White rhino
Indian python	Orang-utan	Asian wild horse
Madagascan ground boa	Asian small-clawed otter	Grevy's zebra
	Maned wolf	Chacoan peccary
Bali mynah	Red wolf	Barasingha
White-naped crane	Red panda	Okapi
Andean condor	Siberian tiger	Gaur
Humboldt's penguin	Asian lion	Arabian oryx
	Snow leopard	Scimitar-horned oryx
	Cheetah	

TABLE IV. Primate Taxa Covered by the U.K. Anthropoid Ape Advisory Panel and the Joint Management of Species Scheme

Anthropoid Ape Advisory Panel	Joint Management of Species Schemes
Gorilla	Lemurs
Orangs	Slender loris
Chimpanzees	Silvery marmoset
Gibbons	Pigmy marmoset
	Golden monkey
	White-faced saki
	Lion-tailed macaque
	Spider monkeys
	Celebes black ape
	Mandrill
	Diana monkey
	DeBrassa's monkey
	Colobus monkeys
	Langurs

between the zoo world and the IUCN SSC Primate Specialist Group, which has recognized the significance of captive propagation by organizing special subgroups in this area.

SPECIES SURVIVAL PLANS

Formulation of plans for both genetic and demographic management entails prescriptions on which animals should mate with each other, how many times

TABLE V. Studbooks for Primates

International	Regional
Black lemur	Geoffroy's tamarin
Ruffed lemur	Woolly monkeys
Pygmy marmoset	
Golden lion tamarin	
Goeldi's monkey	
Lion-tailed macaque	
Douc langur	
Orang-utan	
Bonobo	
Gorilla	

and at what ages animals should breed, and how many offspring should be produced. Implementation of these scientific plans requires a phenomenal amount of cooperation and also coordination. As was indicated above, the most advanced example of this type of program has developed through the Species Survival Plan (SSP) of the AAZPA [1983].

Each SSP program is organized around a Species Coordinator, a person with particular interest, experience, or expertise with the taxon. The Species Coordinator is assisted by a management committee known as the Propagation Group composed of nine members elected from and by the institutions participating in the program (Table VI). Usually, every major zoo and aquarium in the United States and Canada with specimens of the taxon does participate in the SSP program.

A "Memorandum of Participation" (Fig. 14) documents the commitment of an institution to participate in an SSP program. The "Memorandum of Participation" commits each participant to manage their animals in accordance with the guidelines of the Populational Masterplan and the recommendations of the Species Coordinator and the Propagation Group.

The core of the SSP program is the Masterplan for management and propagation. This masterplan provides institution-by-institution and animal-by-animal recommendations on mate selection, animal relocations (from zoo to zoo to produce better genetic combinations), breeding schedules, and culling programs.

Although the AAZPA SSP is currently the most advanced of the scientific and coordinated programs for captive propagation of endangered species, progress is occurring in other parts of the zoo world. The British Federation of Zoos has initiated programs very similar to the SSP. Eventually, it is envisioned that regional (i.e., continental) SSP-type programs will develop worldwide and be coordinated into an international system for endangered

TABLE VI. The AAZPA SSP Propagation Group and Institutional Representatives for Lowland Gorilla

Propagation group	Participating institutions and representatives	
Lester Fisher, DVM, Lincoln Park Zoo—Species Coordinator	Atlanta Zoo, Terry Maple	Los Angeles Zoo, Lester Fisher
	Audubon Zoo, David Anderson	Los Angeles Zoo, Warren Thomas
Gerald Aquilina, Buffalo Zoo (88)	Birmingham Zoo, Bob Truett	Memphis Zoo, Charles Wilson
Diane Brockman, San Diego Zoo (87)	Buffalo Zoo, Gerald Aquilina	Miami Metrozoo, Bill Zeigler
William Conway, New York Zool Pk (86)	Calgary Zoo, Greg Tarry	Milwaukee Zoo, Gilbert Boese
Don Farst, Gladys Porter Zoo (86)	Cheyenne Mtn Zoo, David Allen	National Zoo, Ben Beck
Kenneth Gould, Yerkes Primate Center (87)	Chicago Zoo, George Rabb	New York Zoo, William Conway
Saul Kitchener, San Francisco Zoo (87)	Cincinnati Zoo, Ed Maruska	North Carolina, Les Schobert
Terry Maple, Atlanta Zoo (88)	Cleveland Zoo, Don Kuenzer	Oklahoma City Zoo, David Ruhter
Ed Maruska, Cincinnati Zoo (88)	Columbus Zoo, Margaret Snyder	Philadelphia Zoo, Keith Hinshaw
Warren Thomas, Los Angeles Zoo (86)	Dallas Zoo, John Wortman	Phoenix Zoo, Wayne Homan
	Denver Zoo, Paul Linger	Rio Grande Zoo, Kent Newton
	Detroit Zoo, Steve Graham	Sacramento Zoo, Steve Taylor
	Fort Worth Zoo, Elvie Turner	St. Paul Como Zoo, Tom Hedlund
Official Advisers	Franklin Pk/Stone Mem'l Zoos,	San Antonio Zoo, Russell Smith
Ben Beck, National Zoo	Deborah Schildkraut	San Diego Zoo, Diane Brockman
Joe Erwin, National Geographic Society	Fresno Zoo, Paul Chaffee	San Francisco Zoo, Saul Kitchener
Ulie Seal, IUCN Captive Breeding Specialist Group	Gladys Porter Zoo, Don Farst	St. Louis Zoo, Roger Birkel
	Henry Doorly Zoo, Lee Simmons	Toledo Zoo, Glenous Favata
	Hogle Zoo, LaMar Farnsworth	Topeka Zoo, Gary Clarke
	Houston Zoo, Victor Camp	(Metro) Toronto Zoo, Lawrence Cahill
	Kansas City Zoo, Mike Blakely	Woodland Park Zoo, Violet English
	Lincoln Park Zoo, Lester Fisher	Yerkes Primate Center, Ken Gould

AAZPA SPECIES SURVIVAL PLAN (SSP)
MEMORANDUM OF PARTICIPATION
FOR
GORILLA
Gorilla gorilla

The________________________________(Institutional Name) will participate in an SSP program on gorilla for a period of 5 years or until notifying the AAZPA Species Coordinator of its withdrawal.

This commitment is to cooperate in a program of populational management of the gorilla under the guidance of the Species Coordinator, designated by the Wildlife Conservation and Management Committee, and the Propagation Group of 9 members elected from and by the participating institutions. The Memorandum does not constitute transfer of ownership or relinquishment of control of animals to the SSP, Species Coordinator, or Propagation Group. However, participating institutions will attempt to manage their animals in accordance with the strategic guidelines and specific recommendations of the Species Coordinator and Propagation Group. Proposals from the Species Coordinator and Propagation Group will include advice on mate selection, animal relocations, breeding schedules, and culling programs with the objective of long-term maintenance of a genetically diverse and demographically stable population.

Individual designated to represent your institution and nominated to serve on the Propagation Group:

____

Name Title Phone

Endorsement of institution's chief executive officer:

____

Signed by Title Date

Please return the signed copy to the AAZPA Conservation Coordinator before *31 December 1986.*

Thomas J. Foose, Ph.D.
AAZPA Conservation Coordinator
AAZPA Conservation Office
Minnesota Zoological Garden
Apple Valley, MN 55124

American Association of Zoological Parks and Aquariums

Fig. 14. "Memorandum of Participation" for the AAZPA SSP program for lowland gorilla.

species conservation (Fig. 15). The Captive Breeding Specialist Group (CBSG) of the Species Survival Commission (SSC) of the IUCN and the International Union of Directors of Zoological Gardens (IUDZG) are appropriate vehicles for trying to facilitate such a system. Further details on operating programs

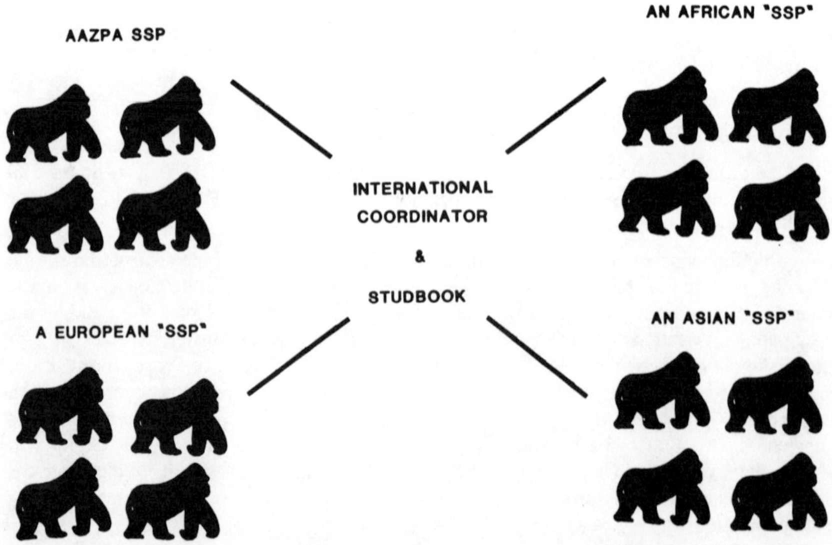

Fig. 15. An International masterplan for gorillas in captivity: a proposed international system of regional SSP-type programs.

for population management and propagation of endangered primates is provided by Kleiman et al. [1982] for golden lion tamarins, Gledhill [1985] for lion-tailed macaques, and Fisher [1984] and Conway [1985] for gorillas.

For at least one species, the chimpanzee, nonzoological primate facilities in the United States are attempting to develop a semblance of a population management program [Seal and Flesness, 1986]. Similar objectives have been articulated by plans for other primates employed in biomedical research (B. Blood, personal communication).

THE PAST, THE PRESENT, THE FUTURE

Species Survival Plans are an encouraging development for the present and the future. Indeed, there have been rapid advances in the captive propagation of primates in the last five years. Hence two previous attempts by Pinder and Barkham [1978] and Stevenson [1985] attempting to survey the success of captive propagation of primates have presented a less sanguine outlook than SSP and other advances inspire. However, it might be worthwhile to reconsider the Pinder and Barkham and Stevenson analyses.

Both surveys were predicated on IZY data, which the authors acknowledge were inadequate. To evaluate success, they applied a number of criteria: 1) population size >100, 2) percentage of captive born >50%, 3) population increase over a period of years, and 4) number of recent births in absolute terms and as a percentage of population. However, any one of these criteria can be misleading. For example, Stevenson notes that lion-tailed macaques are doing well but expresses concern that the number of births is declining. However, this pattern could reflect a saturation of the current carrying capacity for this species. In an unmanaged population, this could still be cause for concern if the response to saturation of the carrying capacity is merely to curtail reproduction without regard for demographic stability [Foose, 1981], but in a managed population such reduction in reproduction would be natural and necessary as part of demographic control.

A combination of population size and percentage captive born might be crude estimations of propagational success. These data for ISIS are presented in Table I. A criterion of 75% rather than 50% captive born seems more consistent with propagational success. On the other hand, since only data from ISIS are being presented, a criterion of population >50 seems logical on the premise that these facilities represent no more than half the world captive population. In considering these results, it must be realized that the failure to propagate primates could be due to 1) an inability to reproduce them or 2) the lack of an attempt to reproduce them. However, analysis of past performance might not be a good indicator of the potential of captive propagation for conservation strategies. Reproductive husbandry of primates is advancing rapidly as the miraculous reversal in the case of the golden lion tamarin indicates (Fig. 16) [Kleiman et al., 1982]. The data that Pinder and Barkham and Stevenson used are already out of date. Conway [1980] has observed that, in every case when a serious effort has occurred to solve problems of reproductive husbandry preventing propagation, zoos have succeeded.

Thus there seems to be evidence that zoos can master reproductive husbandry. What is required to propagate is the kind of commitment and coordination that now seems to be developing among some segments of the zoo community. Table I depicts a gradually improving situation for primates even using the Pinder and Barkham and Stevenson criteria. These results are similar to two more recent analyses by Schmidt [1986] and Flesness [1986] using ISIS data. Once the real effects of SSP management are evident, perhaps in another 5 years, the situation should appear quite encouraging.

Finally, a few promising projects for the future are in progress integrating wild and captive populations for survival of primates. Perhaps the most notable concerns the golden lion tamarin. This species was virtually exter-

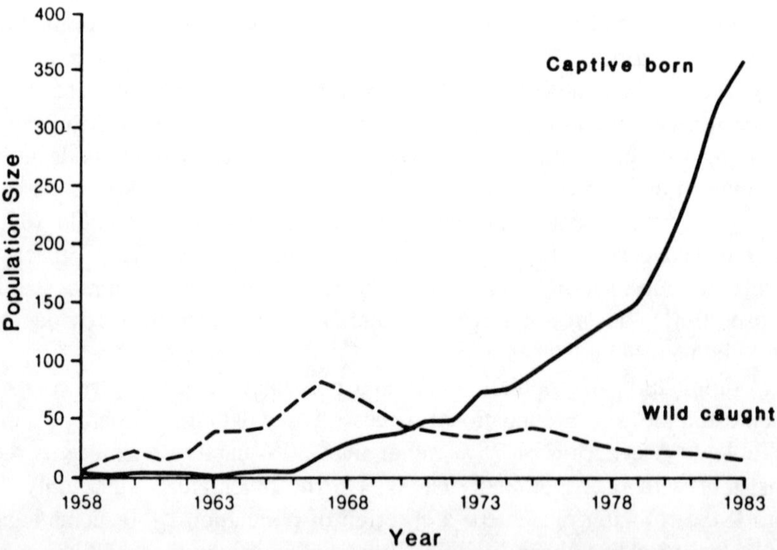

Fig. 16. Growth of the captive golden lion tamarin population since 1958. Curves for both captive-born animals (solid line) and wild-caught animals (dotted line) are shown. The overall population growth rate since 1979 is approximately 26% per year. At the end of 1983, there were 358 captive-born and 10 wild-caught tamarins in the population. From Ballou [1985], with permission.

minated because of habitat destruction and direct killing of the animals, but again, after an uncertain start because of some captive husbandry problems that were solved with concerted efforts of zoo professionals, a captive population of over 400 animals in roughly 20 years has developed [Kleiman et al., 1982]. A very intensive program is now in progress to reintroduce captive-born tamarins into protected patches of forest in their native Brazil [Kleiman and Beck, 1986]. Early results indicate that animals are readapting satisfactorily. A particularly encouraging development is that the Brazilian citizenry are becoming very enamored of and involved in the restoration of this tamarin, and many ranchers are interested in providing sanctuary on their property for reintroduced colonies.

REFERENCES

AAZPA (1983): "Species Survival Plan." Wheeling WV: AAZPA.

Ballou J (1984): Strategies for maintaining genetic diversity in captive populations through reproductive technology. Zoo Biol 3:311–324.

Ballou J (1985): "1983 International Studbook Gold Lion Tamarin." Washington, DC: National Zoological Park.

Bernirschke K (1983): The impact of research on the propagation of endangered species in zoos. In Schoenewald-Cox C, Chambers S, MacBryde B, Thomas L (eds): "Conservation and Genetics." Menlo Park, CA: Benjamin Cummings, pp 402–413.
Chesser RK, Smith MH, Brisbin IL (1980): Management and maintenance of genetic variability in endangered species. Int Zoo Yearbook 20:146–154.
Conway WG (1980): An overview of captive propagation. In Soule ME, Wilcox B (eds): "Conservation Biology." Sunderland, MA: Sinauer Associates, pp 199–208.
Conway WG (1985): Goril 'o' my dreams. In Wagner RO (ed): "Proceedings of the 1985 AAZPA Annual Conference." Wheeling, WV: AAZPA.
Conway WG (1986): The practical difficulties and financial implications of endangered species breeding programs. In Int. Zoo Yearbook (in press).
Crow JF, Kimura M (1970): "An Introduction to Population Genetics Theory." New York: Harper and Row.
Falconer DS (1981): "Introduction to Quantitative Genetics, 2nd Ed." New York: Longman.
Fisher LE (1984): Gorilla SSP report. In Wagner RO (ed): "Proceedings of the 1984 AAZPA Annual Conference." Wheeling, WV: AAZPA, p 264.
Flesness NR (1986): Status and genetic considerations of primates in captivity. In Benirschke K, Linburg D, (eds): "Primate Conservation—The Road to Self-Sustaining Populations." In press.
Flesness NR, Garnatz PG, Seal US (1984): ISIS—An international species inventory system. In Allkin R, Bisby FA (eds): "Databases in Systematics." Orlando: Academic Press, pp 103–112.
Foose T (1980): Demographic management of endangered species in captivity. Int Zoo Yearbook, 20:154–166.
Foose T (1981): Demographic problems and management in captive populations. In Wagner RO (ed): "Proceedings AAZPA 1980 Annual Conference." Wheeling, WV: AAZPA, pp 46–68.
Foose T (1983): The relevance of captive populations to strategies for conservation of biotic diversity. In Schonewald-Cox C, Chambers S, MacBryde B, Thomas L (eds): "Genetics and Conservation." Menlo Park, CA: Benjamin Cummings, pp 374–401.
Foose T, Lande R, Flesness N, Rabb G, Read B (1986): Propagation plans. Zoobiology, 5(2): 139–146.
Foose T, Seal U (1982): Species survival plans for large cats in North American zoos. In: "Proceedings of the International Cat Symposium, Texas." Washington, DC: National Wildlife Federation, in press.
Foose TJ, Conway WG (1985): Models for population management of lion-tailed macaque resources in captivity (a working paper). In Heltne P (ed): "The Lion-Tailed Macaque: Status and Conservation." New York: Alan R. Liss, pp 329–341.
Frankel OH, Soulé ME (1981): "Conservation and Evolution." Cambridge: Cambridge University Press.
Futuyma D (1979): "Evolutionary Biology." Sunderland, MA: Sinauer Associates.
Gledhill L (1985): Progress toward a masterplan of population management for lion-tailed macaque. In Heltne, P (ed): "The Lion-Tailed Macaque: Status and Conservation." New York: Alan R. Liss.
Goodman D (1980): Demographic intervention for closely managed populations. In Soule ME, Wilcox BA (eds): "Conservation Biology." Sunderland, MA: Sinauer Associates, pp 171–195.
Goodman D (1986): The minimum viable population problem. I. The demography of chance extinction. In Soule ME (ed): "Viable Populations." In press.

Gould E (1984): Proposed wildlife park in Sepilok Forest, Sabah, Malaysia with emphasis on proboscis monkeys. In Wayne RO (ed): "Proceedings AAZPA 1984 Annual Conference." Wheeling, WV: AAZPA, pp. 186–189.

Hershkovitz P (1977): "Living New World monkeys (Platyrrhini), Vol 1." Chicago: University of Chicago Press.

Hornacki JH, Kinman KE, Koeppl JW (1982): "Mammal Species of the World." Lawrence, KS: Allen Press, Inc., and the Association of Systematics Collections.

Hutchinson GE (1965): "The Ecological Theater and The Evolutionary Play." New Haven: Yale University Press.

IUCN (1980–85): "Red Data Book." Cambridge, UK: IUCN.

Jacquard A (1975): Inbreeding: One word, several meanings. Theor Pop Biol 7:338–363.

Kleiman DG, Ballou JD, Evans RF (1982): An analysis of reproductive trends in captive golden lion tamarin with comments on their future demographic management. Int Zoo Yearbook 22:94–100.

Kleiman DG, Beck B (1986): The golden lion tamarin. In Benirschke K, Lindburg D (eds): "Primate Conservation—The Road to Self-Sustaining Populations." In press.

Lande R, Barrowclough GF (1986): Effective population size, genetic variation and their use in population management. In Soule ME (ed): "Viable Populations." In press.

MacCluer JW, VandeBerg JL, Read B, Ryder OA (1986): Pedigree analysis by computer simulation. Zoobiology 5(2): 147–160.

Maguire LA (1986): The role of decision analysis in population management for endangered species. Submitted for publication.

Maguire LA, Seal US, Brussard PF (1986): Managing critically endangered species: The Sumatran Rhino as a case study. In Soule ME (ed): "Viable Populations." In press.

Mittermeier RA (1982): The world's endangered primates: An introduction and a case study—the monkeys of Brazil's Atlantic forests. In Mittermeier RA, Plotkin MJ (ed): "Primates and the Tropical Forest." Pasadena, CA, and Washington, DC: The L.S.B. Leakey Foundation and World Wildlife Fund, pp 11–22.

Olney PJS (1984): "1983 International Zoo Yearbook, Vol 23." London: Zoological Society of London.

Pinder NJ, Barkham JP (1978): An assessment of the contribution of captive breeding to the conservation of rare mammals. Biol Conserv 13:187–245.

Ralls K, Ballou J (1982): Effects of inbreeding on infant mortality in captive primates. Int J Primatol 3:491–505.

Roughgarden J (1979): "Theory of Population Genetics and Evolutionary Ecology: An Introduction." New York: Macmillan.

Ryder OA, Wedemeyer EA (1982): A cooperative breeding programme for the Mongolian wild horse *Equus przewalski* in the United States. Biol Conserv 22:259–272.

Schoenewald-Cox C, Chambers S, MacBryde B, Thomas L (1983): "Genetics and Conservation." Menlo Park, CA: Benjamin Cummings.

Schmidt C (1986): A review of zoo breeding programs for primates. Int Zoo Yearbook (in press).

Seal US, Flesness NR (1979): Noah's ark—sex and survival. In Wagner RO (ed): "Proceedings of AAZPA 1978 Annual Conference." Wheeling, WV: AAZPA, pp 214–228.

Seal US, Flesness NR (1986): Chimpanzees in captivity. In Benirschke K, Lindberg D (ed): "Primate Conservation—The Road to Self-Sustaining Populations." In press.

Seal US, Foose TJ, Flesness NR (1986): Gorilla demography. Zoo biol (in prep).

Senner JW (1980): Inbreeding depression and the survival of zoo populations. In Soule ME, Wilcox BA (eds): "Conservation Biology." Sunderland, MA: Sinauer Associates, pp 209–224.

Soulé M, Gilpin M, Conway W, Foose T (1986): The millennium ark: How long the voyage, how many staterooms, how many passengers? Zoobiology, 5(2): 101–114.

Soulé M, Wilcox B (1980): "Conservation Biology." Sunderland, MA: Sinauer Associates.

Stevenson M (1985): Effectiveness of primate captive breeding. In Harper D (ed): "Symposium on the Conservation of Primates and their Habitats." Leicester, UK: University of Leicester, pp 202–230.

Wildlife Preservation Trust International (1985): In the Brazilian jungle, it rises. On the Edge: 26:1. Philadelphia: Wildlife Preservation Trust International.

SECTION III: PRIORITIES

Primate Conservation in the Tropical Rain Forest, pages 303–304

Introduction to Section on Regional Frameworks

The goal of the IUCN/SSC Primate Specialist Group is to maintain the current diversity of the Order Primates, with dual emphasis on the following:

1. Ensuring the survival of endangered and vulnerable species wherever they occur
2. Providing effective protection for large numbers of primates in areas of high primate diversity, endemism and/or abundance

To achieve these goals, the Primate Specialist Group has been preparing regional "Action Plans for Primate Conservation" for the period 1986–1990. These plans are intended to update the Global Strategy for Primate Conservation prepared in 1978, and will determine the highest priority species, protected areas, countries, and biogeographical regions for primate conservation activities in the remainder of this decade. The four articles that follow—on the Neotropical region, Asia, Africa and Madagascar—are excerpted from drafts of the action plans for the different regions. They are not intended to be comprehensive; rather, they aim to give the reader an introduction to each of the regions and the endangered species that occur within them. Far more detail will be included in the action plans themselves, all of which will be produced by IUCN and the World Wildlife Fund in 1986. Inquiries about the availability of these plans should be made to Russell A. Mittermeier, in his capacity as Chairman of the IUCN/SSC Primate Specialist Group.

In the four chapters of this section we briefly discuss both the most endangered species and the most important regions, countries, and protected areas for those species, together with recommendations for continued action or for new initiatives to ensure the survival of all primates species.

There are no easy solutions, and the best approach is often one that employs a variety of different conservation techniques: surveys to determine the status of poorly known species, autoecological and synecological studies of particular primate species or communities, development of endangered species management programs, assessment of existing protected areas and establishment of new ones, development of management plans for parks and reserves, creation of education programs and public-awareness campaigns, training of conservationists from the tropical countries, and efforts to assist in the development of public-policy initiatives for conservation.

Only by taking a multifaceted, multidisciplinary approach can we hope to have all primate species alive today still with us as we enter the next century.

Primate Conservation in the Tropical Rain Forest, pages 305–320

11

Framework for Primate Conservation in the Neotropical Region

Russell A. Mittermeier

World Wildlife Fund, The Conservation Foundation, Washington, D.C. 20037

INTRODUCTION

Neotropical primates are found from southern Mexico through Central America and northern South America south as far as southern Brazil and northern Argentina and, with the exception of Chile and Uruguay, occur in all mainland South and Central American countries and on a few Caribbean islands as well (Fig. 1). Sixteen genera and some 65 species are currently recognized, which means that approximately one-third of all living primate species occur in this part of the world. Of these 16 genera and 65 species, we consider two genera, 12 species, and a total of 28 taxa to be already endangered, and several of these are literally on the verge of extinction (Table I). In this chapter, we analyze the primate conservation priorities in the Neotropical Region by both major phytogeographical region and by country and then discuss a few of the endangered species in more detail. For the purposes of this paper, we will not consider either *vulnerable* or *rare* species (as defined in the IUCN Red Data Book) but only those that are already considered endangered. These are the animals of most immediate concern, and they serve quite well to focus conservation activities over the next 5 years. However, for the convenience of the reader, we do list vulnerable, rare, and indeterminate species as they appear in the Red Data Book [Thornback and Jenkins, 1982] in Table II.

TARGET REGIONS, COUNTRIES, SPECIES, AND POPULATIONS

Priority Phytogeographic Regions

To determine the highest-priority regions for Neotropical primate conservation, we have analyzed distribution of these animals by major *phytogeographical region.* These regions are based on recent work by Gentry [1982]

Fig. 1. Distribution of nonhuman primates in the Neotropical region (map by Stephen D. Nash).

(also personal communication) and include Amazonia (in the broad Hylaean sense that covers the Guianas as well as the Amazon drainage); the Atlantic forest region of eastern Brazil; the Cerrado/Caatinga/Chaco regions of central Brazil, Paraguay, and Bolivia; the northern Andes (north of the Peru-Ecuador border and including the Chocó); the southern Andes; northern Colombia/Venezuela; and Middle America, including all of Central America and Mexico (Fig. 2).

TABLE I. The Most Endangered Neotropical Primate Genera, Species, and Subspecies

Endangered genera	Endangered species	Endangered subspecies
Brachyteles	*Callithrix aurita*	*Callithrix argentata leucippe*
Brachyteles arachnoides	*Callithrix flaviceps*	*Saguinus bicolor bicolor*
Leontopithecus	*Callithrix geoffroyi*	*Chiropotes satanas satanas*
Leontopithecus rosalia	*Saguinus oedipus*	*Cacajao calvus calvus*
Leontopithecus chrysomelas	*Callicebus personatus*	*Cebus apella xanthosternos*
Leontopithecus chrysopygus	*C. p. personatus*	*Alouatta belzebul ululata*
	C. p. nigrifrons	*Ateles geoffroyi azuerensis*
	C. p. melanochir	*Ateles geoffroyi panamensis*
	Saimiri oerstedii	*Ateles geoffroyi grisescens*
	S. o. oerstedii	*Ateles fusciceps fusciceps*
	S. o. citrinellus	*Ateles belzebuth hybridus*
	Alouatta fusca	*Ateles belzebuth marginatus*
	A. f. fusca	
	A. f. clamitans	
	Lagothrix flavicauda	

The results of our analysis show that the highest primate diversity occurs in Amazonia, which has 12 genera and 39 species, followed by the Atlantic forest region of eastern Brazil, with six genera and 15 species (Fig. 3). The relatively high numbers for the northern and southern Andes are somewhat misleading; they result mainly from the extension of certain Amazonian primate species a short way into the Andean foothills above the 1000 m contour line that we have taken as the boundary between the two Andean regions and Amazonia. Above 1000 m in the Andes proper, there are very few primates, and there appears to be only one primate species endemic to the Andes, the Peruvian yellow-tailed woolly monkey (*Lagothrix flavicauda*).

Endemic species serve to highlight the importance of Amazonia and the Atlantic forest even more (Fig. 3). Although most of the regions have few, if any, endemic species, the Atlantic forest has two endemic genera and nine endemic species, and Amazonia has five endemic genera and 26 endemic species.

Priority Countries

We have also analyzed primate distribution for each *country* in the Neotropics, since most conservation programs are developed by country and not by broad biogeographical region. As is indicated in Figure 4, Brazil is by far the richest country in primates in the Neotropics, with 16 genera and 51 species; it is followed by Colombia and Peru, each with 12 genera and 27 species. These three countries also are at the top of the world list of primate

TABLE II. Neotropical Primates in the IUCN Red Data Book [Thornback and Jenkins, 1982][a]

		Genus and species	Common name
Callitrichidae			
81	E	*Callithrix flaviceps*	Buffy-headed marmoset
81	V	*Callithrix argentata leucippe*	White marmoset
81	E	*Callithrix aurita*	Buffy-tufted-ear marmoset
81	V	*Callithrix humeralifer*	Tassel-eared marmoset
81	E	*Saguinus oedipus oedipus*	Cotton-top tamarin
81	I	*Saguinus bicolor*	Bare-face tamarin
81	I	*Saguinus imperator*	Emperor tamarin
81	V	*Saguinus leucopus*	White-footed tamarin
81	E	*Leontopithecus rosalia*	Golden lion tamarin
81	E	*Leontopithecus chrysopygus*	Golden-rumped lion tamarin
81	E	*Leontopithecus chrysomelas*	Golden-headed lion tamarin
81	R	*Callimico goeldii*	Goeldi's monkey
Cebidae			
81	E	*Saimiri oerstedii*	Central American squirrel monkey
81	V	*Callicebus personatus*	Masked titi
81	V	*Chiropotes albinasus*	White-nosed saki
81	E	*Chiropotes satanas satanas*	Southern bearded saki
81	V	*Cacajao calvus*	Uakari (red and white)
81	V	*Cacajao melanocephalus*	Black-headed uakari
81	I	*Alouatta fusca*	Brown howler monkey
81	I	*Alouatta villosa*	Guatemalan howler monkey
81	E	*Lagothrix flavicauda*	Yellow-tailed woolly monkey
81	V	*Lagothrix lagotricha*	Woolly monkey
81	E	*Brachyteles arachnoides*	Woolly spider monkey
81	I	*Ateles belzebuth*	Long-haired spider monkey
81	I	*Ateles fusciceps*	Brown-headed spider monkey
81	V	*Ateles geoffroyi*	Geoffroy's spider monkey
81	V	*Ateles paniscus*	Black spider monkey

[a]The letter preceding each taxon is the category: E, endangered; V, vulnerable; R, rare; I, indeterminate.

diversity, with Brazil being the single richest country on earth in primates and Peru and Colombia being fifth and sixth (Table III).

Brazil stands out in primate endemism as well, with one in every three species being found there and nowhere else (Fig. 4). Peru and Colombia have five endemic species between them, but no other South or Central country has any endemic species (although several have endemics at the subspecific level). Again, on a global basis, Brazil's level of endemism is quite high, being surpassed only by Madagascar, with close to total primate endemism, and Indonesia, with approximately 50% endemism (Table IV).

Fig. 2. Major phytogeographic regions in the Neotropics [Gentry, 1982] (map by Stephen D. Nash).

Endangered Neotropical Primates by Region and by Country

Table I lists the 28 Neotropical primate genera, species, and subspecies that we presently consider endangered and updates the list in the IUCN Red Data Book [Thornback and Jenkins, 1982], but it is important to note that this list represents nothing more than a *minimum* estimate of endangered Neotropical primate taxa. Only those for which some data are available have been listed, and as field data on other little known or unknown species are gathered it will undoubtedly be necessary to add to the endangered list.

On a regional basis, by far the largest number (two genera, nine species) of endangered primates occurs in the Atlantic forest region of Brazil, followed by Amazonia and Middle America (Fig. 5). By country, Brazil, with 19 endangered taxa, 18 of them endemic, once again heads the list (Fig. 6). A brief discussion of the highest-priority regions and species follows.

Atlantic forest region of eastern Brazil. Since 1979, the World Wildlife Fund has carried out a broad conservation program in the Atlantic forest

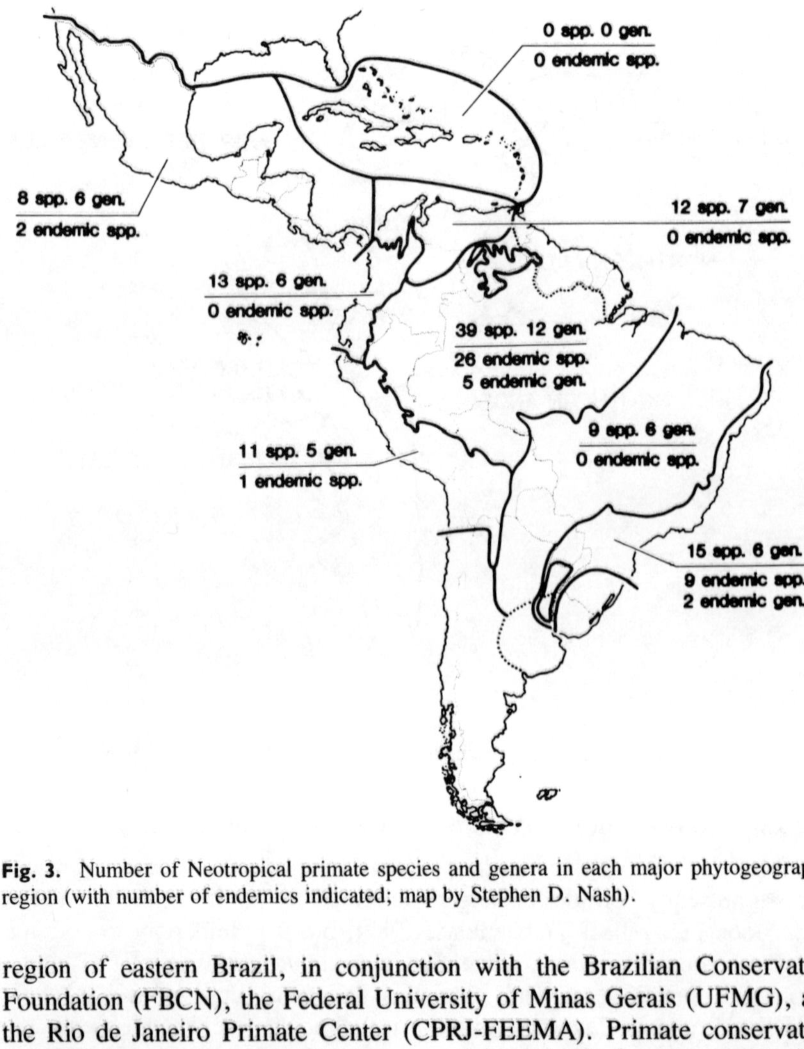

Fig. 3. Number of Neotropical primate species and genera in each major phytogeographic region (with number of endemics indicated; map by Stephen D. Nash).

region of eastern Brazil, in conjunction with the Brazilian Conservation Foundation (FBCN), the Federal University of Minas Gerais (UFMG), and the Rio de Janeiro Primate Center (CPRJ-FEEMA). Primate conservation activities have been a major component of the Atlantic forest program since its inception, and the primate survey project within this program has now investigated some 35 different parks, reserves, and privately protected forests in this part of Brazil. Although this project is still underway and will continue for some time to come, it has already given a good indication of the status and distribution of the primates of the region.

Twenty-one species and subspecies of primates occur in the Atlantic forest, and it appears that 13 of the 15 that are endemic are already endangered

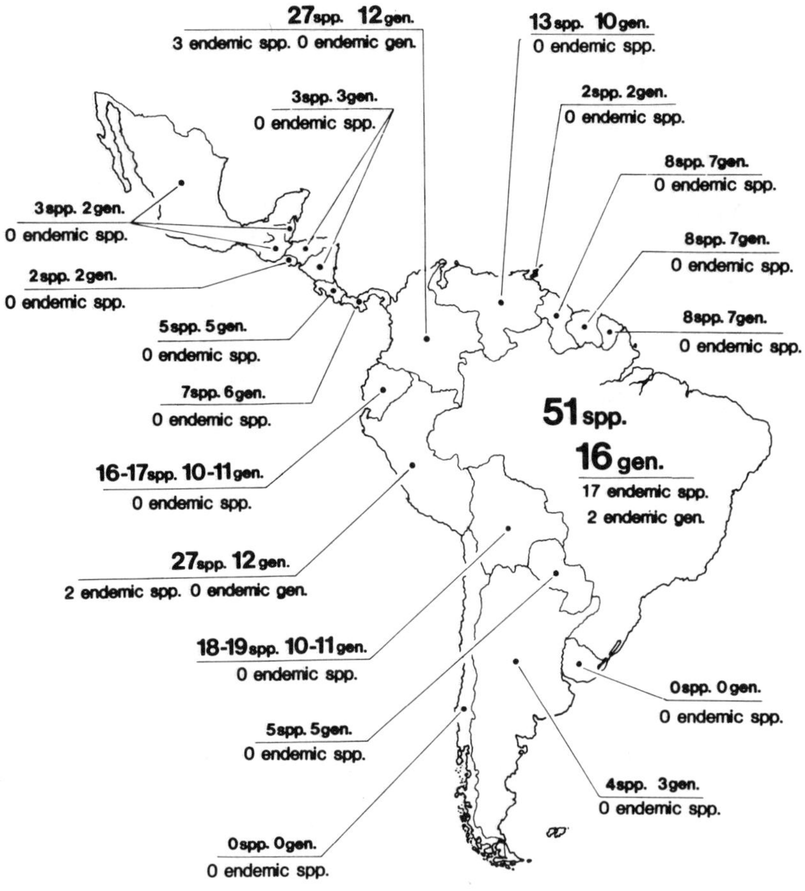

Fig. 4. Number of Neotropical primate species and genera by country (with number of endemics indicated; map by Stephen D. Nash).

(Table V). Using the criterion of taxonomic uniqueness, two endangered genera of the Atlantic forest must be considered the highest primate conservation priorities in the Neotropics, since they are the only primates in the entire region that are actually in danger of becoming extinct *as genera*. These are *Brachyteles*, a monotypic genus with just one species, the muriqui (*Brachyteles arachnoides*), and *Leontopithecus,* the lion tamarins, with three species, the golden lion tamarin (*Leontopithecus rosalia*), the golden-headed lion tamarin (*Leontopithecus chrysomelas*), and the black or golden-rumped lion tamarin (*Leontopithecus chrysopygus*).

TABLE III. Top Countries in the World for Primate Diversity

Country	No. of species	No. of genera
Brazil	51	16
Zaire	29–32	13–15
Cameroon	28–29	14
Madagascar	28	13
Peru	27	12
Colombia	27	12
Indonesia	27–30	8
Nigeria	23	13
Congo	22	14
Equatorial Guinea	21–22	12
Central African Republic	19–20	11–12
Gabon	19	11
Uganda	19	11
Bolivia	17–18	11–12
Angola	18–19	10–11

TABLE IV. Primate Endemism in the World's Top 15 Countries for Primates

Country	No. species (No. endemic species)	No. genera (No. endemic genera)	Endemic species (%)	Endemic genera (%)
Madagascar	28 (26)	13 (12)	93	92
Indonesia	27–30 (12–15)	8 (1)	44–50	12.5
Brazil	51 (18)	16 (2)	35	12.5
Colombia	27 (3)	12 (0)	11	0
Peru	27 (2)	12 (0)	7	0
Zaire	29–32 (2)	13–15 (0)	6–7	0
Nigeria	23 (1)	13 (0)	4	0
Cameroon	28–29 (0)	14 (0)	0	0
Congo	22 (0)	14 (0)	0	0
Equatorial Guinea	21–22 (0)	12 (0)	0	0
Central African Republic	19–20 (0)	11–12 (0)	0	0
Gabon	19 (0)	11 (0)	0	0
Uganda	19 (0)	11 (0)	0	0
Bolivia	17–18 (0)	11–12 (0)	0	0
Angola	18–19 (0)	10–11 (0)	0	0

Fig. 5. Distribution of endangered Neotropical primates by major phytogeographic region (map by Stephen D. Nash).

The muriqui is worthy of special mention in that it is the largest of the New World monkeys and also one of the most endangered. It was once found through much of southeastern Brazil, and the original population has been estimated at about 400,000 individuals [Aguirre, 1971]. However, it has suffered tremendously from habitat destruction and has always been a prime target for hunters, to the point that it must now be considered on the verge of extinction. Surveys conducted by the University of Minas Gerais and the World Wildlife Fund since 1979 have succeeded in accounting for only about 300 muriquis at ten widely separated sites. More might still exist, and surveys are underway to locate them, but these 300 are all that have been documented with certainty thus far.

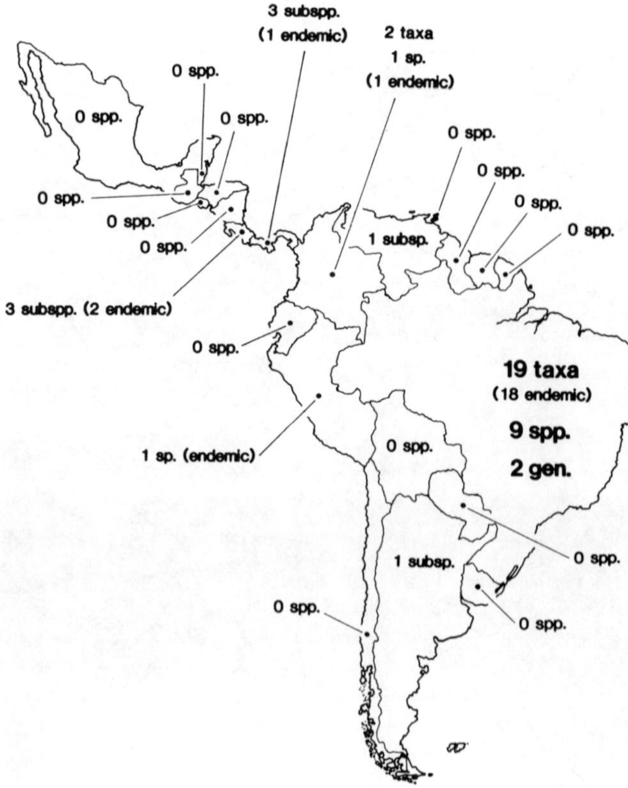

Fig. 6. Distribution of endangered Neotropical primates by country (map by Stephen D. Nash).

TABLE V. Monkeys of the Atlantic Forest Region

Species and subspecies	21
Endemic	15
Endangered	14
Endemic	13
Vulnerable	2
Not yet in danger	5

The muriqui is an important "flagship species" for the Atlantic forest region as well and provides an excellent example of how primates can be used to sell the cause of tropical forest conservation as a whole. Not only the largest of all neotropical monkeys, the muriqui is also the largest mammal endemic to Brazil and, as such, is most appropriate as a symbol of Atlantic forest conservation and of Brazilian conservation in general. Indeed, it is well on its way to becoming for Brazil what the giant panda (*Ailuropoda melanoleuca*) is for China.

The three lion tamarin species are also restricted to the Atlantic forest, and their original range is similar to, but less extensive than, that of the muriqui. The ranges of all three species were already quite small when the first Europeans arrived in Brazil, and they have now been reduced to tiny isolated fragments by widespread forest destruction.

The golden lion tamarin is found only in the state of Rio de Janeiro, and its hope for survival in the wild rests mainly with a single, small (5,000 ha) protected area, the Poço das Antas Biological Reserve. A major international program is now underway to improve the condition of this reserve and to ensure that the lion tamarin population is sufficiently large and well protected for the survival of the species. It includes ecological studies, restoration of the habitat, reintroduction of captive-born animals, and an outstanding education project [Dietz, 1985]. As a result of these efforts, carried out jointly by the National Zoo in Washington, D.C., the Rio de Janeiro Primate Center (CPRJ-FEEMA), the Brazilian Forestry Development Institute (IBDF), and the World Wildlife Fund, the prospects for the survival of wild populations of this spectacular and uniquely Brazilian species are better than they have been at any time during the last 15 years. Fortunately, the golden lion tamarin appears to be secure in captivity, with a large captive population thriving in American and European zoos (D.G. Kleiman, personal communication).

The golden-headed lion tamarin is found only in the southern part of the state of Bahia in eastern Brazil, and although its range is also quite small, it still has more habitat left than either of its two congeners. Unfortunately, forest destruction has been especially heavy within its range over the past 5 years, and the single protected area that exists in this region, the 5,000 ha Una Biological reserve, remains totally inadequate in its present state. An education campaign is now planned for this region as well, as is a field program modelled after the Golden Lion Tamarin Project in Rio de Janeiro.

The area in which the golden-headed lion tamarin occurs is also home to several other endangered primates that are found nowhere else, among them the buff-headed tufted capuchin (*Cebus apella xanthosternos*), the northern masked titi (*Callicebus personatus melanochir*), and Wied's marmoset (*Cal-*

lithrix kuhli), and it may still harbor a few groups of muriqui and northern brown howler monkeys (*Alouatta fusca fusca*). Action is needed there as soon as possible, since the remaining forest is being destroyed very rapidly, and an illegal trade, especially in the golden-headed lion tamarin, is developing there as well [Mallinson, 1984]. Only one legal breeding colony exists at this time at the Rio de Janeiro Primate Center.

The third of the lion tamarins, the black or golden-rumped lion tamarin, is certainly the rarest and probably the most endangered of all New World monkeys. Always restricted to the interior of the state of São Paulo, Brazil's most developed state, it now occurs with certainty in only two widely separated reserves, the 37,157 ha Morro do Diabo State Reserve and the 2,178 ha Caitetus Reserve. It is unlikely that more than 100 individuals remain in the wild, and a project is now underway to determine exactly how many still exist and how they can best be conserved. The state of São Paulo has taken an interest in the future of this species, and a collaborative effort involving the State Forestry Institute (IF), the Energy Company of São Paulo (CESP), the Brazilian Conservation Foundation (FBCN), the Rio de Janeiro Primate Center (CPRJ-FEEMA), the University of Minas Gerais, and the World Wildlife Fund, is now underway. Again, there is only one captive breeding colony in the world, at the Rio de Janeiro Primate Center.

Amazonia. At the generic level, the primates of greatest concern in Amazonia are *Ateles* and *Lagothrix*, which are the next most endangered New World monkeys at the generic level. Neither of these is in nearly as critical a situation as *Brachyteles* and *Leontopithecus*, which are in immediate danger of becoming extinct as genera, but both *Ateles* and *Lagothrix* are disappearing rapidly or are already gone from a good portion of their former ranges. They are hunted for food almost everywhere that they occur, with *Lagothrix* usually being both the preferred monkey food species and the most popular pet monkey. Both genera also reproduce slowly and are thus unable to bounce back from exploitation in the way that smaller monkeys (e.g., callitrichids) can. Finally, they are more demanding in choice of habitat and do not adapt well to man-modified environments. Consequently, the spider monkeys and woolly monkeys are usually the first to disappear in the face of human encroachment on previously undisturbed forest areas, and they have already disappeared from large areas of Amazonia where they once occurred. Market hunting has been an especially serious problem and has resulted in the disappearance of these animals even in areas of otherwise pristine primary forest habitat [Hernandez-Camacho and Cooper, 1976; Mittermeier and Coimbra-Filho, 1977]. The World Wildlife Fund has a special program underway aimed at assessing the status of *Ateles* and *Lagothrix* in key parts of their remaining ranges and also at improving the situation for these animals in captivity [Konstant et al., 1985]. Hopefully, these efforts will help to stimulate further conservation action on behalf of these animals.

Aside from the spider monkeys and woolly monkeys, the most endangered Amazonian monkeys are several subspecies with very restricted ranges in areas of heavy development. These include the southern bearded saki (*Chiropotes satanas satanas*) from the lower Amazonian region of Brazil, south of the city of Belem, the white uakari (*Cacajao calvus calvus*) from a very restricted areas between the Rio Solimões and the Rio Japurá in upper Amazonia, the pied tamarin (*Saguinus bicolor bicolor*) from the vicinity of the city of Manaus, and the white marmoset (*Callithrix argentata leucippe*), found only between two small Rio Tapajos tributaries in an area cut by the Transamazonian Highway. All these monkeys are endemic to Brazilian Amazonia.

Endangered primates from other regions. Although most of the endangered Neotropical primates are found in the Atlantic forest region of Brazil and Amazonia, several are endemic to other phytogeographic regions. Three of the best examples are the yellow-tailed woolly monkey (*Lagothrix flavicauda*), the Central American squirrel monkey (*Saimiri oerstedii*), and the cotton-top tamarin (*Saguinus oedipus*).

The yellow-tailed woolly monkey is the only truly Andean primate and is restricted to a small area in the Andes of northern Peru. Rediscovered in 1974 after not having been seen for almost 50 years [Mittermeier et al., 1975, 1977], it is threatened by habitat destruction and hunting in its once isolated cloud forest range. It is the most endangered Peruvian primate and, like the muriqui in Brazil, is also the largest mammal endemic to its country. An education and public awareness campaign focusing on this species is now underway, as are efforts to establish two new protected areas for this monkey and its habitat [Leo Luna, 1982] (also personal communication).

The Central American squirrel monkey (*Saimiri oerstedii*) is found only in western Panama and adjacent parts of southeastern Costa Rica. Although squirrel monkeys are generally among the most abundant primates in the South American portion of their range, the opposite is true of the isolated remnant populations of the Central American species. *S. oerstedii* is highly endangered, and it has now been separated into two distinct subspecies, *S. o. oerstedii* and *S. o. citrinellus* [Hershkovitz, 1984]. A recent population survey in Costa Rica indicated that there are no more than 3,000 *S. o. oerstedii* in Costa Rica and less than 300 *S. o. citrinellus* [Boinski, 1985]. Since the species is sure to be highly depleted and perhaps already extinct in Panama, it is possible that these Costa Rican individuals are all that remain. Fortunately, each subspecies is found in one of Costa Rica's national parks, Corcovado for *S. o. oerstedii* and Manuel Antonio for *S. o. citrinellus*.

The cotton-top tamarin from northwestern Colombia is another highly endangered Neotropical species. Its range has been subjected to heavy forest destruction over the past 2 decades, and this monkey was also heavily traded in the 1960s and early 1970s, with an estimated 20,000–30,000 exported during the peak years of the primate trade in Colombia [Hernandez-Camacho and Cooper, 1976]. Fortunately, the species is well represented in captivity, with 1,000–1,300 in zoos and research laboratories in the U.S. and Europe [Tardiff, 1986]. Very little information exists on the current status of this species in the wild, but it was already considered endangered when the last survey was undertaken in 1975 [Neyman, 1977], and the situation is certain to have deteriorated since that time.

RECOMMENDED ACTIONS

To ensure the survival of representative cross-section of all Neotropical primate species and subspecies in their natural habitats, the following actions are recommended.

1) Continue the ongoing primate conservation program in the Atlantic forest region of eastern Brazil in order to stimulate still greater interest in the 21 species and subspecies found there, and especially in the two highly endangered, endemic genera, *Brachyteles* and *Leontopithecus*.

2) Develop a similar, major primate conservation program for Amazonia, and especially Brazilian Amazonia, to provide a solid data base on the primates and protected areas of this vast region and to stimulate further conservation action on their behalf.

3) Continue ongoing efforts for key endangered primates and their habitats in priority areas outside Brazil (e.g., *Lagothrix flavicauda* in Peru; *Saimiri oerstedii* in Costa Rica and Panama, etc.)

4) Develop projects for key endangered species that have not yet received adequate attention (e.g., *Saguinus oedipus* in northern Colombia; Central American *Ateles,* etc.).

5) Continue ongoing public awareness campaigns focusing on key endangered primates and develop new campaigns whenever feasible to ensure that primate conservation activities remain in the public eye in the developing countries in which these animals occur *and* in the developed world from which much of the support for primate conservation activities comes.

6) Train a strong cadre of primate-oriented conservation biologists and tropical biologists from the developing countries, both by bringing key

individuals to the United States and Europe for higher-level training and by developing in-country training programs to reach large numbers of future conservationists.

Brazil clearly emerges as the key country for Neotropical primates, being home to 78% of all species from this region and having within its borders more tropical forest habitat than all the other South and Central American countries combined. Colombia and Peru follow closely behind, with highly diverse primate communities and several endemic species each. These are the countries where the greatest conservation effort should be made during the remainder of this century, but certainly not to the exclusion of developing appropriate conservation programs in other countries in which Neotropical primates occur. It should be possible to ensure both the survival of representative populations of *all Neotropical primate species* and the survival of representative primate communities *in all of the countries* of this region. Projects dealing with these recommendations in all Neotropical countries harboring nonhuman primate populations will be included in the *"Action Plan for Primate Conservation in the Neotropical Region*, 1986–1990," soon to be published by the IUCN/SSC Primate Specialist Group and the World Wildlife Fund.

REFERENCES

Aguirre AC (1971): "O Mono *Brachyteles arachnoides*." Rio de Janeiro: Brazilian Academy of Sciences.

Boinski S (1985): Evaluation of the status of the squirrel monkey, *Saimiri oerstedii*, in Costa Rica. Primate Conservation 6:15–16.

Dietz LA (1985): Captive-born lion tamarins released into the wild: A report from the field. Primate Cons. 6:21–27.

Gentry A (1982): Neotropical floristic diversity: Phytogeographical connections between Central and South America, Pleistocene climatic fluctuations, or an accident of the Andean orogeny? Ann Missouri Bot Gard 69:557–593.

Hernandez-Camacho J, Cooper RW (1976): The nonhuman primates of Colombia. In Thorington RW, Heltne PG (eds): "Neotropical Primates: Field Studies and Conservation." Washington, DC: National Academy of Sciences, pp 32–69.

Hershkovitz P (1984): Taxonomy of squirrel monkeys genus *Saimiri* (Cebidae, Platyrrhini): A preliminary report with description of a hitherto unnamed form. Am J Primatol 7:155–210.

Konstant WR, Mittermeier RA, Nash SD (1985): Spider monkeys in captivity and in the wild. Primate Conservation 5:82–109.

Leo Luna M (1982): Conservation of the yellow-tailed woolly monkey, *Lagothrix flavicauda*, in Peru. Int. Zoo Yearbook 22:47–52.

Mallinson JJC (1984): Golden-headed lion tamarin contraband a major conservation problem. IUCN/SSC Primate Specialist Group Newsletter 4:23–25.

Mittermeier RA, Coimbra-Filho AF (1977): Primate conservation in Brazilian Amazonia. In Prince Rainier of Monaco, Bourne GH (eds): "Primate Conservation." New York: Academic Press, pp 117–166.

Mittermeier RA, Macedo H de, Luscombe BA (1975): A woolly monkey rediscovered in Peru. Oryx. 13:41–46.

Mittermeier RA, Macedo H de, Luscombe BA, Cassidy J (1977): Rediscovery and conservation of the Peruvian yellow-tailed woolly monkey (*Lagothrix flavicauda*). In Prince Rainier of Monaco, Bourne GH (eds): "Primate Conservation." New York: Academic Press, pp 95–115.

Mittermeier RA, Oates JF (1985): Primate diversity: The world's top countries. Primate Conservation 5:41–48.

Neyman PF (1977): Aspects of the ecology and social organization of free-ranging cotton-top tamarins (*Saguinus oedipus*) and the conservation status of the species. In Kleiman DG (ed): "The Biology and Conservation of the Callitrichidae." Washington, DC: Smithsonian Institution, pp 39–71.

Tardif SD (1986): Status of the cotton-top tamarin (*Saguinus oedipus*) in captivity. Primate Conservation 6:38–39.

Thornback J, Jenkins M (1982): "The IUCN Mammal Red Data Book, Part 1: Threatened Mammalian Taxa of the Americas and the Australasian Zoogeographic Region (excluding Cetacea)." Gland, Switzerland: IUCN.

Primate Conservation in the Tropical Rain Forest, pages 321–327

12

A Framework for African Rain Forest Primate Conservation

J.F. Oates, J.S. Gartlan, and T.T. Struhsaker

Department of Anthropology, Hunter College of the City University of New York, New York, New York 10021 (J.F.O.), Wisconsin Regional Primate Research Center, University of Wisconsin, Madison, Wisconsin 53706 (J.S.G.), and Wildlife Conservation International, New York Zoological Society, Bronx, New York 10460 (T.T.S.)

INTRODUCTION

It is unlikely that more than a small percentage of the remaining areas of African rain forest can be protected from exploitation and conversion, and many existing primate populations are inevitably going to become extinct. There is thus an urgent need to formulate plans that will 1) ensure that no single species or highly distinct form of primate becomes extinct and 2) ensure the protection of examples of distinct rain forest communities.

In formulating these plans we must aim to protect areas large enough to give populations long-term viability. We must also recognize that local political, social, and economic conditions will play a major role in the success of any project and that projects need to be integrated into local development plans.

TARGET POPULATIONS, SPECIES, AND REGIONS

In selecting rain forest areas for special attention, we need to identify both currently endangered species of primates and distinctive threatened commu-

This chapter is based largely on a planning paper written in 1981 when the International Union for Conservation of Nature and Natural Resources (IUCN) and the World Wildlife Fund (WWF) were drawing up a major program for tropical rain forest and primate conservation. We believed that this program, and other plans being formulated at that time, lacked an appropriate conceptual framework for the identification of conservation priorities. Our paper attempted to lay out a framework for planning the conservation of African rain forest primates, an issue with which each of us is closely involved. The paper was originally published in the March, 1982, edition of the International Primatological Society Newsletter. It should be regarded as a provisional statement. A more detailed Action Plan for African Primate Conservation was recently published by the IUCN/SSC Primate Specialist Group [Oates, 1986].

nities. One problem in identifying endangered species is that there is no universal agreement on the species-level classification of African primates. For the purposes of this analysis, we will therefore regard as species both those populations recognized as such in all widely used classifications and those populations that vary in more than just minor details of coloration and size from others and have been seriously considered for species status in recent years, even if that status is not widely accepted. In the latter category are some of the 14 forms of red colobus monkeys, among which at least four distinct groups can be recognized on the basis of coat pattern and vocalizations: "*badius*" of West Africa (including *temminckii* and *waldroni*); "*pennantii*" of Cameroon, Fernando Po, and the Congo Republic (including *preussi* and *bouvieri*); "*rufomitratus*" of Zaire and East Africa (including *ellioti*, *foai*, *oustaleti*, *tholloni*, and *tephrosceles*); and "*kirkii*" of Tanzania (including *gordonorum*). If these four groups are counted as separate species, then there are at least 45 species of African primate largely restricted to forest areas.[1] Of these species, four can be considered as threatened with complete extinction because they occur in a limited geographical area and are nowhere either abundant or well protected. These are: drill, *Mandrillus leucophaeus*; white-throated guenon, *Cercopithecus erythrogaster*; red colobus, "*pennantii*" group; and red colobus, "*kirkii*" group.

A further seventeen species can be considered as particularly vulnerable in that they are either limited in distribution but at present abundant in a few areas or widespread but threatened in many places. These are: mandrill, *Mandrillus sphinx*; black mangabey, *Cercocebus aterrimus*; crested mangabey, *C. galeritus*; red-capped mangabey, *C. torquatus* (not including *atys* of West Africa); Diana monkey, *Cercopithecus diana*; owl-faced monkey, *C. hamlyni*; red-eared guenon, *C. erythrotis*; l'Hoest's guenon, *C. lhoesti*; Allen's swamp monkey, *Allenopithecus nigroviridis*; black colobus, *Colobus satanas*; black-and-white colobus, *C. polykomos*; black-and-white colobus, *C. vellerosus*; red colobus, "*badius*" group; red colobus, "*rufomitratus*" group; chimpanzee, *Pan troglodytes*; bonobo, *P. paniscus;* and gorilla, *Gorilla gorilla*.

Of these 21 threatened species of African forest primate, the drill, the black colobus, the chimpanzee, the bonobo or pygmy chimpanzee, and the gorilla are listed in the current IUCN Mammal Red Data Book together with the Tana River form of crested mangabey and some populations in the *kirkii*, *pennantii*, and *rufomitratus* groups of red colobus. The olive colobus monkey

[1]In our analysis, we are not considering populations of wide-ranging savanna species secondarily inhabiting rain forest.

(Procolobus verus) is also included as a "rare" form. The Red Data Book is being updated, and a new version should include all the species that we have listed.

Turning now to distinct communities, six different lowland forest primate communities can be recognized in Africa. Each community occurs in what was, until recently, a large forest block, and each displays both high diversity and a high level of endemism at the species or subspecies level. These communities are: 1) western Guinea (Sierra Leone, Liberia, and western Ivory Coast); 2) eastern Guinea (eastern Ivory Coast and western Ghana); 3) Cameroon (area around Mt. Cameroon, including S.W. Cameroon and E. Nigeria; 4) western equatorial (Gabon, Rio Muni, Congo Republic, and Cameroon south of the Sanaga River); 5) Congo Basin (Zaire south of the River Congo); and 6) Ituri (northeastern Zaire and adjacent parts of W. Rwanda and W. Uganda).

The locations of these communities are shown in Figure 1 and their primates are listed in Table I. The areas in which the communities occur are distinctive in their flora and fauna generally, not just in their primates. Much evidence suggests that this distinctiveness is the result of rain forest fragmentation during Pleistocene glacial maxima and that each of today's areas of high endemism probably contained its own forest refuge (or refuges) at some time in the Pleistocene.

Each of these areas is under threat, but a majority of Africa's rain forest primate species could still be protected if effective conservation measures were taken in suitable locations in each area. However, several distinct forms do not occur in the main forest blocks but on islands, both the actual islands of Fernando Po and Zanzibar and the forest "islands" surrounded by other types of vegetation on the East African mainland. Some of these are islands of montane vegetation (such as the Virunga Volcanoes), some are at midaltitude (such as Uganda's Kibale Forest), and some are low-lying (such as the East African coastal forests). The unique primate communities of these islands are listed in Table II.

Each of the species that we regard as threatened with extinction is a member of one of the communities listed in Tables I and II, with a single exception. This is the white-throated guenon, *Cercopithecus erythrogaster*, which occurs only in a limited area of southwestern Nigeria that might have been a secondary Pleistocene refuge.

RECOMMMENDED ACTIONS

To prevent the extinction of any species or distinct community of African forest primates, action must be taken to conserve the following effectively:

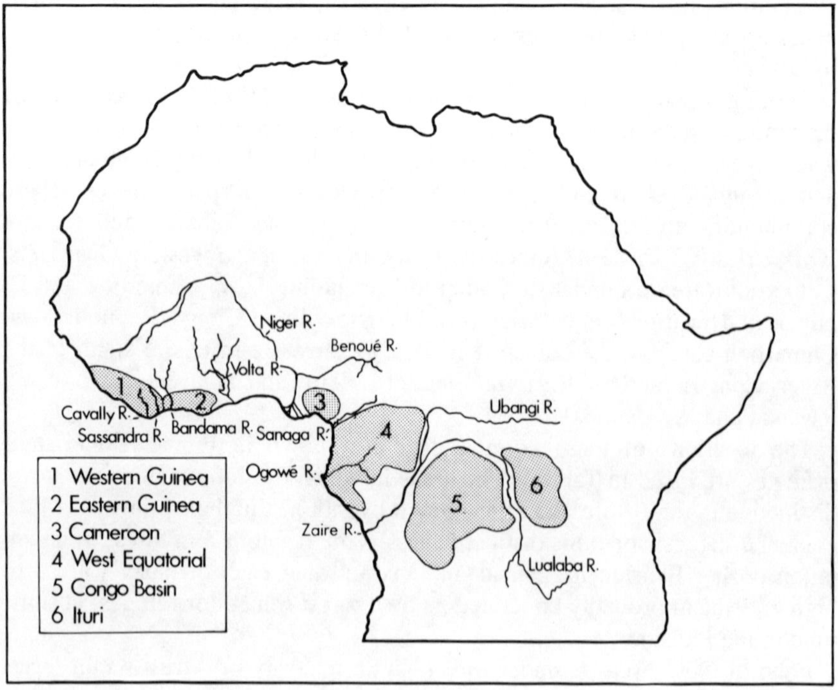

Fig. 1. The six lowland forest primate communities.

1) each of the six major lowland rain forest communities, 2) a representative set of East African forest island communities, 3) the primate communities of Zanzibar and Fernando Po, and 4) a viable population of *Cercopithecus erythrogaster*.

In taking these actions, many factors must be considered, and different considerations will apply in each case. In some areas, national parks or game reserves already exist but need to be made more effective. In other areas, potential reserves have been identified, but action is required to have them formally declared and effectively protected. In still other areas, basic surveys are needed to census primate populations, assess threats to them, and identify reserve sites. Within Africa, conservation education programs are required. Outside Africa, there is need for a greater awareness of the fact that an African reserve on paper is different from a reserve in practice. Many existing national parks are by no means inviolate, and some are almost completely unprotected from exploitation. We would like to stress that "forest reserves" are not equivalent to national parks; they are areas under govern-

TABLE I. Primate Communities of Six Major Lowland Forest Blocks in Tropical Africa

1. Western Guinea

Perodicticus potto potto
Galago demidovii demidovii
Cercopithecus petaurista buettikoferi
Cercopithecus diana diana
Cercopithecus campbelli campbelli
Cercocebus atys atys
Cercopithecus nictitans stampflii
Colobus polykomos
Procolobus "badius" (*badius* form)
Procolobus verus
Pan troglodytes verus

2. Eastern Guinea

Perodicticus potto potto
Galago demidovii demidovii
Cercopithecus petaurista petaurista
Cercopithecus diana roloway
Cercopithecus campbelli lowei
Cercocebus atys lunulatus
Colobus vellerosus
Procolobus "badius" (*waldroni* form)
Procolobus verus
Pan troglodytes verus

3. Cameroon

Arctocebus calabarensis calabarensis
Perodicticus potto edwardsi
Galago demidovii murinus
Galago alleni
Galago elegantulus pallidus
Cercopithecus erythrotis camerunensis
Cercopithecus lhoesti preussi
Cercopithecus pogonias pogonias
Cercopithecus mona
Cercopithecus nictitans martini
Mandrillus leucophaeus
Cercocebus albigena zenkeri
Cerocebus torquatus torquatus
Colobus guereza occidentalis
Procolobus "pennantii" (*preussi* form)
Pan troglodytes troglodytes
Gorilla gorilla gorilla

4. West Equatorial

Arctocebus calabarensis aureus
Perodicticus potto edwardsi
Galago demidovii murinus
Galago alleni
Galago elegantulus elegantulus
Cercopithecus cephus cephus
Cercopithecus pogonias subspp.
Cercopithecus neglectus
Cercopithecus nictitans nictitans
Miopithecus talapoin
Mandrillus sphinx
Cercocebus galeritus agilis
Cercocebus torquatus torquatus
Cercocebus albigena zenkeri
Colobus guereza occidentalis
Colobus satanas
Pan troglodytes troglodytes
Gorilla gorilla gorilla

5. Congo Basin

Perodicticus potto faustus
Galago demidovii phasma
Cercopithecus ascanius whitesidei
Cercopithecus pogonias wolfi
Cercopithecus neglectus
Cercopithecus mitis maesi
Allenopithecus nigroviridis
Cercocebus galeritus chrysogaster
Cercocebus aterrimus
Colobus angolensis angolensis
Procolobus "rufomitratus" (*tholloni* form)
Pan paniscus

6. Ituri

Perodicticus potto ibeanus
Galago thomasi
Galago inustus
Cercopithecus ascanius schmidti
Cercopithecus hamlyni
Cercopithecus neglectus
Cercopithecus lhoesti lhoesti
Cercopithecus mitis stuhlmanni
Cercopithecus pogonias denti
Cercocebus albigena johnstoni
Cercocebus galeritus agilis
Procolobus "rufomitratus" (*ellioti* and *oustaleti* forms)
Colobus guereza occidentalis
Colobus angolensis cottoni
Pan troglodytes schweinfurthi

TABLE II. Primate Communities of African Islands (Forest Species Only)

1. Fernando Po

   Galago demidovii murinus
   Galago alleni
   Galago elegantulus pallidus
   Cercopithecus erythrotis erythrotis
   Cercopithecus nictitans martini
   Cercopithecus lhoesti preussi
   Cercopithecus pogonias pogonias
   Mandrillus leucophaeus
   Colobus satanas
   Procolobus "pennantii" (*pennantii* form)

2. Zanzibar

   Galago garnettii panganiensis
   Galago zanzibaricus
   Cercopithecus mitis albogularis
   Procolobus "kirkii" (*kirkii* form)

3. East African forest "islands" (several of these species occur in only a small number of sites)

   Perodicticus potto ibeanus
   Galago thomasi
   Galago inustus
   Galago zanzibaricus
   Galago crassicaudatus and *garnettii*
   Cercocebus albigena johnstoni
   Cercocebus galeritus (*galeritus* and *sanjei*)
   Cercopithecus ascanius schmidti
   Cercopithecus mitis (several forms)
   Cercopithecus lhoesti lhoesti
   Cercopithecus neglectus
   Procolobus "kirkii" (*gordonorum* form)
   Procolobus "rufomitratus" (*rufomitratus* and *tephrosceles* forms)
   Colobus angolensis (several forms)
   Colobus guereza (several forms)
   Pan troglodytes schweinfurthi
   Gorilla gorilla beringei

ment control in which extensive exploitation (especially lumbering) is usually permitted. The precise meaning of "reserved" varies from country to country as do hunting practices and attitudes toward forest protection.

Bearing these considerations in mind, what known sites could protect the populations and communities listed above? We have identified the following as particularly significant on the basis of their representativeness, potential effectiveness, and/or their unique primates:

1) Gola Forest, Sierra Leone; Sapo National Park, Liberia; and Tai National Park, Ivory Coast (contain all members of the Western Guinea primate community);
2) Nini-Souhien National Park plus contiguous Ankasa Game Production Reserve, Ghana (contains all members of the Eastern Guinea primate community);
3) Okumu Forest Reserve, Nigeria (contains populations of *Cercocebus torquatus* and *Cercopithecus erythrogaster*);
4) Korup National Park, Cameroon (contains most of the Cameroon primate community);
5) Dja and Douala-Edea Reserves, Cameroon, and the Lopé Reserve, Gabon (together contain all members of the West Equatorial community);

6) The Lomako Forest, Zaire (contains a major *Pan paniscus* population, together with *Cercocebus aterrimus* and *Allenopithecus nigroviridis*);
7) The Ituri Forest, Zaire (contains all members of the Ituri community); and
8) an East African network consisting of the Parc National des Volcans (Rwanda), Parc National de Virunga (Zaire), Kigezi Gorilla Sanctuary (Uganda), Nyungwe Forest Reserve (Rwanda), Kibale Forest Reserve (Uganda), Bwindi Forest Reserve (Uganda), Jozani Forest Reserve, Zanzibar (Tanzania), Uzungwa Mountains (Tanzania), and Tana River Primate Game Reserve (Kenya; together these sites contain most of the East African forest island forms).

Efforts must be made to give more effective long-term protection to all these areas, even those that are already wildlife sanctuaries or national parks.

Second, where is investigative work needed to identify the best sites for protection of populations and communities not found in reserves that already exist or are planned? Among a wide range of high priority projects, two of particular importance are: 1) a survey of the area between Mamfe, Cameroon, and Obudu, Nigeria, where an isolated gorilla population may still occur, and 2) a survey of Fernando Po to determine whether any significant populations of endemic primates survive and can be effectively protected.[2]

We believe that all the sites and projects we have listed require attention, and the order in which they are listed does not imply a priority rating. Questions of priority are addressed in the recently published Action Plan for African Primate Conservation [Oates, 1986].

REFERENCE

Oates JF (1986): Action Plan for African Primate Conservation: 1986–90. Stony Brook, NY: IUCN/SSC Primate Specialist Group.

[2]After this paper was written, Thomas M. Butynski completed a survey of Fernando Po (Bioko) and found evidence of an increase in primate populations. His recommendations for conservation are being prepared for publication (Butynski, personal communication, 1986).

Primate Conservation in the Tropical Rain Forest, pages 329–341

13

Framework for Primate Conservation in Madagascar

Alison F. Richard and Robert W. Sussman

Department of Anthropology, Yale University, New Haven, Connecticut 06520 (A.F.R.); Department of Anthropology, Washington University, St. Louis, Missouri 63130 (R.W.S.)

INTRODUCTION

Madagascar is the fourth largest island in the world, with a surface area of about 587,044 km^2. Most of its rainfall is carried on easterly winds and falls on and to the east of the steep escarpment that runs the length of the eastern coast. The climate becomes progressively more arid moving west and south, with near desert conditions prevailing in the southwest. The topographic and climate variations present in Madagascar have given rise to a wide range of ecosystems, and, as a result of the island's long isolation from other land masses, many of the biotic components of these ecosystems are uniquely different from those in analogous communities elsewhere in the world (Fig. 1). The date of Madagascar's separation is now estimated at 200 million years [Rabinowitz et al., 1983], and its modern plants and animals can be classified as 1) descendants of the inhabitants of Gondwana, present before Madagascar drifted into isolation, and 2) descendants of plants or animals that floated or flew by chance to Madagascar after its separation. Many major taxonomic groups are absent or poorly represented. Among the endemic mammals, for example, there are today no perissodactyls, artiodactyls, or lagomorphs, and only one family of carnivores. However, those groups that did reach the island underwent major adapative radiations and in many instances assumed ecological roles occupied by later-evolving forms in other parts of the world.

The primates of Madagascar provide an outstanding example of the way in which evolutionary processes have worked on what must have been one or a handful of small founding populations to produce a unique and diverse array of modern species. Currently, there are a total of 13 genera and 22 species (some authors say as many as 28). All are endemic to Madagascar (although

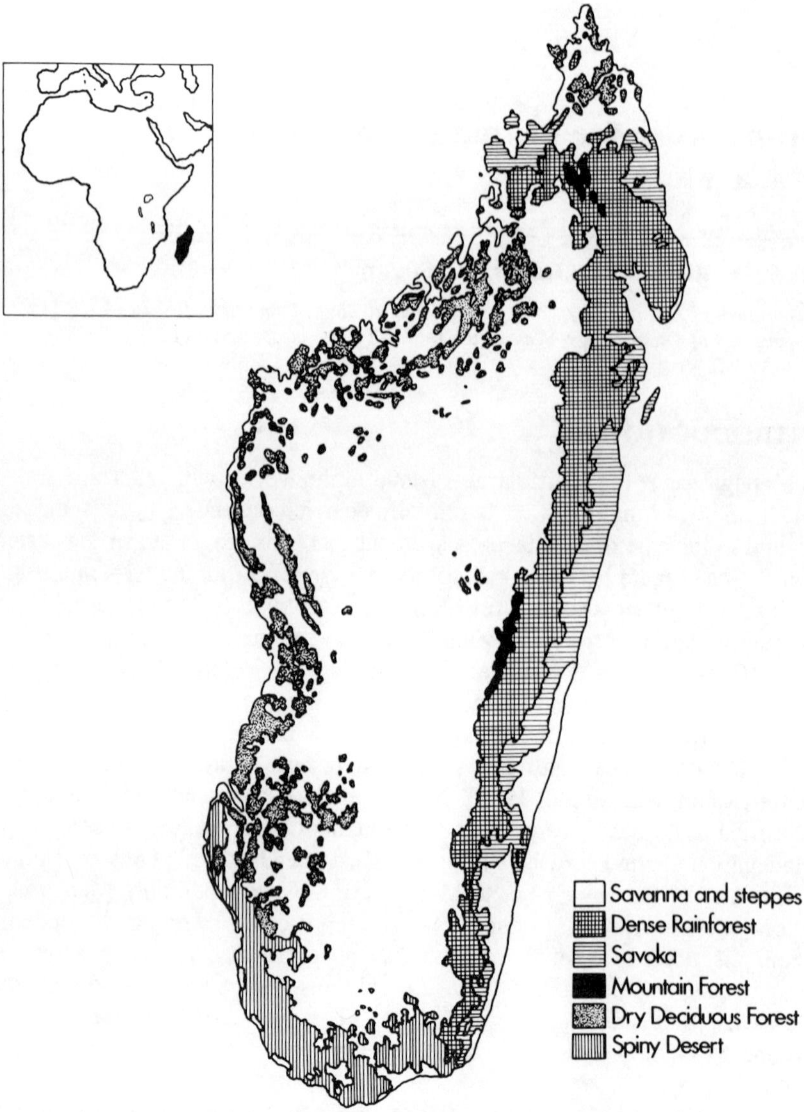

Fig. 1. Forests of Madagascar. Map shows the distribution of existing vegetation formations. Modified from Humbert and Cours Darne [1965].

Lemur fulvus and *L. mongoz* were introduced to two of the Comoro Islands several hundred years ago); all are in danger of extinction, and for some the danger is particularly acute (Table I.) Moreover, at least 14 lemur species, all of them larger than those living today, have become extinct since people arrived on the island about 1,500 years ago (Table II). We do not know what caused these extinctions, although it is likely that habitat destruction, hunting, and competition with introduced species all played a role [Dewar, 1984]. What is clear is that, today, over 90% of the island has lost its original vegetation and that continuing habitat destruction is the greatest threat to the survival of remaining species.

Madagascar is among the world's poorest nations: in 1982, only 21 nations had lower per capita GNPs. In 1986, there are about 10 million people on Madagascar, but World Bank projections indicate that by the year 2000 there might be as many as 16 million. The island exports little and as yet has produced no oil or mineable coal. Forest woods provide a substantial part of the domestic fuel, and forest clearing continues, especially in the east, to make fields for the cultivation of cash and subsistence crops. Despite these depressing facts, the situation is not without hope. The Malagasy are generally proud of their natural heritage, and there is a serious commitment at the level of the national government to incorporate conservation goals into the national development strategy. Moreover, an extensive network of parks and reserves already exists legislatively and, to a limited extent, on the ground (Figs. 2, 3). The central argument of this chapter is that a major goal of any conservation initiative in Madagascar must be to assess the adequacy of the existing reserve system and to reinforce and expand or modify it where necessary.

TARGET POPULATIONS, SPECIES, AND REGIONS

The selection of target populations, species, and regions for special attention is hampered by the lack of recent survey work in much of Madagascar, particularly in the eastern rain forests. Some of the targets proposed here have been selected, therefore, not because we know their problems to be particularly acute but because we know so little about them.

The six target species we have selected are all occupants of the eastern rain forest: *Varecia variegeta, Propithecus diadema, Lemur rubriventer, Indri indri, Allocebus trichotis, Hapalemur simus,* and *Daubentonia madagascariensis.* We selected these six because 1) all occur at low density and are particularly susceptible to the effects of habitat fragmentation, 2) all are threatened by the particularly high rate of habitat destruction in the east, and 3) we have little information about their current geographical ranges; this is

TABLE I. Data Relevant to the Survival of Malagasy Primates[a]

Genus and species	Geographical range			Population density per ha	Mean group size	Home range size (ha)	Population		Major threats	
	Wide	Local	Specialized				Abundant (A) or scarce (S)	Declining	Habitat destruction	Hunting
Lemur catta	x			1.6–2.15	17 (Range 5–30)	5.7–23	A	Probably	x	?
Lemur mongoz		x	?	?	2 (+ Offspring)	>1.15	?	Yes	x	
Lemur macaco		x		0.58	10 (Range 4–15)	?	?	Probably	x	
Lemur fulvus	x			0.4–12	9 (Range 4–17)	0.75–1	A	Probably	x	
Lemur coronatus		x		?	Large, many adult males and females	?	?	Probably	x	
Lemur rubriventer		x	?	?	<5	?	S	Probably	x	
Varecia variegata	?		?	?	2 (+ Offspring)	?	S	Yes	x	?
Lepilemur mustelinus	x			2–8.1	Solitary	0.1–0.3	A	Probably	x	
Hapalemur griseus	x		x	1.1–1.2	3–5 (?)	?	?	Yes	x	x
Hapalemur simus		x	?	Low	5–10	?	S	Yes	x	
Indri indri		x	?	0.09–0.16	2 (+ Offspring)	17.7–18	S	Yes	x	?
Avahi laniger	x			?	2 (+ Offspring)	?	?	Probably	x	
Propithecus diadema	?	?	?	Low	3 (Range 1–6)	20	S	Yes	x	?
Propithecus verreauxi	x			1.1–1.5	5 (Range 3–13)	2–8.5	A	Yes	x	?
Daubentonia madagascariensis		x	x	Low	Solitary	?	S	Yes	x	x
Cheirogaleus major		x	?	?	Solitary	?	?	Yes	x	
Cheirogaleus medius	x		?	3–4	Solitary	?	?	Probably	x	
Microcebus murinus	x		2.5–36	Solitary, nesting groups	0.07–0.2	A	Probably			
Microcebus rufus	x				?	Solitary nesting groups	?	A	Probably	?
Mirza coquereli		x	?	0.3	Solitary	8–10	?	Probably	x	
Allocebus trichotis		x	?	?	?	?	S	Yes	x	
Phaner furcifer		x	?	1–2	2 (+ Offspring)	?	A	Probably	x	

[a]Designations of "wide" distributions and "abundant" populations are used in relation to other lemurs on the island. In relation to most other primates, lemurs have very small geographic ranges and all species must be considered threatened [Wolfheim, 1983]. From Sussman et al. [1985].

particularly true of *Allocebus trichotis* and *Hapalemur simus*. *A. trichotis* is known from just four specimens, the most recent of which was captured south of Maroansetra in 1965; the provenance of the other three, collected during the nineteenth century, is unknown or confused. An effort made to find *A. trichotis* in 1975 was unsuccessful [Petter et al., 1977; Tattersall, 1982]. Our knowledge of *H. simus* is not much better; recent sightings have come only from the humid forest east of Fianarantsoa, although there is some evidence that the species was at one time fairly widespread in the east [Tattersall, 1982]. 4)There are indications that at least some of these species will be able to survive only in climax forest, and data are urgently needed to assess the responses of all six to low or moderate levels of habitat disturbance. We would repeat, however, that we know so little about the distribution and conservation status of almost all the Malagasy primates that it is difficult to identify particular targets.

There are four populations (used here in the sense of subspecies) that we believe may urgently require measures to ensure their immediate survival. These are *Propithecus diadema perrieri, Lemur macaco flavifrons, Hapalemur griseus alaotrensis,* and *Varecia variegeta rubra*. We note that there is a melanistic form of *Propithecus verreauxi* that has sometimes been considered a highly endangered subspecies (*P. v. majori*). It is not included in this list, however, because it is now viewed as a variant that occurs as a polymorphism in populations of *P. v. verreauxi* rather than as a separate subspecies [Tattersall, 1982].

Propithecus diadema perrierei is restricted to a small patch of forest near the northern tip of the island, and there may be as few as 100 or so remaining. Like the other subspecies of *P. diadema*, it lives at very low densities, estimated by Petter et al. [1977] as three or four individuals/km^2, and a study done in 1979 by Ratovomboahangy Bruno suggests that its remaining habitat is rapidly being encroached upon by agriculturalists. Attention has recently been drawn to the presence in northwestern Madagascar of what might be a distinct subspecies of *Lemur macaco*, *L.m. flavifrons* [Koenders et al, 1985]. The validity of this subspecific distinction is a matter of debate [Tattersall, 1982], but the destruction of remaining patches of its habitat is not, and both the taxonomic status and the conservation status of this population urgently require investigation. *Hapalemur griseus* reportedly occurs at low density throughout the eastern forests, but *H.g. alaotrensis*, largest bodied of the three subspecies, is restricted to the reed beds of Lake Alaotra and the surrounding marshes and is subject to intense hunting pressure: Fires are set in the reeds and animals are killed as they flee [Petter and Peyrieras, 1970].

Varecia variegata rubra is found only on the Masoala Peninsula in northeastern Madagascar. A natural reserve on the eastern flank of the peninsula

TABLE II. Characteristics and Distribution of Extinct Lemuriformes[a]

Genus and species	Size[b]	Probable locomotion	Location of finds	Comments
Varecia insignis	Small	Arboreal quadruped (cf. modern forms of *Varecia*)	Southwest	
V. jullyi	Small	Same as *V. insignis*	Central plateau	
Hapalemur gallieni	Small (size of *H. simus*)	Unknown	Central plateau	
Megaladapis madagascariensis	Large (240 mm cranium)	Short-limbed aboreal, possibly modified Koala-like locomation	South, southwest	The animals in this genus probably were aboreal browsers and highly folivorous
M. grandidieri	Large (290 mm cranium)		Central plateau	
M. edwardsi	Largest of three (300 mm cranium donkey size)		South, southwest	
Mesopropithecus pithecoides	Small	Uncertain	Central plateau	Slightly more robust *Propithecus*-type skull
M. globiceps	Small			
Palaeopropithecus ingens	Large	Arboreal hanger and climber (cf. sloth or orang-utan)	Central plateau, south southwest	Long, low skull with upturned nasal bones, generally sloth or orang-like, but dentition like other indriids

Archaeoindris fontoynonti	Large	Unknown	Central plateau	One specimen; dentition of modern indriid, but very large, high, and robust skull
Archaeolemur majori	Medium	Arboreal and terrestrial quadruped (cf. *Papio* baboon)	South, southwest, central plateau	Skull like indriid, but highly derived teeth (cf. ceropithecoid monkey); premolars—continuous shearing blade unlike any other primate; molar dentition highly adapted to grinding probably like gelada baboon—manual grazers; lost tooth comb
A. edwardsi	Medium			
Hadropithecus stenognathus	Medium	Terrestrial quadruped (cr. gelada baboon)	Central plateau, south, southwest	
Daubentonia robusta	Small	As living species but larger	Southwest, possibly central plateau	Represented only by long bones and teeth

[a]All species are probably diurnal except *D. robustus* (for which there is no skull).
[b]Small, equivalent to largest living Malagasy primates or somewhat larger; medium, size of small baboons; large, approximately chimpanzee size or larger.
Source: Tattersall [1982] (also personal communication). Adapted from Sussman et al. [1985].

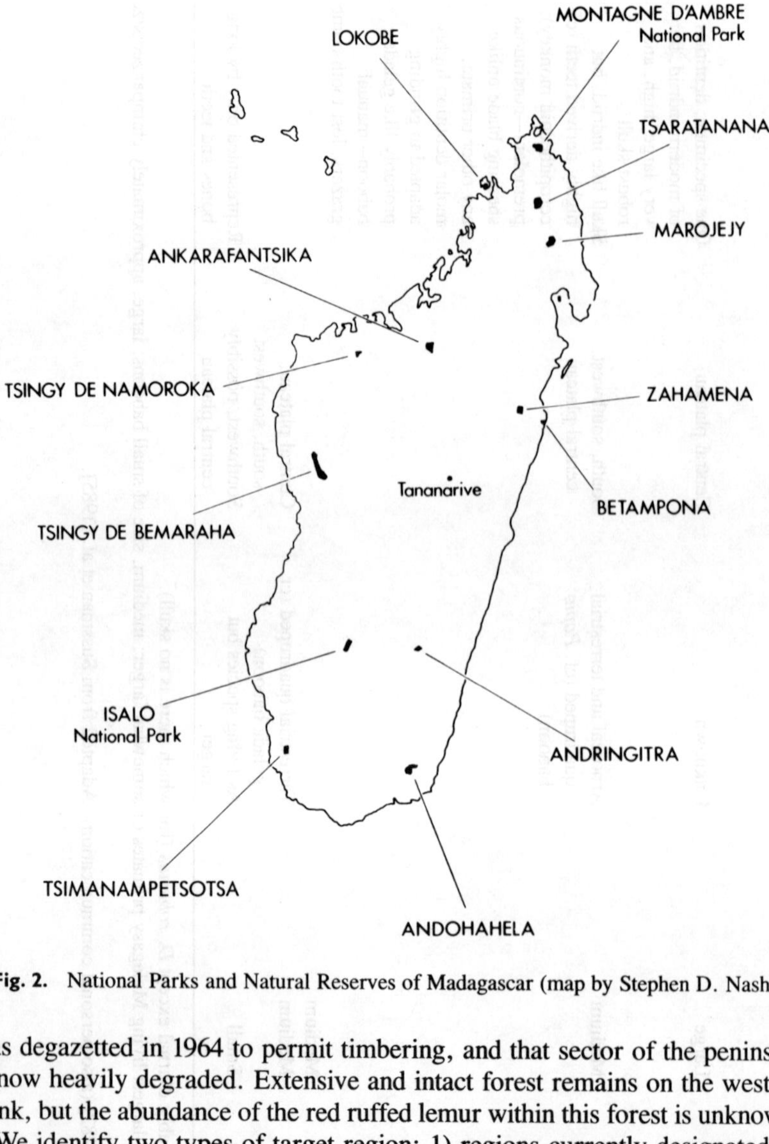

Fig. 2. National Parks and Natural Reserves of Madagascar (map by Stephen D. Nash).

was degazetted in 1964 to permit timbering, and that sector of the peninsula is now heavily degraded. Extensive and intact forest remains on the western flank, but the abundance of the red ruffed lemur within this forest is unknown.

We identify two types of target region: 1) regions currently designated by the government as protected areas and 2) regions containing major habitat types that are poorly represented in or absent from the protected areas. Madagascar's network of natural reserves was established in 1927 and persists with the additon of two reserves in 1939 and 1952 and the abandonment

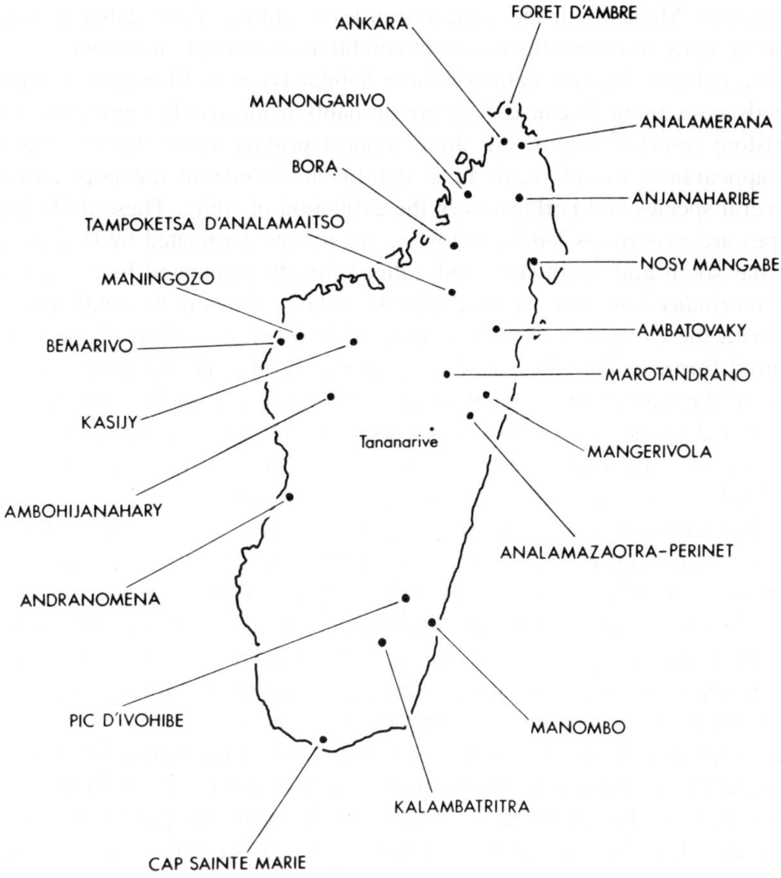

Fig. 3. Special Reserves of Madagascar (map by Stephen D. Nash).

of Reserve Number 2 (the Masoala Peninsula) in 1964, as noted above. In principle, the 11 currently existing reserves are completely protected, and access to them is regulated by a system of permits. Together they represent a surface area of about 569,542 ha, and they encompass a cross-section of most of the main ecosytems of the island. In addition to the reserves, there are two national parks (the Montagne d'Ambre and the Isalo National Park) and 22 special reserves [Andriamampianina, 1984]. These parks and special reserves are regulated by a confusing assortment of laws, but without exception their protection under the law is less complete than that of the natural reserves. Whatever the legal status of the various protected areas is in principle, Madagascar lacks the resources to enforce their status in practice and in many instances their current conditions are simply unknown.

We believe that the status of three habitat types in Madagascar urgently needs assessment, because these are probably inadequately represented in the existing reserve system. All three support primate communities, and their disappearance would result in a significant decline in the populations of several species and perhaps even the extinction of some. These three habitat types are eastern lowland rain forest, dry forests dominated by Didiereaceae in the south and southwest, and gallery forests dominated by the kily tree (*Tamarindus indica*) found on the banks of large rivers in the south and west.

With the exception of forests south of Maroanestra, most of the lowland humid forest of the island is already gone, and it is not well represented in any of the natural reserves. Remaining humid forests in the east are largely restricted to the steep slopes of the escarpment, where practical considerations protect them from encroachment by people, whereas the last vestiges of lowland forest are without this "topographic protection."

None of the existing vegetation maps of Madagascar shows the distribution of forests dominated by Didiereaceae, yet this is a highly distinctive vegetal formation with levels of species endemicity estimated to be as high as 95% [Guillaumet, 1984]. Moreover, preliminary surveys indicate that its distribution is extremely patchy. It occurs at low elevations in the western sector of Reserve Number 11 and in Reserve Number 10 (Tsimanampetsotsa). There has yet to be a thorough inventory of either the fauna or the flora of the latter reserve, located in the very driest part of the island, however, and the extent, diversity, and condition of its forests are unknown. In light of the uniqueness of the Didiereaceae forests, their apparently patchy distribution, and the relatively small areas currently protected, we believe their conservation status urgently requires further consideration.

Gallery forests of the south and west are in some ways the most problematic of the three habitat types targeted, because they occur in ribbons on river banks rather than in extensive tracts, making it practically difficult to protect them and ecologically difficult to ensure their survival even when protected if surrounding areas are degraded. They represent crucial resources for many elements of the southern and western fauna including the lemurs, however, and they are rapidly disappearing. For example, satellite images show that two private reserves, Berenty and Behaloka, contain the last remaining fragments of relatively intact gallery forest over the entire length of the Mandrare, one of the major river systems of the south. This makes the current die-off of *Tamarindus indica* trees at the Berenty Reserve a cause for particularly great concern (O'Connor, in preparation).

In the context of this discussion, it is pertinent to point out that in western Madagascar, where intact and luxuriant gallery forests are still to be found [Sussman, 1977], there are currently no natural reserves between Antsalova in the northwest and Lac Tsimanampetsotsa in the extreme southwest. A private reserve is being established at Analabe, near Morondava, but its surface area is relatively small. We believe that surveys in western Madagascar are a high priority to identify areas where the gradient from gallery forest to adjacent dry forest is still intact, with a view to establishing a natural reserve to protect a representative sample of western ecosystems as well as the gallery forest habitat in particular.

RECOMMENDED ACTIONS

We believe that three kinds of action are urgently needed in Madagascar: 1) systematic and intensive surveys of targeted populations, species, and regions; 2) training, both for the staff of the natural reserve system and, at a more advanced level, for young Malagasy scientists with a serious interest in conservation biology; and 3) development of integrated conservation and development strategies for all the natural reserves.

We envisage two approaches to the survey work. One is to use satellite images to provide rapid answers to questions such as those raised in the previous section of this paper about the extent of particular habitat types or the area of natural reserves still covered with relatively intact vegetation. The other is to carry out on-the-ground assessments of targeted regions and species. The goal of regional studies would be to provide inventories of plants and animals present and to assess their conservation status and current threats to their survival. The goal of species and population studies would be to delineate their geographical ranges, variations in abundance within these ranges, and threats to their survival. It is both necessary and desirable that these surveys be done in close collaboration with the staff responsible for the administration of protected areas and other natural resources in Madagascar. These staff are all employees of the government in the Ministère de la Production Animale (Elevage et Pêche) et des Eaux et Forêts. We also recommend that Malagasy students be included in survey work and given practical training and experience whenever possible.

The impact on primates of the second recommended action, training reserve staff and Malagasy conservation scientists, is less direct but just as essential. There is a growing group of Malagasy who care deeply about the conservation of their natural resources, but there are all too few with sufficient training to translate this concern into action. This second recommendation will take longer to bear fruit than the first, and it is all the more important that immediate steps be taken to initiate appropriate programs. Two heads of natural reserves are already scheduled to attend the School of Forestry in Cameroon next year under the auspices of the WWF-supported Beza Mahafaly Project. This project, based on an international collaboration between the University of Madagascar, Washington University in St. Louis, and Yale University, also provides training for students at the School of Agronomy at the University of Madagascar in the form of a 2 week field school conducted at Beza Mahafaly once or twice each year. These activities represent only a small beginning, however, and we recommend that 1) over the next 5 years all heads of reserves be given the opportunity to receive training in and discuss current ideas in resource conservation and management; an annual, 2 week seminar in Antananarivo might provide an appropriate forum for this undertaking; 2) young Malagasy foresters be recruited and trained as eventual replacements for the current staff of the reserves as they reach retirement; and 3) a serious effort be made to identify young Malagasy scientists at the University with the ability and motivation to go abroad and study for a higher degree in conservation biology and related fields.

The third action we recommend is a natural outgrowth if not, indeed, an integral part of the first. Its impact on primate populations will, like the second, be indirect but equally important. It is both unreasonable and unrealistic to expect that the island's natural reserve system can be strengthened if it is at the expense of people living in the vicinity of the reserves: For the most part, these people are subsistence cultivators and pastoralists who can ill afford to carry the primary, albeit indirect, economic cost. Consequently, we recommend that planning to offer surrounding populations alternatives to the exploitation of protected areas be an integral part of the conservation strategy for each existing reserve and for areas that may be added to the reserve system.

ACKNOWLEDGMENTS

This paper is based not only on our own work and ideas but also on the work and ideas of colleagues and collaborators, and it has benefitted greatly from discussions with them. In particular, we thank Joseph Andriamampianina, Robert Dewar, Al Gentry, Alison Jolly, Russ Mittermeier, Martin Nicoll, Sheila O'Connor, Mark Pidgeon, Jon Pollock, Pothin Rakotomanga, Georges Randrianasolo, Voara and Bodo Randrianasolo, and Gilbert Ravelojaona.

REFERENCES

Andriamampianina J (1984): Nature reserves and nature conservation in Madagascar. In Jolly A, Oberlé P, Albignac R (eds): "Madagascar." New York: Pergamon Press; pp. 219–228.

Dewar RD (1984): Recent extinctions in Madagascar: The loss of the subfossil fauna. In Martin PS, Klein RG (eds): "Quaternary Extinctions." Tucson: University of Arizona Press, pp 57–93.

Guillaumet, J-L (1984): The vegetation: An extraordinary diversity. In Jolly A, Oberlé P, Albignac R (eds): "Madagascar." New York: Pergamon Press, pp 27–54.

Humbert, H and Cours Darne G (1965): Notice de la carte Madagascar, Carte International de Tapis Vegetal. Trav. Sect, Sci. Tech. Inst. Français. Pondichery. Hors Série 6:1–162.

Koenders, L., Rumpler Y, Brun B (1985): Notes on the recently rediscovered Sclater's lemur *Lemur macaco flavifrons* Primate Conservation 6:35.

Petter J -J, Albignac R, Rumpler Y (1977): "Faune de Madagascar 44: Mammifères Lémuriens (Primates Prosimiens)." Paris; ORSTOM/CNRS.

Petter, J -J, Peyrieras A (1970): Observations éco-écologiques sur les lémuriens malgaches du genre *Hapalemur*. Terre Vie 24:356–382.

Rabinowitz PD, Coffin MF, Falvey D (1983): The separation of Madagascar and Africa. Science 220:67–69.

Sussman RW (1977): Distribution of the Malagasy lemurs. Part 2: *Lemur catta* and *Lemur fulvus* in southern and western Madagascar. Ann NY Acad Sci 293:170–184.

Sussman RW, Richard AF, Ravelojaona G (1985): Madagascar: Current projects and problems in conservation. Primate Conservation 6:53–59.

Tattersall, I (1982): "The Primates of Madagascar." New York: Columbia University Press.

Wolfheim, JH (1983): "Primates of the World: Distribution, Abundance, and Conservation." Seattle: University of Washington Press.

Primate Conservation in the Tropical Rain Forest, pages 343–354

14

A Framework for Primate Conservation Priorities in Asian Moist Tropical Forests

Clive W. Marsh

Senior Conservation Officer, Sabah Foundation, Kota Kinabalu, Sabah, Malaysia

INTRODUCTION

As for other tropical regions, the primary goals of Asian primate conservation are to ensure that no species becomes extinct and that representative examples of all distinct communities are protected. Primates can also serve as the "selling point" for protected areas with much wider conservation value. Conversely, funding and political support for reserves that are important for primate conservation can sometimes best be pursued by stressing the conservation of other species or the potential of an area for education and training. Although such practical considerations must often determine which projects are implemented, these decisions should be made in the context of an underlying system of conservation priorities. This requires the identification of endangered species and biogeographically distinct communities and an assessment of what is known of these and the effectiveness of current conservation measures. This paper presents such an outline biogeographical scheme for the region and suggests priorities for status surveys and protection efforts in the field.

Biogeographical divisions vary with the taxa under consideration and the aims of the classification. The system used here is confined to primates and gives particular weight to areas of high species diversity or endemism. Only forms that are generally considered as distinct species are included [following Kavanagh, 1983], but it is recognized that the taxonomy of some groups (e.g., *Presbytis* and Sulawesi macaques) may be subject to revision. More

An Action Plan for Asian Primates is currently being prepared by the IUCN Primate Specialist Group. This chapter is offered as a contribution to the review process and has benefitted from correspondence or discussion with the following individuals: R. Ali, E.L. Bennett, W.Y. Brockelman, D.J. Chivers, A.A. Eudey, M. Kavanagh, and R.E. Salter.

detailed analyses within particular zoogeographic provinces or countries should elaborate at the subspecies level.

TARGET POPULATIONS, SPECIES AND REGIONS

Asian primates are confined principally to those tropical regions south and southeast of the Himalayan ranges, which constitute the Oriental biogeographical region. The only exceptions are three species of the highly adaptable genus *Macaca*, which occur outside the region (*M. fuscata, M. sylvanus,* and, marginally, *M. mulatta),* and a baboon, *Papio hamadryas,* in southern Arabia. Within the Oriental region, four subregions are generally recognized [e.g., Lekagul and McNeely, 1977], and within this framework a scheme of 18 provinces based on the distribution of 51 commonly recognized primate species in the region is presented in Table I. Approximate provincial boundaries are shown in Figure 1. In many cases, boundaries are precisely delineated by coastlines or rivers, but in some they are more diffuse. In general, boundaries have been placed where the limits of at least three species' ranges are approximately coincident.

As elsewhere, the distribution and speciation of primates in Asia has been strongly influenced by climatic changes in the Pleistocene, which have caused periodic regression of forest to montane refugia, especially in the Indochinese subregion [Eudey, 1980]. A related factor, of unique importance in the Sundaic subregion, has been Pleistocene changes in sea level, which have several times created and broken land bridges between the principal islands. Great rivers, such as the Brahmaputra, Salween, and Mekong have also acted as barriers to some species. In total, 24 species of primate occur in the Sundaic subregion, 21 species in the Indochinese subregion, nine species in the Indian subregion, and seven in Wallacea.

Starting in the west, in the Indian subregion, several species thrive outside forest areas and are not immediately threatened (e.g., *Presbytis entellus, Macaca mulatta,* and *M. radiata*), although certain subspecies might be vulnerable (e.g., *M. radiata diluta* [Korup, 1982]). However, two obligate forest species, *Presbytis johnii* and *Macaca silenus,* are confined to relict forest areas in the western Ghats, and both are endangered, the latter species critically so [Green and Minkowski, 1977; Ali, 1985].

In the northeast of the Indian subregion, a line joining the lower Brahmaputra river with the foothills of the Himalayas, in Bhutan, demarcates a boundary with the much richer Indochinese primate fauna to the east. Within the Indochinese subregion, there are six langurs (*Presbytis* spp.), all in the *cristata* species group of this genus, which are largely allopatric in distribu-

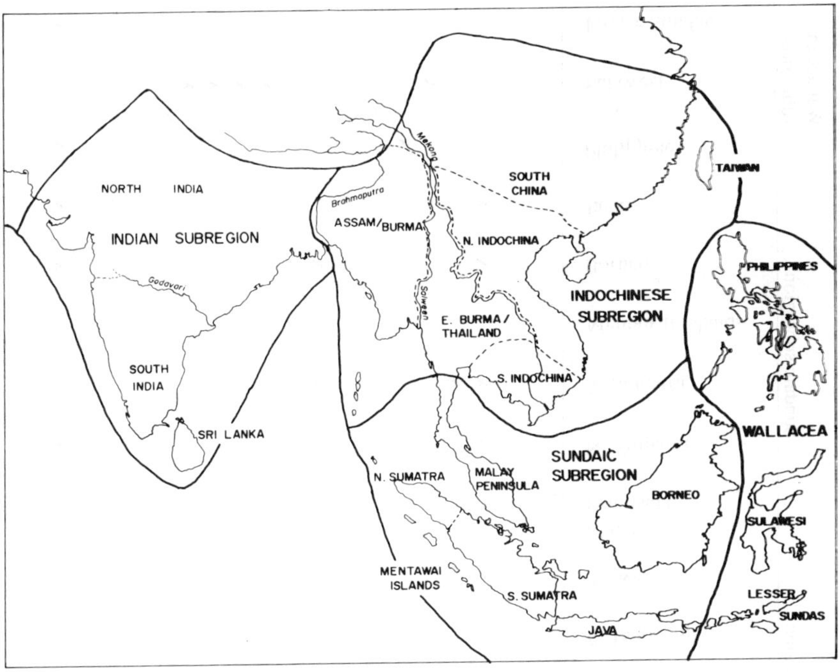

Fig. 1. An outline biogeographical scheme for primate conservation in the Oriental region based on four subregions and 18 provinces.

tion [Fooden, 1976]. The westernmost species, *Presbytis geei,* is confined to a tiny area around Bhutan but appears to be quite common within its range [Mukherjee and Saha, 1974; Mukherjee, 1978]; the status of *P. pileata* is little known in Assam and western Burma [Mukherjee, 1978; Tun Yin, 1967] but is reasonably secure in Bangladesh [Khan, 1981; Gittins and Akonda, 1982]. Farther east, *P. francoisi* has a restricted range in North Indochina and has yet to be adequately surveyed. Also occurring in this area and neighboring South China are two other endangered colobines in different genera, *Rhinopithecus* and *Pygathrix.* Little is known about either of these forms in the wild [see Baldwin et al., 1975], but all three subspecies of *Rhinopithecus roxellanae,* in China, and *R. avunculus,* in Vietnam, appear to be critically endangered [Zhang et al., 1981; Poirier, 1983; Constable, 1982]. Also in Vietnam, and possibly still on the island of Hainan, is *Pygathrix nemaeus,* which is listed as endangered by IUCN [Goodwin and Holloway, 1978]. For Vietnam, there is a recent report that this species might still be quite widely distributed [Constable, 1982].

TABLE I. Biogeography of Primates in the Oriental Region[a]

		Indian subregion			Indochinese subregion						Sundaic subregion						Wallacean subregion		
Species		Sri Lanka	S. India	N. India	Assam/Burma	E. Burma/Thailand	S. China	N. Indochina	S. Indochina	Taiwan	Malay Peninsula	N. Sumatra	S. Sumatra	Mentawai Islands	Borneo	Java	Philippines	Sulawesi	Lesser Sundas
Prosimians (5 spp.)																			
Loris tardigradus		x	x																
Nycticebus coucang				?	x	x		x	x		x	x	x		x	x			
Tarsius bancanus													x		x				
T. spectrum																		x	
T. syrichta																	x		
Cercopithecines (15 spp.)																			
Macaca silenus	*Silenus* group		x																
M. nemestrina					x	x		x			x	x	x		x				
M. pagensis														x					
M. tonkeana																		x	
M. maura																		x	
M. ochreata																		x	
M. nigra																		x	
M. sinica	*Sinica* group	x																	
M. radiata			x																
M. assamensis				x	x	x	x	x											
M. thibetana							x												
M. fascicularis	*Fascicularis* group					x			x		x	x	x		x	x	x		x
M. cyclopis										x									
M. mulatta				x	x	x	x	x											
M. arctoides					x	x	x	x	x		x								

Colobines (21 spp.)															
Presbytis aygula	*Aygula* group														x
P. hosei														x	
P. melalophos										x		x		x	
P. thomasi											x				
P. frontata														x	
P. rubicunda														x	
P. entellus		x	x	x	x										
P. senex	*Senex* group	x													
P. johnii			x												
P. cristata	*Cristata* group					x			x	x	x	x		x	x
P. pileata					x										
P. geei					x										
P. obscura										x					
P. phayrei					x	x		x							
P. francoisi							x	x							
P. potenziani													x		
Nasalis larvatus														x	
Simias concolor													x		
Rhinopithecus roxellanae					?		x								
R. avunculus								x							
Pygathrix nemaeus								x							
Apes (10 spp.)															
Hylobates lar						x				x	x				
H. pileatus									x						
H. moloch															x
H. muelleri														x	
H. agilis										x		x		x	
H. hoolock					x										
H. klossi													x		
H. concolor								x	x						
H. syndactylus										x	x	x			
Pongo pygmaeus											x			x	

[a]Note that provincial names and boundaries refer to those shown in Figure 1 and do not necessarily correspond to political boundaries, e.g., N. Indochina includes part of what is politically China.

Like most of the Indochinese langurs, gibbons in continental Southeast Asia are almost exclusively allopatric (Chivers, 1977). Two species, *Hylobates concolor* and *H. pileatus,* are considered threatened, and *H. hoolock* might also be, but the bulk of its range, in Burma, has yet to be properly surveyed (R. Salter, personal communication).

In contrast to langurs and gibbons, several Indochinese macaques are sympatric, at least at the provincial level, and in some cases at the local (syntopic) level also. For example, Huai Kha Khaeng reserve, in western Thailand, is remarkable for its community of five species—*Macaca mulatta, M. assamensis, M. arctoides, M. nemestrina,* and *M. fascicularis* [Eudey, 1980]. In this case, high species diversity argues for a high conservation priority, and one of the species, *M. arctoides,* might deserve threatened status. A further macaque, *M. thibetana,* is confined to South China but is quite widely distributed in mountainous regions [Tan Bangjie, 1985]. *M. cyclopis* in Taiwan has a smaller range and is apparently threatened [Poirier, 1980].

In the Sundaic subregion to the south, the climate is equatorial rather than monsoonal, and thus less seasonal. Forests are generally taller in structure, richer in species, and more completely evergreen than those to the north. A typical primate community in the Malay Peninsula (which includes southern Thailand), Sumatra, Borneo, or Java includes at least one prosimian (*Nycticebus coucang*), two macaques (*Macaca fascicularis* and *M. nemestrina*), two langurs (usually one in the *cristata* species group and the other in the *aygula* group, except in Borneo), and from one to three apes (*Hylobates* spp. and *Pongo pygmaeus*). This subregional uniformity reflects the conjunction of the Greater Sunda Islands and the Malay Peninsula until land bridges were broken in Pleistocene times; since then sea barriers and island extinctions have given rise to many differences at the species and subspecies level. Thus each of the major islands plus the Malay Peninsula constitutes a separate zoogeographic province. On Sumatra, consideration of ape and langur distributions suggests a marginal further division between the northern one-quarter of the island and the remainder [Wilson and Wilson, 1977; Rijksen, 1978], and the Mentawai islands off the continental shelf of western Sumatra are highly distinctive. These islands have been isolated for at least 500,000 years and support four endemic species, one of which is also a monotypic genus (*Simias concolor*). This is an unmatched level of endemism in relation to area for primates anywhere and demands the highest priority. As a result of both intensive traditional hunting and recent exploitation of timber on the main island of Siberut, all four Mentawai species are endangered [Whitten et al., 1979; WWF, 1980].

TABLE II. Asian Primates Currently Listed in the IUCN Red Data Book [Goodwin and Holloway, 1978, and updated]

Species	Status
Tarsius bancanus	Indeterminate
Tarsius syrichta	Endangered
Macaca silenus	Endangered
Macaca pagensis	Indeterminate
Presbytis johnii	Vulnerable
Presbytis geei	Rare
Presbytis potenziani	Indeterminate
Nasalis larvatus	Vulnerable
Simias concolor	Endangered
Rhinopithecus roxellanae	Rare
Pygathrix nemaeus	Endangered
Hylobates pileatus	Endangered
Hylobates moloch	Endangered
Hylobates klossi	Vulnerable
Hylobates concolor	Indeterminate
Pongo pygmaeus	Endangered

Borneo supports another monotypic colobine genus (*Nasalis*) and four other endemic species (three *Presbytis* spp. and *Hylobates muelleri*), but since the island is very large these forms are probably less vulnerable. However, the two largest Bornean primates, *Pongo pygmaeus* and *Nasalis larvatus,* both appear to be rather intolerant of habitat disturbance and are considered threatened [Davies, 1983]. The status of *Presbytis frontata* is uncertain (M. Kavanagh, personal communication).

Primate diversity is poorer on Java than on the other Greater Sunda Islands, and the forests are under great pressure from human population. Hence the Javan gibbon, *Hylobates moloch,* is endangered, and the langur *Presbytis aygula* is threatened to a lesser degree (E. Brotoisworo, personal communication). Both are confined to the western one-third of the island.

The Wallacean subregion is really less a discrete zoogeographic entity than three separate areas east of the Sunda Shelf that support some relicts of the ancestral Asian primate community together with more recent immigrants. *Macaca fascicularis,* for instance, is widespread in the Philippines and most of the Lesser Sunda Islands, where it is probably a recent arrival. However, it is now rare or extinct on some Philippine islands, probably because of excessive trapping (personal observation). Of greater concern for conservation are two species of tarsier (*Tarsius syrichta* in the Philippines and *T. spectrum* in Sulawesi) and a relict assemblage of allopatric macaque forms in

Sulawesi, which are variously considered as from one to seven species. Four species are recognized here, following Groves [1980], and at least one of these, *M. maura,* on the southwestern arm of the island is threatened [Brotoisworo, 1982]. Sulawesi, like the Mentawai Islands, is thus a province of threatened endemism within a relatively small area [Mackinnon, 1980].

Currently, 16 species of Asian primate are listed in the Red Data Book in four categories (Table II), but the list is in need of review. On the basis of published or informal evidence, or lack of it, species that appear candidates for addition to the list include the following: *Macaca maura (V), M. thibetana (I), M. cyclopis (R), M. arctoides (V), Presbytis aygula (V), P. frontata (I), P. francoisi (I),* and *Rhinopithecus avunculus (E).* On the other hand, it is not clear that *Tarsius bancanus* still merits indeterminate status following recent evidence of wide distribution and habitat tolerance in Sabah [Davies and Payne, 1982] and Sarawak [Aken and Kavanagh, 1982].

RECOMMENDED ACTION

For several of the species listed in the Red Data Book or which might merit inclusion, there exists very little recent field information. A first recommendation is therefore to list provinces and species most in need of more thorough survey from the standpoint of primate conservation. These are given in Table III. Note, however, that many species that are not threatened overall nonetheless urgently require surveys within particular parts of their range (e.g., *Macaca fascicularis* in the Philippines).

Five out of the ten provinces listed as needing better surveys lie partly or wholly within the communist or socialist countries of continental Southeast Asia. Although most of these present severe difficulties for foreign scientists to visit or to liase with local scientific institutions, some, such as China, are currently more open to foreign contact. A valuable goal of scientific diplomacy should be to cultivate such links and encourage or assist local scientists to carry out surveys and make their findings available internationally.

For other species listed in Table II, improved information on status or taxonomy is considered less important than the gazettement or improved management of national parks and other protected areas. Table IV lists a selection of reserves within each province that are believed to be particularly significant on the basis of their representativeness, potential effectiveness, and/or their unique primate species. However, the list is certainly not comprehensive, nor is it meant to imply that the establishment of one or two large reserves within each province is a sufficient commitment even for the

TABLE III. Zoogeographic Provinces and Species Most in Need of Status Surveys

Province	Countries	Species
Assam/Burma	India, Bangladesh, and especially Burma	*Nycticebus coucang* *Macaca assamensis* *Presbytis pileata* *Presbytis phayrei* *Hylobates hoolock*
E. Burma/Thailand	Burma, Thailand	Macaca spp. especially *Macaca arctoides* *Presbytis phayrei*
South China	China	*Rhinopithecus roxellanae* *Pygathrix nemaeus* *Macaca thibetana*
N. Indochina	China, Laos, Vietnam	*Macaca arctoides* *Rhinopithecus avunculus* *Pygathrix nemaeus* *Presbytis francoisi* *Hylobates concolor*
S. Indochina	Especially Kampuchea	*Hylobates pileatus*
Borneo	Especially Indonesia	*Nasalis larvatus*, *Hylobates muelleri/agilis* border zone *Presbytis frontata* *Pongo pygmaeus*
Java	Indonesia	*Presbytis aygula* *Hylobates moloch*
Philippines	Philippines	*Tarsius syrichta* *Macaca fascicularis*
Mentawai Islands (other than Siberut)	Indonesia	*Macaca pagenesis* *Presbytis potenziani* *Simias concolor* *Hylobates klossi*
Sulawesi	Indonesia	*Macaca maura*

limited purpose of primate conservation. Since species occurrence, ecology, and behavior vary greatly with habitat, the principal dimensions of habitat variation within each country must be recognized and protected [Marsh and Wilson, 1981]. Concern for primates should therefore be set within the broader context of a detailed national conservation strategy for each of the countries in the region.

TABLE IV. Some Important Reserves for Conservation of Asian Primates in Moist Tropical Forests

Province	Country	Reserve	Remarks
Sri Lanka	Sri Lanka	Sinharaje	Last lowland rainforest
S. India	India	Agastyamalai	S. Indian community
		Silent Valley/ New Amarambalam	S. Indian community Esp. *Macaca silenus*
		Anaimalai	*Presbytis johnii*
Assam/Burma	Burma	Alaungdaw Kathapa	*Presbytis pileata*
		Tamanthi	*Hylobates hoolock*
South China	China	Wolong and others	*Rhinopithecus roxellanae* *Macaca* spp.
E. Burma/ Thailand	Thailand	Huay Kha Khaeng	Five sympatric *Macaca* spp.
		Khao Yai	*Hylobates lar*/*H. pileatus* border zone
		Khao Soi Dao	*Hylobates pileatus*
Malay Peninsula	Malaysia	Taman Negara	Malayan community
		Krau	Malayan community
N. Sumatra	Indonesia	Gunong Leuser	Three sympatric apes and *Presbytis thomasi*
S. Sumatra	Indonesia	Kerinci-Seblat	South Sumatran community
Mentawai Island	Indonesia	Teitei Batti	Four endemic spp.
Borneo	Malaysia	Lanjak-Entimau	*Pongo pygmaeus* *Presbytis frontata*
		Samunsam	*Nasalis larvatus*
		Bako	*Nasalis larvatus*
		Danum Valley	N. Bornean community
		Sepilok	*Pongo pygmaeus*
	Indonesia	Tanjung Puting	*Pongo pygmaeus* *Nasalis larvatus*
Java	Indonesia	Gunung Halimun	*Presbytis aygula* *Hylobates moloch*
Sulawesi	Indonesia	Dumoga-Bone	*Macaca nigra* *Tarsius spectrum*
		Lore-Lindu	*Macaca tonkeana* *Tarsius spectrum*
		Bulusaraung	*Macaca maura* *Tarsius spectrum*

REFERENCES

Aken KM, Kavanagh K (1982): Species conservation priorities in the tropical forests of Sarawak, Malaysia. In Mittermeier RA (ed): "Species Conservation Priorities in the Tropical Forests of Southeast Asia." Gland, Switzerland: IUCN Species Survival Commission.

Ali R (1985): An overview of the status and distribution of the lion-tailed macaque. In Heltne PG (ed): "The Lion-tailed Macaque: Status and Conservation." New York: Alan R. Liss, Inc., pp 13–25.

Baldwin LA, Kavanagh M, Teleki G (1975): Field research on langur and proboscis monkeys: An historical, geographic and bibliographical listing. Primates 16:351–363.

Brotoisworo E (1982): Conservation of Sulawesi macaques. Kyoto University Overseas Research Reports on Studies of Asian Non-human Primates 2:39–50.

Chivers DJ (1977): The lesser apes. In Prince Rainier III, Bourne GH (eds): "Primate Conservation". New York: Academic, pp 539–598.

Constable JD (1982): Visit to Vietnam. Oryx 16:249–254.

Davies G (1983): Distribution, abundance and conservation of Simian primates in Borneo. In Harper D (ed): "Symposium on the Conservation of Primates and their Habitats: Vol. I, Primate Conservation in the Wild." Leicester: University of Leicester (Vaughan Paper No. 31), pp 122–148.

Davies G, Payne J (1982): "A Faunal Survey of Sabah." Kuala Lumpur: World Wildlife Fund Malaysia.

Eudey A (1980): Pleistocene glacial phenomena and the evolution of Asian macaques. In Lindberg DG (ed): "The Macaques: Studies in Ecology, Behavior and Evolution." New York: Van Nostrand Reinhold, pp 52–83.

Fooden J (1976): Primates obtained in Peninsular Thailand June–July, 1973, with notes on the distribution of continental Southeast Asian leaf-monkeys (*Presbytis*). Primates 17:95–118.

Gittins SP, Akonda AW (1982): What survives in Bangladesh. Oryx 16:275–281.

Goodwin HA, Holloway CW (1978): "The IUCN Red Data Book, Vol. I, Mammals." Revised by Thornback J. Morges, Switzerland,: IUCN.

Green S, Minkowski K (1977): The lion-tailed monkey and its South Indian rain forest habitat. In Prince Rainier III, Bourne GH (eds): "Primate Conservation." New York: Academic, pp 290–337.

Groves CP (1980): Speciation in Macaca: the view from Sulawesi. In Lindberg DG (ed): "The Macaques: Studies in Ecology, Behavior, and Evolution." New York: Van Nostrand Reinhold, pp 84–124.

Kavanagh M (1983): "A complete Guide to Monkeys, Apes and other Primates." London: Jonathon Cape.

Khan MAR (1981): The non-human primates of Bangladesh. Tigerpaper 8:12–15.

Korup GU (1982): Survey of non-forest populations of monkeys in South India. New Delhi: Report to India Dept. of Environment.

Lekagul B, McNeely JA (1977): "The Mammals of Thailand." Bangkok: Association for the Conservation of Wildlife.

Marsh CW, Wilson WL (1981): "A Survey of Primates in Peninsular Malaysian Forests." Kuala Lumpur: University Kebangsaan Malaysia.

Mackinnon JR (1980): A glimmer of hope for Sulawesi. Oryx 15:55–59.

Mukherjee RP (1978): Further observations on the golden langur (*Presbytis geei* Khajuria 1956) with a note to capped langur (*Presbytis pileatus* Blyth 1943) of Assam. Primates 19:737–749.

Mukherjee RP, Saha SS (1974): The golden langurs (*Presbytis geei* Khajuria 1956) of Assam. Primates 15:327–340.

Poirier F (1980): Taiwanese macaques: Ecology, population dynamics, conservation needs. Paper presented at VIIIth Congress Int. Primat. Soc., Florence.

Poirier F (1983): The golden monkey in the People's Republic of China. IUCN/SSC Primate Specialist Group Newsletter 3:31–32.

Rijksen HD (1978): "A Field Study on Sumatran Orang Utans (*Pongo pygmaeus abelli* Lesson 1827): Ecology, Behaviour and Conservation." Wageningen, The Netherlands: Veemand, Zonen.

Tan Bangjie (1985): The status of primates in China. Primate Conservation 5:63–81.

Tun Yin (1967): "Wild Animals of Burma." Rangoon: Rangoon Gazette.

Westing AH, Westing CE (1981): Endangered species and habitats of Vietnam. Environmental Conservation 8:59–62.

Whitten T, Whitten J, House A (1979): Solution for Siberut? Oryx 15:166–169.

Wilson CC, Wilson WL (1977): Behavioural and morphological variation among primate populations in Sumatra. Yearbook Phys. Anthropol. 20:207–233.

World Wildlife Fund (1980): "Saving Siberut: A Conservation Masterplan." Bogor, Indonesia: World Wildlife Fund.

Zhang YZ, Wang S, Quan GQ (1981): Geographical distribution of primates in China. J Hum Evol 10:215–266.

Index